PENGUIN BOOKS

ALTERED HARVEST

Jack Doyle is Director of the Agricultural Resources Project at the Environmental Policy Institute, where he has worked since 1974, specializing in agricultural and rural energy issues. Born in Philadelphia, he holds degrees from Millersville University and the Pennsylvania State University. He is the author of *Lines Across the Land* and numerous articles and essays, which have appeared in *The New York Times*, the *Boston Globe*, *The Progressive,* and other newspapers and magazines. He has served as an adviser and consultant to government agencies and has appeared as an expert witness before congressional committees.

ALTERED HARVEST

Agriculture, Genetics, and the Fate of the World's Food Supply

by JACK DOYLE

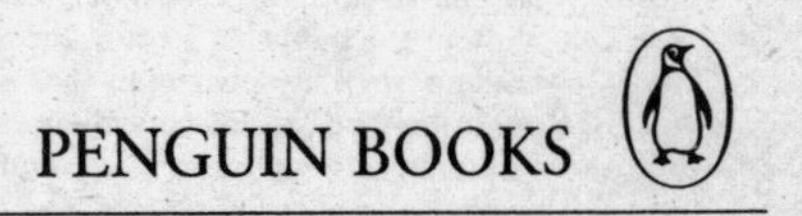

PENGUIN BOOKS
Viking Penguin Inc., 40 West 23rd Street,
New York, New York 10010, U.S.A.
Penguin Books Ltd, Harmondsworth,
Middlesex, England
Penguin Books Australia Ltd, Ringwood,
Victoria, Australia
Penguin Books Canada Limited, 2801 John Street,
Markham, Ontario, Canada L3R 1B4
Penguin Books (N.Z.) Ltd, 182–190 Wairau Road,
Auckland 10, New Zealand

First published in the United States of America by
Viking Penguin Inc. 1985
Published in Penguin Books 1986

LIBRARY OF CONGRESS CATALOGING IN PUBLICATION DATA
Doyle, Jack, 1947–
Altered harvest.
Bibliography: p.
Includes index.
1. Food industry and trade—Technological innovations —Government policy—United States. 2. Agricultural innovations—Government policy—United States. 3. Genetic engineering industry—Government policy—United States. 4. Biotechnology industries—Government policy—United States. 5. Drug trade—Government policy —United States. 6. Seed industry and trade—Technological innovations—Government policy—United States. 7. Agricultural chemicals industry—Government policy—United States. 8. Food adulteration and inspection—Government policy—United States. I. Title.
HD9006.D65 1986 338.1'9'73 86-17082
ISBN 0 14 00.9696 5 (pbk.)

Printed in the United States of America by
Offset Paperback Mfrs., Inc., Dallas, Pennsylvania
Set in Sabon

To those who have gone before me—
the Doyles, the McCuens,
the Rowans, & the O'Briens
. . . and for Dana

ACKNOWLEDGMENTS

This book could not have been completed, nor could the nearly three years of research behind it have begun, without the support and encouragement of a number of individuals and organizations. For financial support to the Agricultural Resources Project of the Environmental Policy Institute while I worked on this book, I am indebted to the Arca Foundation, the Carbonnel Foundation, the CS Fund, Joint Foundation Support, the Joyce Foundation, Rodale Press, the Ruth Mott Fund, and the Tortuga Foundation. Individuals at these organizations who also deserve a word of thanks for their encouragement and support include: Smith W. Bagley, James Ketler, John Haberern, Patricia Hewitt, Craig Kennedy, Kristin McKendall, Maryanne Mott, Bob Rodale, Ron Stevens, Marge Tabankin, Marty Teitel, and Herman Warsh.

Among those who reviewed and commented upon one or more parts of the draft manuscript for this book—and to whom much thanks is due for helping improve the book's final quality—are the following: Dr. Martin Alexander, Paul Bauman, Joe Browder, Bees Butler, Pete Carlson, Kate Clancy, Dr. Robert N. Ford, Cary Fowler, Tim Galvin, Charles Goodall, Joan Gussow, John Haberern, Dick

Harwood, Maureen Hinkle, Dr. Virgil A. Johnson, Mark Kramer, Carolyn Jabs, Star Lawrence, Dr. Morris Levin, Mark Messing, Jim Miller, Bob Mullins, Chuck Murphy, Bob Nicholas, Frances Sharples, Skip Stiles, Marty Strange, Kent Whealy, Nancy Weiman, Cathy Woteki, and Keith Zary.

I am also indebted to those individuals in industry and government who, for the most part, gave freely of their time, consenting to be interviewed or explaining in some detail, the nuances of the seed, chemical, and/or biotechnology industries. Others in the business community are also owed thanks for taking the author on production tours at seed factories and walk-throughs of biotechnology laboratories. All in all, I found the people in industry with whom I visited—those at the largest chemical companies as well as the smallest start-up biotechnology firms—to be hardworking and well-meaning. The industry criticism that follows in these pages is leveled more at process and institutional decision-making than it is at any individual.

During the three years of work on this project, I have also been helped enormously by several able and dedicated research assistants who did everything from run down obscure leads in the Library of Congress—when not battling finicky word processors or temperamental Xerox machines—to making insightful observations and suggestions that helped improve the author's view of his subject. This book, in its better parts, is as much theirs as it is mine. For their assistance, hard work, and moral support, I am indebted to Jeff Hoover, Mary Black, Kathy Backer, Cheryl Valdivia, and Jackie Williams.

And finally, there has been a special group of individuals who have served as professional confidants and friends at crucial junctures during the course of this project, listening to one idea after another, reading draft after draft, challenging points of possible contention, and generally trying to make me think things through. These people have helped sustain my energy level and confidence in the project while I pursued occasional dead ends and waged the writer's war of keeping at it, and I am indebted to them in a special way. They are: my mother, Eleanor M. Doyle, and father, John C. Doyle; my sisters Ellen and Teresa and brothers Mike, Tim, and Tom—each of whom read various draft chapters and offered varying opinions but always confidence; Louise Dunlap, president of the Environmental Policy Institute, who believed in an idea, and stood by the project in its earliest and then, darkest moments; Sara Ebenreck, Cynthia Lenhart, Dick Schneider, and David Weiman, four very important friends who each read much of three books' worth of material and offered some of the most rigorous and insightful reviews, as well as encouragement and moral support; and Jeanne Richards, a very able research assistant who became an indispensible editor, reviewer, and confidant, and helped sustain me in the final months of the manuscript's preparation.

To all of these people, and to Bill Strachan and Walt Bode, my editors at Viking, and the good people at the Environmental Policy Institute who lent much moral support, I say thank you for your encouragement and assistance.

I must also add that much of this book was written in 1983 and 1984, and a few sources and discussions with industry people date to August 1981. Since then, and almost on a monthly basis, the genetic sciences have been exploding with change and new information, and there is great flux in the biotechnology industry. Several people used as case examples in this book have moved on to new opportunities, and some companies have been changed in their ownership and/or orientation. Nevertheless, the philosophies and strategies conveyed in these pages, as well as the overall direction of the technology, commerce, and polity, remain on course as we approach mid-year 1985. Moreover, the stories, concerns and information imparted in the following pages cover a time frame of more than sixty years, and leave lessons, I believe, that are more enduring than temporal.

Jack Doyle
Washington, D.C.
May 1985

CONTENTS

INTRODUCTION

It is a rainy December afternoon in New York City. Ronald Reagan is in town pinning medals on people for valor—traffic is a mess. I hail a cab and head for a place called Zabar's. I'm going there to learn why a New Yorker I met said she doesn't have to worry about food. "Why should I be concerned about food," she says, "when I can buy any exotic food I want from anywhere in the world right here in New York City?"

Zabar's, it turns out, is a New York institution. Located on the city's Upper West Side, Zabar's is a family-owned deli grown large by reputation and luminary patrons such as Neil Simon, Woody Allen, and Ted Kennedy. Presidential candidates go there to campaign. In 1984, just before the New York primary, Walter Mondale was there amid the hanging sausage.

Zabar's is a cornucopia of good things to eat, a true gourmet's delight. "What Tiffany is to diamonds or F.A.O. Schwarz to toys, Zabar's is to food," says the *Zabar's Deli Book*. Over 20,000 customers pass through Zabar's each week, more than one million each year, and they spend nearly $10 million on what they buy there.

Standing in the middle of Zabar's cheese section—which stocks more than 400 different varieties—I notice a Zabar's worker, a white-coated woman in her thirties, carrying a stack of cheddar cheese blocks so high she can barely see over it. She is headed for a "low spot" in one of the cheese counters. I wonder if the shelves of Zabar's have ever been bare for more than a few minutes.

Zabar's sells a lot of cheese—over 12,000 pounds each week, with Jarlsberg and Brie leading the list at 1,200 pounds and 1,000 pounds, respectively. And that's not all.

Toward the rear of the store, near the stairway to the second floor, three Zabar's workers are surrounded by a dozen wooden barrels of coffee beans, scooping away and grinding selections as customers call them out. Hanging from the rafters are links of sausages and bags of onions. At the end of the meat counter, in white plastic pails, are the olives—"oil-cured 'hot' Moroccan," Gaeta, Sicilian green, French Hyon, and others. At the smoked fish counter you can buy Alaskan lox, smoked eel, smoked cod roe, and more. Scotch Salmon is selling for $5.98 per quarter pound; Scotch Kippers for $4.98 a pound. Matthew Walker's Plum Pudding is stacked in small red boxes on the stairway leading to the second floor, and you can also buy Swiss Vegetarian Pâté with herbs, or Talko Texas Brand crisp okra pickles in a jar, or Zabar's "pasta in pesto" or its tortellini salad with broccoli and cauliflower.

Zabar's is the epitome of food opulence in America, and for some New Yorkers, the proof that the American cornucopia will never run dry. Yet Zabar's and countless other delis, supermarkets, and restaurants throughout America do not *make* food, they only sell it.

Making food is another matter.

As a highly urbanized society, America is a nation far removed from the sources and origins of its daily bread. In a mere hundred years, we have traveled from being a nation of farmers with our hands in the soil to a nation of consumers lining up in supermarkets to purchase an extravagant array of fresh, processed, and imported foods. In the course of this progress, we have come to take abundance for granted, and we put a good deal of trust in our farmers, businesses, and government agencies who produce, process, and assure the purity of the substances we eat. As a matter of practical necessity and modern living, few of us really know much about the food that comes to our dinner tables. We assume for the most part that it still "comes from" farmers. Yet that is changing.

Today, America is on the leading edge of a powerful new era in food making; an era in which all of agriculture and much of food manufacturing will be reshaped by genetic technologies—by gene splicing in crops, embryo engineering in livestock, and powerful new microchemicals that will "turn-on" the genes of entire fields of wheat, potatoes, and other crops.

In this new era, food will be more finely shaped at its point of origin—at the level of the seed, the gene, and the molecule. Food will begin in the laboratory rather than the farmer's field. Crops will still grow in fields, but they will be given their biological marching orders by scientists who design them in the laboratory.

For what is hurtling toward all of us—and the age-old occupation of farming worldwide—is an unprecedented revolution in agricultural genetics; a revolution that began slowly, thousands of years ago, with the first domestication of plants and animals, progressing ever since with man's attempts at controlled crossbreeding. Today, with genetic engineering and biotechnology, that revolution is moving at astronomical speed, and with it, the practice and control of agriculture and food making are shifting dramatically from farmers to scientists, and most importantly, to those who own science.

This book traces that transformation; it is a story about the economic race to own the biological and genetic ingredients of agriculture; a story about political and economic change-in-the-making; a story about who will wield the new genetic ingredients of food power in billion-dollar world markets.

My reasons for writing this book are rooted in experiences that range from picking blackberries in open fields as a young boy, to following then Secretary of Agriculture Bob Bergland around the country one November as he conducted his "structure of agriculture" tour.

During the time I grew up in Delaware County, Pennsylvania, my younger brother and sister and I would occasionally venture into the "wilderness" across the Conchester Highway to a plot of land where blackberries seemed to exist solely for our morning cereal. We regarded that patch of land as common property, never thinking that one day it, or the blackberries on it, could be owned.

Property rights—at least those pertaining to open fields of blackberries—never crossed our minds. Years later, working as a lobbyist for the Environmental Policy Institute in Washington, D.C., I learned about plant patenting laws, and that seed and food crops could, indeed, be owned. I soon realized how few people knew that, and that such laws would increasingly determine the kinds of crops used by farmers. In 1980, while working as a consultant to the U.S. Secretary of Agriculture's Office, I became involved with the political and economic nuances of federal farm policy. I learned then that the issues of farm income, farmland loss, and soil erosion—while serious and unyielding—were not the only problems threatening farmers or the nation's food system. A particularly foreboding trend, it seemed to me, was an evolving shift in the control of agriculture; a shift that was inexorably moving farmers and natural resources out of the picture, and nonfarm corporations, technology, and international politics into the picture.

Biotechnology and "super genetics" are now becoming the driving engine behind a more rapid phase in that power shift; one that is moving all of us farther away from the sources of our food while moving farmers into a new kind of technological-genetic dependence that may reduce our agricultural options in the future and diminish ecological diversity in the process. Such a shift—one that is centered on corporate decision-making and high technology—suggests a certain kind of ecological and economic vulnerability, and that is part of what this book is all about.

On Wall Street, meanwhile, lots of money has already changed hands. Wealthy investors like Henry Ford II, William Weyerhaeuser, and the European Rothschilds are putting their money into the new genetics of food-making. And so are corporations. Seed companies in places such as Iowa, Illinois, and Missouri continue to be bought up by corporations like Monsanto and Ciba-Geigy. Universities are making million-dollar research deals with big business; patent applications are flying everywhere. Clearly, big changes are in the offing, and established interests are scrambling.

New business strategies are visible too. Chemical companies now plan to move from a clumsy era of indiscriminant and toxic pesticide development to a more sophisticated age of designer chemistry—a time when agricultural microbes and chemical molecules will be made to work together through the wizardry of computer modeling and gene machines. A few energy corporations are shifting surplus capital into the business of food and plant biology—the molecular bases of energy and food are not that far apart. Agribusiness and food-processing corporations are also moving into the seed-and-molecular ends of the food pipeline, eyeing field-to-table possibilities. And hundreds of new science-based biotechnology companies—or "research boutiques," as some call them—have formed almost overnight, many seeking "fast-buck" agricultural breakthroughs.

All of these players view the possibilities for economic gain through the prism of world food growth—slated to double in the next thirty-five years. In that period, more food has to be produced than in all prior history. And whoever holds the genes will wield extraordinary power.

As we enter the age of biogenetic food production, much is being promised for a new Horn of Plenty. And much is being promised that is potentially good and benign. We are told that a "pesticide free" agriculture, self-fertilizing crops, and nutritionally improved foods are just over the horizon. Productivity will rise, food will be cheaper, and all men will be fed. Harvests everywhere—even in the most hostile and parched regions of the globe—will be bountiful and abundant. Yet there are reasons for concern.

A series of thorny questions now confront societies everywhere faced with the advance of the agrigenetic revolution. Despite the promises of a better day in agriculture, there are concerns about how this new technology will affect food security, environmental quality, and biological diversity. The genetic manipulation of crops, farm animals, chemicals, and microorganisms for use in food production will not come without risk. The variables involved in the biological and chemical realms are inordinately numerous, and most scientists do not rule out the possibility of ecological or economic problems with the use of genetically altered substances in agriculture. One very persistent microbe or errant gene that wrongly colonizes an important ecological niche could touch off a chain reaction of economic and political events that could seriously disrupt a country's food system. We need only look to the recent havoc created by the Mediterranean fruit fly in California, the Avian influenza outbreak in the Middle Atlantic poultry belt, and the new strain of citrus canker in Florida to be reminded of how quickly small insects and microbes can bring modern agriculture to its knees.

Already, many of our crops and livestock are "genetically uniform," just waiting for the right disease or pest to strike. Even as genetic uniformity becomes more widespread in the modern agricultural systems of advanced nations, it is also being exported to developing countries. And in those countries themselves, genetic diversity is being destroyed by growth and agricultural development.

There are also nagging concerns about the influence of chemical and pharmaceutical corporations in the development of agricultural biotechnology, the place and use of patents, and what should be regulated. Consumers want to know how food and food prices will be affected. Farmers and consumers both want to know whether there will be more or less competition in the food sector, and how antitrust rules will apply to corporations manufacturing the new products of agricultural genetics. Farmers also wonder if big farms and small farms will be equally served by this new technology.

For others, there are deeper ethical and social questions pertaining to genetic alteration of food-producing resources, the potential genetic mixing of species, and the private ownership of plants, seeds, and microbes. Some scientists and citizens are concerned about the commercialization of academe and the eroding neutral ground of public universities and nonprofit research institutes. Still others want to know how Third World countries will be helped by the powers of agricultural biotechnology. But perhaps the most important questions facing society today are political: how will the raw power in the food production process now made possible with genetic engineering be used? and who will use it?

Change in the food-production system is nothing new, of course, but it is always rooted in the conduct of science, commerce, and politics. Now, with the new levels of scientific sophistication that exist in the mid 1980s, our technologies are capable of moving with extraordinary speed, followed closely by commercial investment and political support to sustain them. Yet this ability to move quickly in one direction around a revolutionary technology—with the full weight and momentum of our most powerful institutions—ought to give us pause as we ply the "new fire" of genetic technology to the biology that sustains us. Clearly, with gene splicing and bioengineering, there is revolutionary power at hand. But there is also a concern that these new tools of productivity be used with measured application and some benefit beyond concentrating wealth or enhancing political power.

It is in this spirit of proceeding with caution—ever mindful of the social dislocation and economic vulnerability that such a new, high-powered technology may bring—that this book is offered.

ALTERED HARVEST

CHAPTER 1

IT CAN HAPPEN HERE

> The corn fell victim to the epidemic
> because of a quirk in the technology . . .
> —*National Academy of Sciences,*
> *1972*

In the summer of 1968, when the nation was preoccupied with the Vietnam War, the assassinations of Martin Luther King, Jr., and Robert F. Kennedy, and a divisive presidential election campaign, the first signs of trouble went almost unnoticed. Out in the heartland, on a few isolated seed farms in Illinois and Iowa, a mysterious disease was producing "ear rot" on corn plants. At the time, scientists thought the strange disease might be a combination of two familiar diseases called "yellow leaf blight" and "charcoal rot," but they were wrong. Yet only a tiny amount of hybrid corn seed was lost to the new disease that summer, so no alarms were sounded. Whatever it was, the new malady was probably a freak occurrence that would most likely die off over the winter. Diseases like that were one of the "normal" consequences of doing business with nature.

But in 1969, a few farmers and scientists noticed the same problem recurring in midwestern seed fields and hybrid corn test plots. One account noted: "In the late summer and early fall of 1969, a few corn fields in southern Iowa began behaving erratically. Ears rotted inside husks. Stalks fell to the ground. Shortly, the same thing happened in isolated fields in Illinois and Indiana." This time scientists noticed that only certain hybrid corn varieties were susceptible to the disease. In Florida, too, a few seedsmen found that hybrid corn varieties growing there were particularly vulnerable. Yet there was no adequate scientific explanation for the new disease. Scientists knew it was a fungus, but they didn't know what kind or how it worked.

In 1970, the disease was first reported in February from southern Florida, near Belle Glade. Between May 5 and May 20, heavy infestations were cited in southern Alabama and Mississippi. By June 18, the disease covered the entire state of Florida, lower Alabama, and most of Mississippi. The lower third of Louisiana and coastal Texas were also infected.

Reproducing rapidly in the unusually warm and moist weather of 1970, its spores carried on the wind, the new disease began moving northward toward a full-scale invasion of America's vast corn empire. Later to be identified as "race T" of the fungus *Helminthosporium maydis,* it soon became known as the Southern Corn Leaf Blight.

The new fungus moved like wildfire through one corn field after another. In some cases it would wipe out an entire stand of corn in ten days. Moisture was a key factor; a thin film on leaves, stalks, or husks was all the organism needed to gain entry to the plant. Within twenty-four hours it would start making tan, spindle-shaped lesions about an inch long on plant leaves, and in advanced form would attack the stalk, ear shank, husk, kernels, and cob. In extreme infections, whole ears of corn would fall to the ground and crumble at the touch.

The fungus moved swiftly through Georgia, Alabama, and Kentucky, and by June its airborne spores were headed straight for the nation's Corn Belt, where 85 percent of all American corn is grown. By this time, however, unsuspecting Corn Belt farmers had already planted their crops and were largely unaware of the bitter harvest headed their way.

The fungus could begin reproducing within sixty hours of landing on a corn plant—yielding a new generation of its own kind every ten days—and its spores could survive temperatures of 20 degrees below zero and still germinate, which meant they could linger in fields and plant remnants through the winter. In some cases, the fungus could even penetrate corn seed, causing it to fail or produce blighted seedlings.

In its wake, the Southern Corn Leaf Blight left ravaged corn fields with withered plants, broken stalks, and malformed or completely rotten cobs covered with a grayish powder. When farmers harvested what they could, clouds of spores were thrown up into the air behind their combines, spreading the disease even farther.

In just four months—from May to September 1970—the disease had spread as far north as Minnesota and Wisconsin (it later entered Canada), and as far west as Kansas and the Oklahoma panhandle. The nation's corn farmers were facing a full-blown crisis.

UNCERTAINTY, CONFUSION, & PANIC

The U.S. Department of Agriculture (USDA) was caught completely off guard by the blight. On August 1, 1970—a time when millions of acres of corn in the Southeast had already been laid waste by the blight—agriculture officials were confidently predicting a record 4.7-billion-bushel corn crop. A week later, they began revising their estimates downward, suggesting that the disease could cut the corn harvest by 10 percent.

On Sunday morning, August 16, the *Des Moines Register* jolted the Midwest with the banner headline CORN MARKET IN TURMOIL. That account reported steep price rises in corn-futures trading on the Chicago Board of Trade the previous Friday, fueled by Iowa State University reports that the new blight had been found in Iowa. One midwestern trading firm, which normally did about half a million bushels in corn trading on a busy day, did seven million bushels' worth that Friday. "Somebody's trying to manipulate the corn market," said one midwestern trader. "This corn blight thing isn't that serious. Those southern states just don't grow that much corn. Too many people are getting too excited about too little."

Nevertheless, with the opening of business that Monday, panic gripped the commodity and futures markets. Estimates that the blight might wipe out half the nation's corn crop fueled frantic trading and speculation. At the Chicago Board of Trade, the nation's largest commodities market, 193 million bushels of corn changed hands in one day, smashing a trading record that had stood for 122 years. The Dow Jones index for commodity futures hit 145.27, and had its highest one-day advance in nineteen years. The futures prices of corn, wheat, oats, and soybeans all jumped to their allowable one-day limits. Trading in livestock also soared, as prices for live hogs, cattle, and poultry rose in reaction to the prospect of higher-priced feed grains. Between August 17 and 20, the corn blight boosted the future price of corn thirty cents a bushel—a huge increase when measured in the millions of bushels traded. During the frantic August trading, some speculators became wealthy overnight; one corn trader made paper profits of $500,000 that month.*

The August rally in the commodities markets was sparked by newspaper accounts like the report in the August 16 *Des Moines Register*. But officials at the USDA weren't talking, knowing that any statement on the blight from the department could affect the markets. The department's official crop report was due on September 11. Senator Allen J. Ellender of Louisiana, chairman of the Senate Agriculture Committee, was quoted on the commodity wires as saying that no more than 5 percent of the nation's corn crop was affected. An unofficial figure of 4 percent was attributed to U.S. Secretary of Agriculture Clifford M. Hardin. One USDA administrator, James U. Smith, then chief of the Farmers Home Administration, was reprimanded for his agency's leaking a statement about the blight to United Press International, and was told by assistant secretary T. K. Cowden to inform his people, "to make no statements that could be interpreted as a governmental figure regarding the size of the corn crop."

In Chicago, meanwhile, some traders complained of misinformation and exaggeration by the media. "On Sunday before the limit move," said Charles Mattey, who then headed up Bache & Company's commodities department, "all the media had the wrong numbers. The Georgia pathologists were talking about the seven Southern states, not the entire country. Damage to eighty million bushels instead of two and a half billion. . . ." But in reality, the blight had already far surpassed the eighty-million-bushel mark.

BLIGHT IN THE CORN BELT

What really panicked commodity traders and government officials was the blight's penetration of the Corn Belt; just three mid-

*On the afternoon of August 17, in an effort to slow speculation, the directors of the Chicago Board of Trade met in special session and immediately increased the margin requirement—the amount of cash a trader had to have in his account when placing an order to buy or sell a futures contract.

western states—Illinois, Indiana, and Iowa—accounted for half the nation's total corn production. "There has always been blight in the South," said former Chicago Board of Trade Chairman William Mallers, "but when you get [blight] in the Corn Belt, you're really talking." In August 1970, Illinois Secretary of Agriculture John W. Lewis was estimating that 25 percent of his state's corn crop was already lost to the blight. Just one year earlier, Illinois had been the nation's top corn producer, accounting for more than one-fifth of the crop.

When the first reports of the blight's severity hit the newspapers in mid-August, the U.S. Congress was in its traditional summer recess, and political reaction to the blight's damage and the rising prices caused by the blight came in piecemeal fashion, mainly from farm-state congressmen and senators at home in their states and districts. However, 1970 was an election year, and while a few congressmen and senators made inquiries of the USDA's action on the blight, with some calling for emergency farm aid, no congressional hearings were ever held on why the blight occurred. But the growing national scope of the problem, and its potential for fueling food-price inflation, did come to the attention of the White House.

In late August 1970, the USDA began to acknowledge that there was a problem. By August 23, Secretary of Agriculture Clifford M. Hardin had opened up the government's corn reserves to help dampen speculation in the commodities markets. On September 3, 1970, Hardin wrote a two-page brief for President Richard Nixon on the corn-blight situation, saying that the blight was not a new problem, but had become "economically significant." The disease's new strength, Hardin explained, was the result of "an unforeseen mutation."

"Although we are concerned about 1970 damage," wrote Hardin, "we feel that . . . there is ample feed grain for livestock to carry us well into the 1971 harvest period." Hardin also assured Nixon that many of the earlier private reports which projected 1970 losses ranging from 10 to 50 percent of the crop "were exaggerated."*

*Later, in December 1970, when the corn blight was seen as something more of a potential political problem, USDA and White House officials organized a Corn Blight Information Conference at which President Richard Nixon spoke to a group of farmers assembled at USDA's research station in Beltsville, Maryland, just outside of Washington, D.C. The purpose of the meeting was to assure farm leaders

But for many farmers who had already lost entire corn fields to the blight, such figures were no exaggeration. And while major corn processors and other corn-using industries moved quickly to protect their interests by raising prices,* farmers could not raise their prices. "We'll be lucky if we have enough corn to pay our fertilizer bill," said 52-year-old Indiana farmer, Melvin Pflug, after surveying his 600 acres of corn, about half of which was devastated by the disease.

A BREAK IN THE WEATHER

As the official tally of the blight's nationwide toll remained unknown through August 1970, farmers, traders, and USDA officials anxiously looked on to September. Despite the growing and justified fears of continued damage to the nation's corn crop, there was still one favorable possibility: a break in the weather. "It may seem ironic with all the technology at our command today, but everything now hinges on the weather," said Dennis B. Sharpe, then an agricultural economist with the Federal Reserve Bank of Chicago. "If it is fairly cool and dry over the next two weeks," Sharpe told *Business Week* that August, "there is nothing to worry about. On the other hand, if it rains and it is hot and humid, the fungus will spread quite rapidly."

On September 21, corn prices on the Chicago Board of Trade

that USDA was working on the blight, and that the White House was concerned, too. In his speech, Nixon talked about the recently passed farm bill and he praised the American farmer. He also spoke briefly about the blight: "There was a time," said the President, "I suppose not ten, maybe not twenty-five years ago, when corn blight came, we might not have had enough in storage to take up the slack, but beyond that, we might not have developed the capability to deal with the problem.

"But now we not only have the amount in storage to take up the gap, but we also, as I understand, due to the enormous facilities of research and the brains and the overtime and genius that has gone into it, we are finding an answer to this problem.

"And that means that in the future we will be able to deal with it more effectively. It means also we can share this knowledge with other people throughout the world."

*On Wednesday, August 20, following the dramatic increases for corn and other grain contracts in the futures markets, major food processors began raising their prices for certain corn products. CPC International, Inc., an industry leader in the corn-processing business, announced immediate price increases for corn syrup and corn starch—up 45 cents a hundredweight and 75 cents a hundredweight, respectively. Other corn processors followed suit.

dropped sharply on the basis of rumored USDA reports that favorable weather in the Corn Belt could slow the spread of the blight. However, humid weather in the first half of September intensified the disease in the southern portions of Ohio, Indiana and Illinois, although the weather did break in the northeast states and western Corn Belt, sparing a huge portion of the crop. In other words, the nation was lucky.*

As it was, the Southern Corn Leaf Blight devastated 15 percent of America's 1970 corn crop, reducing the average national corn yield from 83.9 to 71.7 bushels per acre, costing farmers about $1 billion in losses. Some southern states lost more than 50 percent of their corn crop. In all, more than 1.02 billion bushels of corn were lost in 1970. But the crisis wasn't over.

THE RIPPLE EFFECT

Although the most immediate effects of the 1970 blight fell on the shoulders of farmers, its ripple effect soon began to reach other parts of the American economy. Small-town bankers and businessmen who had loaned farmers money began to worry about repayment. Washington worried about exports. At that time, the United States was exporting about 600 million bushels of corn annually, and large quantities of corn were also fed to cattle, poultry, and swine. Domestic food processors and distillers also depended upon corn. If losses in the cornfields became severe, a three-way tug-of-war over existing supplies could ensue between food processors, livestock feeders, and grain exporters.

Adding to the reality of the disease itself were rumors that any blighted grain would be toxic to humans and animals. Further questions emerged about "secondary organisms" that might invade the grain, causing still other kinds of toxic problems. In fact, pathologists at the University of Illinois did discover "secondary fungi"—capable of producing the potent poisons known as aflatoxins—

*"It should be recognized," wrote University of Illinois plant pathologist A. L. Hooker in 1972, "that dry weather reduced disease spread in the western Corn Belt and delayed northward spread of the disease on the eastern seaboard. In addition, because of favorable climatic conditions, northern states had above normal yields. Without those two features, national disease losses could have been greater than those estimated."

growing on blighted corn stalks, husks, and ears. But no toxic effects were reported in livestock or humans.

However, it was learned that the blight itself could be transmitted in corn seed. And that fed speculation that the blight was being exported to foreign countries through American corn seed. By early 1971, the corn blight was reported in Japan, the Philippines, Africa, and Latin America, and some importers of corn seed, such as Australia and New Zealand, were wondering if the problem didn't originate with American seed. Addressing the question, *Ramparts* magazine, in a March 1971 editorial, wrote, "There is considerable speculation as to whether through our exports of diseased corn . . . We are spreading the blight around the world." At that time the United States was exporting some 46.8 million pounds of corn seed to all parts of the world, worth about $5 million annually. Yet proving that blight in other countries originated in U.S. seed was difficult when the importing countries weren't looking for it in the imported seed.

A SHORTAGE OF SEED

One concern about the blight that began to haunt USDA officials as early as August 1970, was the question of an adequate supply of seed for 1971. Practically all the nation's hybrid corn seed was then grown in the Midwest, where the fungus was taking its toll. Some farmers and seedsmen meeting in the South at that time were beginning to wonder if there would be any corn seed available for 1971. "If this stuff spreads to the Corn Belt," said Ed Komarek of Georgia's Greenwood Seed Company, ". . . we just won't have any seed . . . I hate to think of next year." Yet, D. D. Walker, President of the American Seed Trade Association, meeting in Washington with Secretary Hardin on August 21, confidently predicted that there would be "ample seed corn supplies for the 1971 crop." Farmers, however, weren't merely concerned with an adequate supply of seed, but with an adequate supply of disease-resistant seed. And that would take time.*

*Offers to help produce new sources of corn seed came from some interesting quarters. In an August 20 telegram to Secretary Hardin, for example, Bernard Steinweg, senior vice-president of the Continental Grain Company, one of the

Although several American seed companies did produce new supplies of seed in locations such as Mexico, Hawaii, and Argentina, there was a shortage of disease-resistant corn seed as the 1971 planting season approached. By spring there was only enough new seed to plant about 23 percent of the nation's corn crop, and much of this seed was diverted to southern states in an attempt to create a "buffer zone" to block the expected northerly progression of the blight again in 1971.

Another strategy seed companies used to stretch their limited supplies of corn seed was to sell stocks of disease-susceptible seed in states where drier and cooler conditions had stymied the blight's spread in 1970. During 1971, susceptible corn seed was sold to farmers in western Corn Belt states such as Nebraska, Kansas, and western Iowa, and northern states such as Michigan, Wisconsin, Minnesota, and the Dakotas.

Meanwhile, corn farmers in the Midwest were provided with "blends" of seed—supposedly 50 percent resistant seed and 50 percent susceptible seed. One midwestern farmer who started spotting the blight on his corn in June 1971, said of the 50/50 arrangement, "I can't find the 50 percent of the stalks that don't have blight."*

Other farmers complained of supply problems. "I've only got

largest grain companies in the world, made the USDA an offer of Argentine land and production assistance to help in growing corn for seed. "Our affiliate company in Argentina, which has a hybrid seed company subsidiary," cabled Steinweg, "is willing to assist American hybrid seed corn companies in the production of seed this winter. We not only have lined up acreage for this purpose, we consider we have the technical ability to handle the production in an efficient manner . . . We would appreciate being able to cooperate with you and any American seed companies not now aware of our capabilities and interest."

More than a month later, after the USDA had engaged the cooperation of the Mexican government in allowing American seed companies to grow seed there, U.S. Assistant Secretary of Agriculture Ned Bayley wrote in reply to Steinweg, "Your offer has been brought to the attention of the U.S. seed trade. We could not, however, find any firm that is able to take advantage of the offer. We understand that it would be very difficult at this late date to produce seed corn in Argentina for return to the U.S. for planting next spring." In one sense, Bayley's reply to Continental was a polite way of saying that American seed companies were not very enthusiastic about one of the world's major grain corporations getting into their business at a time of shortage.

*In at least one case, a group of farmers in Iowa brought a class action suit against some sixty seed companies which allegedly sold hybrid corn seed to Iowa

about 25 percent normal [blight-resistant] seed," reported Illinois farmer Carl Thompson in May 1971, "and even that is more than some of my neighbors have." A black market in resistant seed developed, with some farmers paying two to three times the going market price. And in at least one case, a truckload of resistant seed was hijacked.

The few seed companies that managed to produce blight-resistant corn seed didn't waste any time in raising their prices. For some reason, the Funk Brothers Seed Company of Bloomington, Illinois, had noticed as early as 1968 that the popular corn hybrids were becoming increasingly vulnerable to insects and some milder midwestern strains of blight, and had switched the company's seed production operations back to an older kind of hybrid. By late 1970, when other seed companies were struggling to come up with a plan to produce some new seed on an emergency basis, Funk Brothers was sitting pretty. On October 12, 1970, the company announced it would increase seed prices on its new hybrid by 17 percent, selling its best blight resistant line at nearly thirty dollars per fifty-pound bag. In 1969, for example, before the blight, the average U.S. price of hybrid corn seed was $13.70 a bushel. Thereafter, the price of hybrid corn seed continued to spiral upward due to the difficulty in producing blight-resistant seed. By 1974, the average U.S. price had jumped to twenty-five dollars a bushel, an 84 percent increase over pre-blight prices.

farmers in 1970 with prior knowledge that the seed was susceptible to blight, and failed to warn the farmers of that susceptibility. That suit, *Lucas et al. v. Pioneer, Inc., et al.* alleged that seed-company officials had knowledge of the disease susceptibility of their hybrid corn seed "prior to 1969, and perhaps as early as 1962." The suit also alleged that the officials knew in February 1970 that the blight had reached epidemic proportions in Florida and was moving north, but failed to warn farmers of the potential disaster, even though many of the companies had sold susceptible seed to Iowa farmers during 1970. Besides this, the suit charged that seed-company officials did not instruct Iowa farmers about any precautionary measures to protect themselves from the impending disaster, though they knew of such measures. The farmers sued for damages and losses of 100 million bushels of corn, then valued at roughly $100 million. The suit, however, was not resolved until the late 1970s, having been dismissed by the Iowa Supreme Court as an improperly brought class action, after which it was refiled by individuals in separate actions, with settlements of court costs awarded to some farmers. Similar suits were also filed by farmers in several other states, but their outcomes are either still in the balance or otherwise undetermined.

CORN BLIGHT, PART II

As farmers began planting their fields in the spring of 1971, no one knew for sure what the prognosis for the corn blight would be that year. "Hope is mixed with fear as we go into the 1971 corn growing season," wrote *Successful Farming* editor Charles E. Sommers in March 1971—"hope that the new southern corn leaf blight disease epidemic won't hit again [and] fear that it probably will . . . few people in the know are making predictions as to what to expect." In May 1971, George F. Sprague, a USDA scientist from Illinois who was coordinating the fight against the blight, admitted "a considerable degree of uncertainty and speculation" about its outcome in 1971. "We shall have to wait for a final answer," he said.

The prospect of rising food prices and food-based inflation caused by the possibility of two successive years of blight began to surface in the press. "Corn accounts for 70 percent of all grain fed to beef and dairy cattle, hogs and poultry," commented *U.S. News & World Report* in a May 1971 story. "If this year's crop is severely cut by blight, there will inevitably be shortages—and soaring prices—in beef, pork, milk, eggs and chicken." One Wall Street analyst following the blight remarked later that year, "the biggest question mark overhanging the near-term outlook for inflation does not concern the steel-wage negotiations but the progress of the corn-leaf blight."

On May 2, 1971, in a nationally broadcast speech on agriculture, President Richard Nixon ordered more money for research to fight the corn blight, noting that the disease had created "major problems for corn farmers." The National Aeronautics and Space Administration (NASA) and the U.S. Air Force had also been enlisted in the effort to monitor the blight's progress using satellite and remote-sensing technologies.

Generally though, the infestations of 1971 were regarded as light compared with the previous year. As in 1970, weather again was an important factor, with a cool, dry spring slowing the blight's progress initially, and somewhat drier summer conditions prevailing in the Corn Belt. Nevertheless, the blight was still spotted in 581 counties in 28 states by July, and in parts of the Midwest, some severe outbreaks were reported. Overall, however, the nation sustained only minimal losses in 1971. By 1972 enough blight-resistant seed had been produced by seed companies, and farmers throughout the country were adequately supplied. The crisis was over.

THE AFTERMATH

The corn blight epidemic of 1970–71 was not a crisis for most Americans at the time. Although many were no doubt aware of it, few were directly affected. Had the billion bushels of corn that were lost to the blight been fed to cattle, they would have produced over 7 billion one-pound steaks, or more than 30 billion quarter-pound hamburgers. And while some food prices did rise slightly, corn on the cob, chicken, and hamburger were still on the dinner table.

Corn surpluses from previous years and substitutions of other grains helped to ease the blight's impact. In fact, the nation's grain reserves probably could have absorbed two very bad years of blight before things would have become really tight. However, a few weeks of "blight weather," coming in the late harvest weeks of 1970 and 1971, might have put the nation to that test very quickly. As it was, man and science won this round.*

By 1972, American scientists and seedsmen were congratulating themselves for their "heroic" actions, now reassured that the system worked and could respond to an unforeseen disease in a relatively brief span of time. However, beneath the self-congratulations and public confidence, there were some reservations. Some plant pathologists were taken by surprise by the strength of the Southern Corn Leaf Blight and the speed with which it spread, and a few were privately shaken when they learned why this new mutant strain of fungus spread so quickly. While plant-disease epidemics had occurred in the United States before and were a regular fact of life in agriculture, scientists discovered something new about crop diseases in 1970; something they did not know before this particular corn blight occurred.

At the beginning of the epidemic, there was no defense against the Southern Corn Leaf Blight because the new strain of fungus had found a "genetic window" that made its infestation rapid and wide-

*However, as biologist H. Garrison Wilkes has pointed out, "Such a crop failure in countries such as Guatemala or Kenya, where people obtain half of their calories from corn, would have been disastrous."

spread. The genetic window in this case was a gene found in the cytoplasm, the watery material that surrounds the cell nucleus and makes up the bulk of most living cells. In terms of crop disease, that was a new twist.

Commenting on that discovery in 1971, pathologist A. L. Hooker noted that it was "most unusual" that the cytoplasm of corn plant cells played a major role in determining the disease reaction, since in almost all other diseases, genetic factors in the nucleus of the cell determined disease resistance or susceptibility. Because of this, explained Hooker, corn breeders and seedsmen had no reason to suspect that uniformity in the corn crop would pose any problem. But it did.

The cytoplasm found common in most hybrid corn at that time was called "Texas male-sterile cytoplasm," or "T-cytoplasm," after a Texas variety of corn in which it was discovered. For twenty years preceding the blight, T-cytoplasm was used by plant breeders and seed companies to simplify the process of hybrid corn seed production. Male-sterile cytoplasm produced tassels on corn plants that bore impotent pollen, which—in combination with a fertility-restoring gene in the hybrid cross—enabled scientists to cross-breed and pollinate large numbers of plants more easily. T-cytoplasm thus eliminated the time-consuming, labor-intensive, and economically expensive step of hand detasseling corn plants. It was a revolutionary invention in plant breeding. But what scientists didn't know then about T-cytoplasm was that it also carried a gene in the mitochondria (an organelle of the cell that produces chemical energy for the cell) which enabled the new strain of the corn blight fungus to do its damage.

T-cytoplasm was a man-made change in corn plants used to foster the quick and profitable production of high-yielding, hybrid corn seed. It was a change accomplished and advanced by science and commerce without full knowledge of the potential consequences. The new strain of corn blight fungus, *Helminthosporium maydis,* was a mutation perfectly keyed to a gene in that cytoplasm.*

*Interestingly, two Philippine plant breeders had reported in the scientific literature of 1962 and 1965 that they had observed *Helminthosporium maydis* wreaking havoc on some of their hybrid corn lines as early as 1957. The inbred lines used to develop these hybrids were from the United States, and contained T-cytoplasm. Yet in 1972, a study by the National Academy of Sciences (NAS)

At least 80 percent of the hybrid corn in America in 1970 contained T-cytoplasm, which is why "race T" of *Helminthosporium maydis* laid waste to 15 percent of the nation's corn crop. "The USA in 1970 had 46 million acres of corn with Texas male sterile cytoplasm," wrote Iowa State University Pathologist J. Artie Browning in 1972. "Such an extensive, homogenous acreage of plants . . . is like a tinder-dry prairie waiting for a spark to ignite it. Race T was the spark. . . ."

The official scientific response to the corn blight came in August 1972, with the release of the National Academy of Sciences study

discounted these reports almost casually, noting that in neither of the reports did the scientists warn of a possible epidemic. Perhaps the Filipinos did not warn that the fungus could be damaging to all varieties having T-cytoplasm, said the NAS, "because scientists are disciplined to avoid extrapolation. They probably reasoned, too, that they were working in a tropical environment not at all typical of the world's major corn lands."

In America, meanwhile, two scientists, Donald Duvick of Pioneer Hi-Bred International (the largest hybrid corn seed company in the United States) and A. L. Hooker, a plant pathologist with the University of Illinois, did check the Philippine report. Duvick reported in 1965 that to his knowledge, no differences between T-cytoplasm and normal cytoplasm had been reported or noted in the United States. Duvick charged the increased susceptibility in the Philippines to a possible "secondary effect of reduced plant vigor, accentuated by the Philippine environment."

Hooker, who would later become one of the authors of the NAS study reporting on the corn blight, tested Illinois corn varieties to see if they were especially vulnerable to *H. maydis*. In his tests, Hooker used the same inbred lines found vulnerable in the Philippines, containing both normal and T-cytoplasm. He and his colleagues tested these lines in 1963, but they did not use "race T" of *H. maydis*, and so found no differences. Therefore, the results of their tests were not published.

By September 1969, however, Hooker and his colleagues had isolated some of the "race T" fungus from an Illinois cornfield, and officially identified it as a new strain in early 1970. Hooker's paper describing the new strain was not published until August 1970, when he reported: "A majority of the acreage of America's most valuable crop is now uniformly susceptible and exposed to a pathogen capable of developing in [epidemic] proportions."

When did the seed companies first know of the new race T, and the fact that most of the hybrid seed they were selling in 1970 would be highly susceptible to the new disease?

About a year later, in August 1971, Hooker provided the following observation in a paper presented before the American Society of Agronomy in New York: "Seed companies were unaware of the potential susceptibility of hybrids containing T-cytoplasm in the commercial crop in time to inform their customers of the advantages of hybrids containing all or a portion of plants with normal cytoplasm, or in time to make significant changes in their seed production methods during the 1970 season."

Genetic Vulnerability of Major Crops. "The corn crop fell victim to the epidemic," said the Academy's report, "because of a quirk in the technology that had redesigned the corn plants of America until, in one sense, they had become as alike as identical twins. Whatever made one plant susceptible made them all susceptible." The Southern Corn Leaf Blight, said the NAS study, was genetically based—a key finding.* Looking beyond corn, the Academy also warned that most other crops were "impressively uniform genetically and impressively vulnerable." Moreover, the study added, "this uniformity derives from powerful economic and legislative forces," such as food company preferences for one kind of crop and government marketing orders requiring specific kinds of fruits and vegetables. But despite these warnings, not much has changed since 1972. Corn is less vulnerable, but 43 percent of the nation's corn acreage is planted to varieties derived from 6 inbred lines. Other crops are even more vulnerable. And cytoplasmic breeding systems are still being used in a number of crops, including corn.†

What has changed since 1970–72 is the emergence of something called "biotechnology"—an all-powerful genetic technology that will increasingly be at the center of agriculture and food production worldwide. And at the hub of this new technology, more than was ever imagined in 1970, is the gene.

GENES, FOOD, & POWER

Unseen by most of us, and familiar only to those who peer into the arcane world of plant and animal cells, genes are the building blocks of our food supply. They can determine everything from the

*In a 1976 paper entitled "An Evaluation of Special Grant Research on Southern Corn Leaf Blight," the USDA also acknowledged the genetic uniformity in the nation's corn crop as one of the primary causes of the 1970 Corn Leaf Blight. "In the [1960s], it became clear that relatively few corn breeding parents were being used to produce the bulk of American hybrid corn varieties," said the report. "This narrowness of germplasm set the stage for potential vulnerability to diseases, insects and other stresses. In early 1970, environmental conditions in Southern and North-central corn producing regions were favorable for easy disease establishment and spread among vast plantings of highly uniform varieties. The [Southern Corn Leaf Blight] epidemic became of national and international significance."

†For more details on the issue of genetic uniformity in agriculture, see Chapter 10.

protein content in a slice of bread to how much milk a dairy cow produces. Genes, and the configurations in which they occur inside plant and animal cells, hold key instructions of growth that govern cell and organism; instructions that determine the enzymes and biochemical reactions that build proteins and other materials inside the organism, as well as governing its interactions with the outside environment. By tinkering with genes, these carefully choreographed instructions of growth and environmental give-and-take can be altered, and through such changes, a nation's food system can be altered for better or worse.

America's food system—one of the largest, most productive, most sophisticated such systems in the world today—is an incredibly far-reaching system, permeating vast areas of modern society and everyday life. The business of agriculture and its related industries account for approximately one-fourth of the nation's gross national product. Agricultural exports alone add more than $25 billion annually to the nation's balance-of-trade ledger. In terms of employment, one out of every five jobs in America involves the storing, transporting, processing, or merchandising of farm commodities. In producing crops and livestock, over one-half of the nation's land mass, roughly 1 billion acres, is normally used for agricultural purposes.

In any given July, there are billions of corn plants growing in the rich and warm soils of Illinois and Iowa; thousands of dairy cows roaming the hillsides of Vermont and Wisconsin; and millions of chickens, hogs, and turkeys being fattened from Maine to Missouri. Yet, underlying this huge and continuous American empire of food production are genes; the millions of genes carried in the cells of all plants and animals; genes governing microbes in the soil, fungi in the wind, and insects on the move—genes which are the ultimate foundation of all living things that grow and move. To understand and control the function of these genes is to wield whole systems of power. Even a single genetic alteration to one crop line in one subpart of America's huge agricultural system can have ramifications touching millions of people—alterations which are also worth millions of dollars. Multiply such alterations many times over throughout world agriculture, and there are innumerable revolutions made possible; revolutions of food production and polity, and of fundamental economics.

In the United States, large sums of capital have already been invested in the agrigenetic revolution. New industries have formed and major corporate realignments have occurred. The scientific establishment is poised for change and politicians of all stripes are eager to help. For what is happening "backstage" in America's food system today—in the halls of government, in university laboratories, and in corporate boardrooms—is the beginning of the genetic centralization of food production.

In one sense, the new agrigenetic technologies will "transistorize" the food-making process, reducing it to a compact set of genetic components which will be, for the most part, out of public view and increasingly held by governments and corporations. While this technological reductionism is occurring, world food needs, of course, will be expanding. The pie will be bigger, in other words, as "chip-like" power accrues to those who own food genes. In this situation, risks of all kinds will escalate.

In food production systems as we know them today, the variables involved—technological, economic, and ecological—are numerous and wide ranging. Mistakes, unforeseen consequences, and miscalculations giving rise to damaged or failed harvests are not infrequent occurrences. Add to this now the new dimensions of biotechnological food-making—with its near instant ability to screen millions of cells at a laboratory workbench, produce millions of specifically designed microbes, or to leap species barriers in the making of new crops and livestock—and the prospects for mistake or calamity swell geometrically.

Today, we may be moving toward a high-tech, house-of-cards agriculture worldwide, with genetic engineering at its base; a system in which one monkey wrench or one unforeseen mutation can create enormous problems. Just as the technology of hybrid corn production "went wrong" in 1970, aiding the advance of the corn blight, the agricultural biotechnologies of genes, microbes, and molecules might "go wrong" on a much grander scale in the future. Despite what its proponents may claim for it, this is not an invincible or fail-safe technology.

Yet clearly, this technology *does* have the potential to be safely and beneficially applied, improving food production, environmental quality, and agricultural diversity in the process. But making sure that happens could well be the challenge of the mid-1980s and beyond.

CHAPTER 2

GENETIC DISTANCE

To understand the source of one's next meal
is to understand one's own political vulnerability.
—*Mark Kramer,* Three Farms

It was unseasonably warm in Vermont for late November. We had just landed at the Barre airport in a small New England Airlines plane that had been buffeted by high winds on a flight from Boston. Three of us—needing a ride to Montpelier's Tavern Inn, where then–U.S. Secretary of Agriculture Bob Bergland would hold his first "Structure of Agriculture" hearing the next day—decided to share a cab. Montpelier was the site of the first of ten hearings Bergland would hold across the country that fall, listening to hundreds of farmers, consumers, and businessmen express their concerns about the nation's food and farm system.

En route to the hotel, I asked our cab driver if he knew what the number-one industry was in Vermont. The young, long-haired cabbie said he thought it was lumbering or slate quarrying. "Don't you know that agriculture is the leading industry in Vermont?" boomed a big, middle-aged man named John York who was riding in the front seat. York, then an executive with the Le-Hi Dairy Cooperative, cited some figures about the number of dairy farms in the state to make his point. "Oh, yeah, agriculture," replied the cabbie, "I forgot about agriculture."

The cabbie's response typifies that of many Americans who neither think of agriculture as an industry nor one of the nation's leading economic activities. Yet it is both. And on the following day, some of the people participating in Bergland's "Structure of Agriculture" hearing would explain why many Americans, like the cabbie, tend to "forget about agriculture."

THE WAY WE PRODUCE OUR FOOD

Sally Taylor was one of about seventy-five people who came before Bob Bergland that day in November 1979 at the Tavern Inn. A local activist in her thirties, Taylor was then working with farmers and consumers in a direct marketing project in Hartford, Connecticut, and what she had to say to Bergland was as important as anything the former U.S. Secretary of Agriculture would here on his entire tour. "The way we produce, process, and distribute our food [puts] the maximum distance between consumers and producers," said Taylor. "And that's physical distance," she explained, "in terms of concentrating production in certain parts of the country, and it's also . . . distance in terms of the numbers of steps in between food as it comes out of the ground, and us, when we eat it."

Sally Taylor was, first of all, talking about the miles separating consumers in New York City from the farmers in the Corn Belt; the literal distance between Boston and the vegetable fields of California. She was also talking about specialization in agriculture; about the fact that corn is produced in one region, wheat in another, and dairy products in another. And she was concerned too about the "distance" inherent in the modern process of food manufacturing, between raw and finished product—between, for example, corn and cornflakes. Here, she lamented the fact that in modern food manufacturing, consumers are left out of the process until it's time to buy the finished product. Consumers can't see this process of food production; they are far removed from it and must take a lot on faith. In short, they must trust the system; trust it for handling, taste, nutrition, additives, and pricing.

Farmers, too, are in the same sort of situation, although at the other end of the process. With the sophisticated manufacturing of modern farm supplies—fertilizers, pesticides, antibiotics, medicated feeds, hybrid seed, and livestock hormones—farmers are as much removed from the content and substance of their farm goods and farming as consumers are from the making of cornflakes and milk. Farmers, like consumers, must trust the system; trust it for the supplies that provide them with their livelihood.

And while we all trust the system for all kinds of goods—from automobiles to television sets—food, and the substances that go into its making, are among the most vital, most perishable, and most sacrosanct of all economic provisions—they are life necessities. There is, therefore, a fundamental reason for being concerned about them—about where they come from, how they are made, and who produces them—in a uniquely different way than we are about most other economic goods.

Today, even as farmers and consumers are "distanced" from what goes into the making of food on one end of the system and what finally comes out of it on the other end, a new increment of distance is about to be added to the food-production process—genetic distance.

FROM THE INSIDE OUT

For thousands of years, man* has fed himself by being an astute observer. He has participated in food production by more or less standing *outside* of the process, watching and reacting to the natural growth of plants and animals. Even during the industrial and scientific revolutions of the last one hundred years or so, man has improved his crop and animal lines by carefully observing and recording how and what they produced, and then breeding the best of them to obtain selected characteristics. Now, however, scientific man is about to make an extraordinary leap in food production power; he is about to begin manufacturing his food from the *inside out*—from within the confines of the cell and the inner sanctum of heredity.

*The word "man" is used here in the generic sense of "all mankind" and is not intended to slight or exclude the achievements of women in the domestication of plants or animals. Indeed, Norman Borlaug, Nobel Prize winner for his work on Green Revolution wheats, has said the Neolithic woman ". . . is responsible for all the bread and cereal products that we eat today. Neolithic woman was the greatest plant biologist in history. She understood that she had to domesticate wild plants and improve them. No one knows how she accomplished these feats. Meanwhile, scientific man has never created an important new grain of any magnitude. We have worked only with what neolithic woman gave us. . . ."

But in this process, man and his new-found genetic power will largely be out of view, made remote and distant by the nature of the technology itself. Biotechnology and genetic engineering will shift the centers of food production from the traditional resources of land, water, and farmers to the remote reaches of the electron microscope and the tiny world of molecular biology. These technologies will present a new and special kind of "distance" in food production, and will consign their economic and political powers to those who understand and use them. But some say this is nothing new.

To be technically correct, "genetic engineering" has been practiced in agriculture for thousands of years. Early hunters making their kills and early farmers selecting certain crops were genetic engineers of a sort, affecting the size and composition of particular plant and animal gene pools.* But today's plant breeder, molecular biologist, and genetic engineer are in a more powerful position, relying less on instinct and serendipity than did their hunting and farming ancestors. They practice their alterations and selections much closer to the living hub of the genetic command. No Neolithic hunter or farmer could move genes from one species to another or propagate thousands of copies of a new microbe in a matter of days. Nor could he photograph a chromosome or recombine bits of the all-powerful molecule called DNA.

ON THE WAY TO DNA

To understand how the new innovations of genetic science will affect crops, farm animals, and food production, one must journey inside plants and animals, into the cell nucleus, and finally to some of the smallest known components of heredity—the chromosome and the gene.

All living things are made up of cells, and all cells, even those without a nucleus, contain chromosomes, the main component of which is the complex molecule known as DNA, or deoxyribonucleic acid. Each chromosome is actually a double-stranded coil of DNA, structured like a twisted zipper, with the two halves of the zipper bound together by a sequence of special chemical "crosslinks" or

*A "gene pool" is the collective of all the genes of a population or species.

base pairs. These pairs—adenine-thymine and cytosine-guanine—comprise the "teeth" of the zipper. One or the other of each pair is located on each strand of DNA, and they always match up the same: adenine with thymine, cytosine with guanine. It is these pairs which "unzip" down the middle of their chemical bonds to replicate DNA prior to cell division.

DNA is the chemical constant in all things made up of cells, from corn to cats, bacteria to elephants, and plant cells to brain cells. Each strand of DNA is segmented into genes—thousands of them—which control specific traits and functions, such as eye color in a man or the protein content in a particular type of wheat. When individual genes or groups of them are identified as controlling one specific function or characteristic, they can sometimes be "snipped out" and spliced into other DNA, thus the terms "gene splicing" and "recombinant DNA."

During the 1970s a whole host of laboratory acrobatics once consigned to the far-off Buck Rogers future began to increase man's ability to alter the genetic composition of plants and animals. These various laboratory techniques, involving both direct and indirect genetic methods are today's new "biotechnologies," and they embrace everything from growing genetically altered bacteria in fermentation tanks to gene splicing in tomato plants. Broadly defined, biotechnology is any practice controlled by man that involves modifying the genetic material in a living thing. And today, the realm of biotechnology extends over virtually all of biology, with its most powerful practice occurring at the cellular, subcellular, and molecular levels.

Our knowledge about the inner world of cells and their genetic powers is owed, in large part, to some prime movers and turning-point discoveries. For example, modern genetics began in the 1860s as a theory of one man, Gregor Mendel, a nineteenth-century monk and contemporary of Charles Darwin who worked with pea plants in a small garden of an Austrian monastery. Mendel theorized the existence of hereditary "elements," later to be known as genes, that carried specific traits from one generation to the next. He was the first to prove, entirely by observation and careful record keeping, that plants inherited their characteristics—such as whether a pea was green or yellow, wrinkled or smooth—from their parents. More importantly, Mendel reasoned that it took two "elements"—one from each parent (although only one would be expressed)—to determine the final outcome of any trait. Much of what Mendel discovered about heredity—published in an obscure journal in 1866—went unnoticed until 1900, when his work was rediscovered. Today, Mendel's theories about heredity have been validated by further scientific proof.

In the last few decades of the nineteenth century, scientists had only the crudest of early microscopes with which to observe cells, but by the late 1870s they began seeing "colored bodies" (because of the dyes they used in experiments) in the nuclei of cells. From these rare sightings of unknown "colored bodies" within the cell came the term "chromosome."

Meanwhile, a few scientists in Switzerland and Germany began to determine that what was in the nuclei of the germ cells of living organisms—the sperm and egg—had something to do with heredity. They noted that the embryo developed after the nuclei of the sperm and egg had fused. Another scientist, Edouard Van Beneden, who was studying parasitic worms in horses, noted that the sperm contributed the same number of chromosomes to the embryo as did the egg. He also discovered meiosis, the cell-division process in which the chromosomes of the parents' germ cells split in half, with only one strand of DNA going into each sperm or egg cell.

In 1903, chromosomes were conclusively linked to hereditary traits by cytologist Walter Sutton, and in 1909 the term "gene" was coined. During the years 1910–15, Thomas Hunt Morgan, working with fruit flies at Columbia University, greatly advanced the gene theory of heredity and the knowledge of how individual traits were governed. By 1933 the "mapping" of genes on chromosomes was in practice, and in 1944, Oswald Avery and a team of scientists at Harvard University proved that genes were part of DNA. By 1947, Edwin Chargaff had measured the amounts of cytosine, guanine, adenine, and thymine in DNA. Then in 1953, James Watson and Francis Crick made their seminal "double-helix" discovery, revealing the two-stranded, zipper-like structure of the DNA molecule. By 1958, scientists had proven that after the chromosome has split down the middle, the DNA molecule duplicates itself by recombining with a complementary second strand.

During the 1960s, the "genetic code" along the DNA chain was beginning to be deciphered, and enzymes were discovered (later to be called "restriction enzymes") that could "cut" DNA segments at specific locations. These enzymes became a key tool for transferring individual genes and larger DNA segments to other segments in gene-splicing experiments. By 1970, with the aid of an electron microscope, DNA was photographed for the first time in all its double-helical splendor. Then came the watershed years for the science that would propel gene splicing.

In 1971, biochemist Paul Berg of Stanford University recombined genetic material from two different kinds of viruses in the first successful recombinant DNA experiment. In 1972, Ananda Chakrabarty, a scientist then working with the General Electric Company, filed an application with the U.S. Patent Office for a laboratory-produced microorganism that would, eight years later, become the subject of a landmark Supreme Court case testing the United States patent laws. And in 1973, biochemist Stanley Cohen of Stanford University and bacteriologist Herbert Boyer of the University of California hit gold; they discovered a direct method of ferrying foreign genes into the DNA of bacteria that would then multiply into millions of identical offspring, carrying and reproducing the inserted genetic trait. The Cohen-Boyer discovery incorporated Boyer's work with restriction enzymes—the surgical "scissors" that enable the cutting of DNA and the splicing of genes—and Cohen's work with plasmids—small cellular pieces of DNA, three-to-four genes in length, that could be used to piggy-back new genetic material into reproducing bacteria. The Cohen-Boyer discovery meant that substances such as interferon and insulin, whose production could be controlled by genes spliced into multiplying bacteria such as *Escherichia coli,* could be reproduced in huge quantities. But then came the cold water.

EARLY CONTROVERSY

By 1974, the advance of the emerging DNA technologies was far enough along to raise the specter of accidents, escaping organisms, and runaway science. In that year, American biologists called for a moratorium on recombinant DNA experiments, and during the next four years debated how to regulate this new technology and calm public fears.

On the heels of some public controversy, the National Institutes of Health (NIH) in 1976 adopted a set of guidelines regulating federally sponsored recombinant DNA work, and during 1977, at least sixteen bills were introduced in Congress to regulate recombinant DNA research. Some scientists feared that a new Washington-based bureaucracy would be created to bridle genetic engineering, and that it would crush the creativity and promise of the new science. However, no legislation was passed, and the more exaggerated public fears began to subside.* NIH began relaxing its guidelines; meanwhile, the genetic sciences marched forward.

In 1978, Harvard researchers using genetic techniques produced rat insulin; Genentech, a small genetic-engineering company based in San Francisco, produced human insulin; Stanford University researchers transplanted a mammalian gene, and scientists at Cornell University implanted a gene that produced the amino acid known as leucine into a yeast cell. In 1979, Genentech produced a human growth hormone, and in 1980 two kinds of interferon. Biogen, a Swiss-based biotechnology company, also produced human interferon.

In 1980, the Nobel Prize for chemistry was awarded to three U.S. scientists for their work with recombinant DNA techniques and genetic decoding. But 1980 was also a banner year for the fledgling genetics industry for other reasons. In June, the U.S. Supreme Court narrowly decided, in the *Diamond v. Chakrabarty* case, that man-made microbes could be patented. Then, in October, Wall Street responded in a frenzy to the first public stock offering by a genetic engineering company—Genentech.

BIOTECHNOLOGIES BOOM

At the time, Genentech was an unknown company to the general public, and even to Wall Street. It was a small firm with a few exciting breakthroughs under its belt, but it had not yet produced a single commercial product. Yet the market went wild: Genentech's stock soared from $35 to $89 a share within minutes of the opening

*These fears were prompted by the location of genetic engineering facilities and the conduct of the new science generally, and fears about escaping, disease-causing organisms. A later generation of concerns would arise in the early 1980s with the release of genetically engineered substances into the environment (see Chapter 12).

offer, and the company raised $55 million in a few frantic hours of trading. Not far behind was Cetus, another California-based biotechnology company, which set a Wall Street record in 1981 for the largest amount of money—$115 million—raised in an initial public stock offering. By March 1981, Genentech's scientist-businessman and co-founder, Herbert Boyer, was on the cover of *Time* magazine, and featured in its cover story entitled, "Shaping Life in the Lab—The Boom in Genetic Engineering." At year's end, more than eighty new biotechnology companies were operating in the United States. Investment capital was moving in new directions.

The Supreme Court's 1980 patenting decision in *Diamond v. Chakrabarty*, followed by Genentech's and Cetus's successful outings on Wall Street, sent unmistakable signals to corporate America. Within months of these pivotal events, several giant pharmaceutical manufacturers—Hoffman-La Roche, Abbott Laboratories, Schering-Plough, Upjohn, and Bristol-Myers—all began recombinant DNA programs for interferon. Du Pont stunned its competitors by announcing in 1981 that it would spend $120 million for research and development in the life sciences, with an emphasis on biotechnology.

While commercial investment and public attention focused initially on genetic engineering's promise for producing plentiful supplies of drugs such as insulin and the miracle cures of interferon, a string of genetic breakthroughs with plants and livestock began to stir interest in agriculture.

In March 1981, Monsanto, a chemical company, and Genentech, jointly announced the development of a new "bovine growth hormone"—a genetically produced substance that promised to increase milk and beef yields by as much as 40 percent, and do so with less feed. On June 21, a genetically engineered vaccine for foot-and-mouth disease was developed by a team of USDA scientists. Nine days later, on June 30, in a first-of-its-kind experiment, a USDA scientist, Dr. John Kemp, and a University of Wisconsin scientist, Dr. Timothy Hall, successfully transferred one gene from a French bean plant to a sunflower cell, calling their new creation the "sunbean." U.S. Secretary of Agriculture John Block heralded the sunbean as a "breakthrough achievement" and said that it ushered in "a whole new era" in plant genetics. "It is the first step," he said, "toward the day when scientists will be able to increase the nutritive value of plants, to make plants resistant to disease and environmental stress, and to make them capable of fixing nitrogen from the air."

Genetic experiments and biotechnology breakthroughs of significance to agriculture continued to occur almost monthly. During 1983 and 1984, for example, cells from different plant and animal species were fused together to create entirely new entities such as the "pomato" (the result of protoplast fusion between potato cells and tomato cells) and the "geep" (an animal created by the fusion of goat and sheep cells). In an experiment with ramifications for the livestock industry, rat genes with a growth hormone "rider" were microinjected into the nuclei of mouse embryos, giving rise to adult mice which grew to twice their normal size. In the plant world, Monsanto scientists (and some German scientists at another location), moved bacterial genes into plant cells. Later these genes "expressed" their designated function in the plants and passed that trait on to succeeding plant generations.

All of these developments continued to improve the prospects for lucrative opportunities in agricultural biotechnology. Some reports speculated that the worldwide market for "agrigenetic" products might be ten times that of genetically manufactured medical and pharmaceutical products, and one report placed the agrigenetics market potential at $50 to $100 billion by the year 2000. Corporate America got the message. By 1982, thirty major American corporations had invested in the new biotechnologies with an eye toward agriculture and/or food production. Two years later, more than one hundred companies—both new biotechnology firms and established corporations—were involved in agricultural genetics research. The race was on.

A NARROWING FIELD OF PLAYERS

Today, biotechnology and genetic engineering are entering an American food and agriculture system that is already highly consolidated and incredibly powerful; where fewer and fewer farmers and corporations control most of what is produced, processed, and exported. Over the last fifty years, the trend throughout America's food and agricultural system—driven by technology and government policy—has been to specialize and consolidate; to produce more with less. The aim has been to bring about rising agricultural productivity and increasing economic efficiency. The result has been increasingly fewer farmers, farm suppliers, and food processors. While some efficiencies have been achieved, they have come with the liability of increasing economic concentration in food production.

In the United States today, approximately 7 percent of all farms control 56 percent of the nation's agricultural production; fifteen agribusiness corporations provide 60 percent of all farm supplies; and about sixty companies handle 70 percent of the nation's food processing. When specific markets or sub-markets are examined, such as cold breakfast cereals or hybrid corn seed, even higher ratios of market concentration are found.

The three largest cereal companies—General Foods, General Mills, and Kellogg—control 80 percent of the ready-to-eat breakfast cereal market. Sunkist, one of the nation's largest agricultural cooperatives, markets between 50 and 75 percent of the nation's oranges and lemons. In the grain trade, six companies—Cargill, Continental, Pillsbury, Bunge & Born, Garnoc, and Louis Dreyfus—account for about 95 percent of American wheat and corn exports; and Cargill, Farmland Industries, and Continental own more than half of the nation's grain elevators.

In the area of farm supplies, International Minerals & Chemicals, Occidental Petroleum, Farmland Industries, and CF Industries top the markets for fertilizers. Monsanto, Eli Lilly, and BASF sell 63 percent of all herbicides used on soybeans. Four companies—Pioneer Hi-Bred International, DeKalb AgResearch, Ciba-Geigy, and Cargill—sell more than 60 percent of all hybrid corn seed.

And some of the world's largest corporations move in and out of the agricultural and food sectors anytime they want to. In June 1981, for example, Occidental Petroleum, a giant energy corporation, spent more than $800 million to acquire Iowa Beef Processors, the nation's leading beef packer. After the acquisition of Iowa Beef, A. Robert Abboud, then president of Occidental, said: "Our strategy for the 1990s is to be prominent in the food area. We're going to be running into a food scarcity situation in the 1990s in the same way that we have an energy shortage in the 1980s. We will continue to build in this area." Occidental, in fact, already owned a seed company (acquired in 1978) when it purchased Iowa Beef, and today holds patents on eleven soybean varieties, some of which are used extensively in states such as Missouri and Kentucky. The company's plant breeders and geneticists are also working on a variety of hybrid rice for the People's Republic of China.

Frito-Lay, a company that uses 3 percent of the nation's potatoes, is now studying the genetics of the potato. Unilever, a British corporation with $21 billion in annual sales, has successfully cloned the oil palm, a plant that provides 15 percent of the world's vegetable oil. The Carnation Company, known for its dairy products, is also in the business of genetically manipulating and selling frozen cattle embryos.* And there are others. Du Pont, Dow, and FMC are all studying what crops and microbes do in the nitrogen cycle, hoping to produce crops that might one day make their own fertilizer. Eli Lilly has put money into photosynthesis, and Campbell and Heinz are now pursuing the genetics of the tomato.

"DISTANCE," CONSOLIDATION, & POWER

What is occurring in the United States today, and what will continue to occur here and throughout the world in the remainder of the twentieth century, is a coupling of genetic power with corporate power. For consumers and farmers this means an even greater physical and technological separation from the sources of food and food production; a greater "distancing" from the economic entities which own the genetic ingredients capable of producing food. More foreboding perhaps—given the drive for international supremacy and the lure of burgeoning world markets—is that governments will move to favor and protect those in commerce using biotechnology. This can be done (more than it is already) through tax breaks, research incentives, patents, export policy, and other areas. Too much of that, however, would run counter to some deeply embedded traditions, at least in the United States.

Since the days of Thomas Jefferson, a long line of philosophers, statesmen, and agrarian activists have told us that it is a good idea—

*In September 1984, Nestlé S.A., of Switzerland, the world's largest food company, acquired Carnation for $3 billion. In addition to Carnation's involvement in livestock genetics, as well as plant genetics, Nestlé subsidiaries are also conducting plant genetics research on soybeans and other crops.

both politically and economically—to keep the growing, processing, and selling of food in as many hands as possible; to keep the family farmer going; to keep the resource base widely owned; and to prevent food production from becoming lodged in the hands of a few entities. When America was a nation of farmers and small businesses, this philosophy generally worked, and it worked within the framework of the marketplace. But times have changed, and so has the business of food production—including the technologies that drive agriculture. Clearly, agrigenetic technologies have the ability to consolidate food-production power far beyond anything we have seen in the past. Already, in terms of the ingredients used to produce crops and livestock, agribusiness corporations, chemical companies, and pharmaceutical businesses account for as much, if not more than, farmers and ranchers in the food-making process. This will only be magnified with commercial bio-engineering.

The ability to influence, if not control, the process of food production—whether through the sale of a new kind of hybrid wheat in South Dakota or the price charged for a box of cereal in Boston—is a power nearly as great as that of government itself. That is, the ability to produce food—especially the ability of a few people to produce food for a great many people—is a power equivalent to any political power. Food is, after all, the stuff of lining up; the stuff of riots when prices rise or there isn't enough to go around.

Throughout many parts of the world, food, politics, and government are intertwined as one. Nowhere is this clearer than in those countries where food is scarce and political upheaval is an ongoing process. In the United States, agricultural abundance has, for the most part, kept the connection between food and politics somewhat at arm's length from the average citizen. But that is bound to change. Food is basic, and in a crowded world, it will become even more basic, with political consequences washing up on every doorstep in every nation.

VULNERABILITY

In the long run, the trend toward centralization in food production—a trend which biotechnology is now facilitating at a rapid rate—will mean greater vulnerability in the food system. More of the technological "pieces" essential to food production—seeds, livestock embryos, microbes, and gene-keyed agrichemicals—will be in fewer and fewer hands, and mostly corporate and government hands. Since neither huge corporate structures nor cumbersome government bureaucracies are known for agility or responsiveness in times of crisis, disruptions in the food systems of tomorrow—either natural or man-made—could get out of hand quickly. Moreover, such disruptions will probably be addressed first and most lastingly for their business and political impacts, not their humanitarian or environmental consequences.

Vulnerability of many kinds, then—agricultural, economic, political—will likely increase as agrigenetic technologies and corporate/government "cooperation" move to center stage in the food production process; making that process increasingly "distant" from the people it feeds as well as farther removed from a simpler set of natural resources that once sustained it. But the new genetic technologies of agriculture could be used to create agricultural and economic diversity, reducing the trends toward centralization and a too-heavy reliance on one technological fix. Unfortunately, however, the historical pattern has been to use agricultural technology to concentrate food-making power rather than diversify it.

In the United States, a first step toward such concentration of power, and one of the first signs of "genetic distancing" in the food system, came with the development and introduction of hybrid crops more than sixty years ago. For that story, it is necessary to turn to the American Midwest in the 1920s.

CHAPTER

3

HYBRID POWER

We hear a great deal these days about atomic energy.
Yet I am convinced that historians will rank the
harnessing of hybrid power as equally significant.
—*Henry Agard Wallace (1955)*

In the summer of 1923, President Harding's Secretary of Agriculture, Henry C. Wallace, gave a speech at the annual DeKalb County Farm Bureau picnic in DeKalb, Illinois. After his speech—and with a few hours to kill while he waited for a train—Wallace and Tom Roberts, then chief farm adviser for the DeKalb County Farm Bureau, retired to Roberts' office with a few of Roberts' associates for some informal conversation. Very shortly, the conversation turned to corn, and Roberts soon retrieved one of his Bureau's prize-winning ears of corn to impress Wallace. Admiring Roberts' corn, Wallace said that back in his Washington office he had some small, twisted ears of corn whose seed would produce a bigger crop yield than any of Roberts' good-looking prize winners. Wallace explained that his corn was a new kind of hybrid that would, in time, increase corn yields above anything previously known in the Corn Belt.

Today, more than sixty years later, some of the descendants of that informal meeting in Tom Roberts' office are now principal players in the world hybrid corn seed business. The Wallace and Roberts families are among the chief owners and executives of two of the world's most prominent hybrid corn seed companies—Pioneer Hi-Bred International of Des Moines, Iowa, and DeKalb

AgResearch of DeKalb, Illinois. The story of the ascent of these two families and their companies to the top of the $1-billion-a-year hybrid corn seed market begins well before 1923, however, and is continuing today as biotechnology begins to change the way corn and other crops are used in agriculture.

FIRST, THE SEED

Seed is the primary ingredient of agriculture. With the exception of a few crops that can be propagated by vegetative means—by runners, cuttings, or grafting—most of the food crops on which the world depends are reproduced by seed. Without seed, there is no agriculture—no crops, no livestock, no meat, no vegetables, no food.

Worldwide, the market for agricultural crop seed is nearly $45 billion. In the United States alone, farmers and ranchers spend nearly $4 billion each year for seed, and use millions of bushels of field, range, and crop seed annually. In 1979, for example, farmers used about 16 million bushels of corn seed, 79 million bushels of soybean seed, and 364 million pounds of cotton seed.

Seeds have become the substance of a business only in the last one hundred years, and selling seed to farmers is an even more recent development. In the past, most farmers grew their own seed year after year, as some still do today. Only with the advent of hybrid corn in the 1930s and 1940s did a "seed industry" begin to take shape. (Yet, to this day, the U.S. Department of Commerce has no Standard Industrial Classification [SIC] number to define, or statistically keep track of, seed companies per se.)

The seed industry began as an informal, "over-the-fence" business created by farmers. The first seedsmen were, most likely, the leading local farmers of their day, bartering seed to their neighbors in exchange for hauling crops to market, blacksmithing, or other services. Some of these farmers began to specialize in the growing, collecting, and bagging of seed, and with time their services became accepted as local businesses. Other companies emerged that "handled" seed much as they did grain and feed. The Cargill Company, for example, one of today's dominant grain corporations, began cleaning and bagging grain for seed in the late 1800s. Still other companies formed to sell garden seed through the mail.

The W. Atlee Burpee Company of Warminster, Pennsylvania, for example, formed in 1876. The Ferry-Morse Seed Company, of Mountain View, California, formed in 1892. And the Asgrow Seed Company of Kalamazoo, Michigan, has been in business since 1865. When the American Seed Trade Association first convened in 1883—then composed of 35 companies—"unjust damage claims, excessive tariffs on seeds, and high postage rates" were the most pressing issues.

As early as 1839, however, the U.S. Congress appropriated money for seed research, using revenue raised from U.S. Patent fees to fund the research. And during the 1840s, Congressmen distributed free packets of vegetable, flower, and farm seed to their constituents—a practice that amounted to the distribution of more than 60 million packets of seed by 1923. With the opening of the western frontier and the passage of the Homestead Act in 1862, demand for seed grew, and so did the need for more supportive public policies. In addition to the creation of USDA in 1862—whose mission in part was to "collect . . . propagate . . . and distribute" seed—Congress also established a network of federally supported land-grant universities and agricultural experiment stations. One of the initial responsibilities of the experiment stations was to "increase," or multiply, available seed varieties for agricultural purposes.*

Working with the agricultural experiment stations, farmers served as the first line of seed growers and distributors, but often seed varieties were mixed, renamed, or contaminated under the farmer-grower arrangement, and little reliable information was available on seed varieties and seed quality. This situation led to the movement for seed breeding and seed inspection in the agricultural experiment stations during the 1900-1915 period, and eventually to the establishment of seed certification programs in the states between 1915 and 1930. Meanwhile, the commercial seed industry was also taking form.

In the 1890s, the nation's loose collection of seed farms and seed companies began to evolve into a modest trading and packaging industry. By then, some companies, such as that formed between

*Seed for any new plant variety is always limited at first, derived usually from the few plants used by the plant breeder. In order to have widespread use of new seed, more of it has to be produced by growing more plants and harvesting the yield for seed—thus the term "increase."

Jesse Northrup and Charles Braslan in Minneapolis, were buying grain seed from farmers who had too much for their own use, and selling it to other farmers who didn't have enough. The Northrup/Braslan sales force, for example, journeyed from farm to farm in top hats and Prince Albert coats making their deals. By the early 1900s, a few seed companies began using a "brand name" approach for merchandising and marketing farm seed.

Yet it was hybrid corn, more than any other single development, that created the American seed industry. "Hybrids made it a business rather than a commodity," says George Jones, now president of the Northrup King Company. Prior to the arrival of hybrid corn in the 1930s, there were forty to forty-five seed companies operating in the United States. Forty years later, there were nearly three hundred new seed companies, most in the business of selling hybrid corn seed. But the emergence of hybrid corn in agriculture was not an overnight affair.

THE HYBRID CORN REVOLUTION

Most of the corn used by American farmers until the 1890s consisted of varieties domesticated by the American Indian. In fact, the earliest domestication of corn dates to its extensive use by North and South American Indians. Corn was the cultural and economic foundation of the Inca, Mayan, and Aztec empires of South and Central America. When Columbus landed in Cuba in 1492, his men found large, continuous Indian cornfields stretching for miles across some Caribbean islands. Indian corn sustained the early colonists and spread throughout Europe and the rest of the world after its discovery in the Americas. By 1880, American farmers, using Indian corn varieties, were raising about 34 bushels of corn for every person in the United States. But little was known about the genetics of this plant that would become the most important crop in American agriculture.

Some of the more visible reproductive features of the corn plant, and how they "worked," began to be understood in 1694 when Rudolf Jakob Camerarius, a Dutch botanist, discovered corn pollen and the fertilization process. In 1716, Cotton Mather, the Salem, Massachusetts, preacher, noted that corn pollen could be carried by wind to other plants effecting cross-fertilization. In 1876, Charles

Darwin, also pondering the corn plant, wrote about the differences between self-fertilization and cross-fertilization in a paper entitled, "Cross and Self Fertilization in the Vegetable Kingdom."

All plants, including corn, are generally pollinated in one of two ways: self-pollination or cross-pollination (also called open pollination). Wheat, cotton, oats, soybeans, and peppers are examples of self-pollinated crops; apples, corn, broccoli, figs, and strawberries are examples of cross-pollinated crops. Pollination is the beginning of reproduction in plants, and involves the mixing of genes from the male and female lines. In the plant world, this mixing of genes begins in the flower, where the anther, or male part of the flower, produces pollen, and the stigma, or female part, contains the ovary. In the microscopic ballet of pollination, it is the job of tiny pollen grains to find their way to the stigma, and eventually to the ovary wherein an egg, or ovule, lies waiting to be fertilized. However, it is the location and operation of the plant's flower that determines how a particular plant is pollinated.

Corn's "flower" is found on two different parts of the plant: the tassel, or male part, is at the very top of the plant, while the silk, or visible female part of the corn flower, is located at the end of each "ear." Each strand of cornsilk—of which there are some 800 to 1,000 per ear—is attached to a site on the corn cob where a single kernel of corn forms. If all goes according to plan, 800 cornsilk strands will make 800 kernels on an ear of corn. Yet no one kernel of corn will form on the cob unless pollen—the "sperm" from the tassel—makes its way to the silk.

Pollen grains, dustlike in appearance and brownish in color, are light, dry, and easily blown about by the slightest wind. Each corn tassel produces about 25 million pollen grains, or something on the order of 25,000 grains for every strand of cornsilk, although only one is needed to pollinate the silk. Upon arrival on the sticky silk, pollen grains become mighty players in the act of corn reproduction. Within minutes of their arrival, each begins to sprout a tiny pollen tube which travels the length of the silk to a rendezvous with the female ovule on the corn cob. When that meeting occurs, a kernel is born. After ripening, that kernel, along with its 800 to 1,000 sibling kernels, becomes an ear of corn.

The corn plant, due to its widely separated flower parts and the mobility of its pollen, is both a cross-pollinated and a self-pollinated species. That is, pollen from one corn plant might fertilize its own silks or those of different kinds of corn plants. But it is the accessibility of the corn flower—its easy-to-get-at male and female parts—that has enabled man to play a very active role in corn breeding, developing special lines of corn by catching pollen in bags surrounding the tassel, and using that pollen to selectively fertilize the silks of other purposely chosen corn varieties. Thus, with the help of man, corn became the first plant species to travel the road of selective inbreeding and cross breeding, leading ultimately to controlled hybridization.

American botanist James Beal, of Michigan State University, was the first to use "detasseling" of corn plants to control pollination and produce a hybrid cross. He recorded yield increases of 50 percent in some of his varieties as early as 1877, but his work went largely unnoticed. In 1896, at the Illinois Agricultural Experiment Station, Cyril G. Hopkins developed an early, pure-line breeding system, which laid the groundwork for using inbred lines in hybrid corn development.

Hybrid corn is the product of both inbreeding and cross breeding. An "inbred line" is a variety of corn that has been crossbred with its own kind for several generations, to the point where certain valued characteristics in that line are "fixed" and retained in subsequent generations. Inbred lines are also called "pure lines." Such inbred lines are used in crosses with other inbred lines to develop first-generation, or single-cross, hybrids. In the late 1890s, Hopkins and his co-workers in Illinois were in the early stages of developing stable inbred lines.

Edward Murray East, a chemist who had studied with Hopkins at Illinois, moved to the Connecticut Agricultural Experiment Station in 1905. Discovering that inbreeding depressed corn yields, East set out to overcome this barrier. He and another scientist, George Harrison Shull, who was then working at the Carnegie Institution's Cold Spring Harbor Experiment Station on Long Island, are generally credited with conducting similar research on corn at about the same time, which led to the important discovery that crosses between inbred lines produced great yields, or what is called "hybrid vigor."

However, at this point in the story, the real problem was seed: the inbred corn seed used to produce the "single-cross" hybrids came from small, runty ears of corn like the ones Henry C. Wallace bragged about to Tom Roberts at the DeKalb County farm picnic. Extracting enough seed from these small, malformed ears was just not practical, and would prove to be very costly and inefficient for anyone who tried. While the yields of the single-cross hybrids were good, the seed source was limited and impractical from a commercial standpoint. Then along came Donald Jones.

Jones, a graduate student working with East in Connecticut, took the single-cross hybrid work of East and Shull one step further. In 1917, Jones took the seed of two single-cross hybrids and crossed the resulting plants again—in effect crossing a hybrid with a hybrid—to produce a "double-cross" hybrid. Jones was the first scientist to try this novel approach. The results were as good in every respect as with the single-cross hybrids. The double cross produced corn yields 20 to 25 percent above those of the existing, open-pollinated varieties used by farmers at the time. But more importantly, the large, normally shaped, and plentifully kerneled ears of the double-cross could then be used as a bountiful source of hybrid corn seed.* From that point on, the hybrid corn revolution was on its way.

PIONEER V. DEKALB; THE ROAD TO HYBRID POWER

As Jones and other scientists were unlocking the secrets of hybridization in corn, some farmers and businessmen were already experimenting with corn breeding, each in their own way. Among them were Henry Agard Wallace of Des Moines, Iowa—the son of Henry C. Wallace—and Tom Roberts, the DeKalb County farm advisor.

Henry Wallace, encouraged by his father, had begun experimenting with corn as a high school student in 1904. As a schoolboy, Wallace had discovered that the seed from runty, inbred ears of

*Double-cross hybrid corn seed is made in the following way: A selected variety of corn is inbred for several generations. This inbred (A) is then crossed or hybridized with another inbred (B). The resulting hybrid (AxB) is the single-cross hybrid seed. This hybrid (AB) is then crossed with another single-cross hybrid (CD) to produce double-cross hybrid corn seed, the article of commerce sold to farmers.

corn yielded a vigorous crop when planted. After he graduated from Iowa State College in 1910, Wallace began writing for *Wallaces' Farmer* magazine, and followed the work of scientists such as Shull and East. Wallace experimented for a time with pure-line inbreeding, but concluded in 1913 that the process was "too laborious." However, he continued his cross breeding work with Iowa corn varieties. After Donald Jones's discovery of double-cross hybridization in Connecticut, Wallace began his experimenting anew, using some of the inbred lines developed by East and others at Connecticut, and crossing these with midwestern corn varieties. By 1924, Wallace had produced the Copper Cross hybrid, the first hybrid corn variety developed for sale in the Iowa Corn Belt. In that first year, Wallace sold all the Copper Cross seed he had—fifteen bushels, sold at a dollar a pound.

As an editor of *Wallaces' Farmer,* the senior Wallace, Henry C., wrote often about the promise of hybrid corn. "No seed company, farmer, or experiment station has any inbred seed or cross of inbred seed for sale today," he wrote during the mid-1920s. "The revolution has not come yet, but I am certain that it will come within ten or fifteen years." By 1926, with the help of two friends and his wife's money, his son, Henry A. Wallace, had established the first hybrid corn seed company in the United States. The new firm was first known as the "Hi-Bred Corn Company," later renamed the "Pioneer Hi-Bred Corn Company."

Meanwhile, in northern Illinois, Tom Roberts and his friend Charlie Gunn, who had listened to Agriculture Secretary Henry C. Wallace extol the bountiful future of hybrid corn in 1923, decided to begin their own hybrid corn program. For five years—until the summer of 1928—the program was kept secret from the DeKalb Agriculture Association while Gunn proceeded with developing inbred lines. The secrecy of their effort was chiefly aimed at beating a potential competitor, William Eckhardt, who had also established a seed business in northern Illinois.

After some difficulty in developing hybrid lines of suitable maturity for use in the northern Corn Belt, and a disastrous first-year seed harvest in 1934, the DeKalb Agriculture Association moved ahead full-steam into the hybrid corn business. In 1935, DeKalb sold 15,000 bushels of hybrid seed, some of it through the mail in response to an advertisement in *Prairie Farmer,* a popular farm magazine of that day. Tom Roberts began thinking of expansion. He quickly realized he could sell much more hybrid seed, and broadened DeKalb's breeding program under Charlie Gunn to develop hybrids for farmers beyond the Illinois Corn Belt. Gunn's hybrids were doing 25 to 35 percent better than anything farmers had in their fields.

In 1936, DeKalb produced 90,000 bushels of seed and sold it throughout the northern sections of Illinois, Iowa, and Indiana. By 1937, nearly 250,000 bushels of seed were produced. Seed-production operations were expanded to Nebraska, Iowa, Indiana, and Minnesota. DeKalb production zoomed to 370,000 bushels in 1938, and 500,000 bushels in 1939. By 1940, more than 4 million acres thoughout the Corn Belt were planted with DeKalb hybrids, and fully 10 percent of all corn produced in America was grown from DeKalb hybrid seed. That year, one hybrid alone, known as "DeKalb 404 A" was planted on 2.5 million acres and was credited with moving the Corn Belt two hundred miles north.

While DeKalb and Pioneer were not the only seed companies to prosper during the 1930s, they were two of the earliest, and the two destined to become the predominant, most influential companies in the industry for the next forty years.

SELLING SEED & BUILDING EMPIRES

A crucial challenge facing both DeKalb and Pioneer in their first years of operation was convincing farmers of the value of a "new kind" of seed. By the late 1930's both companies had begun to use a similar and unique form of marketing called the "farmer-dealer" system. Tom Roberts, Jr., recalls how his father went about building DeKalb's farmer-dealer system:

> Until then [1936], grain stores and elevators stocked seed and waited for the farmers to come and get it. Many of these stores had additional suppliers who were in the alfalfa seed business. We reasoned that these suppliers would eventually come out with competitive hybrids, and, because they could offer a full line, these stores would stock their seed. They would want to make ours available along with theirs on a cigarette-machine

> basis. In other words, just another commodity. But my father, who had been a county farm advisor, had a shrewd idea up his sleeve. He went to the county agent and asked, 'Who are the best farmers in this county?' Then he went out and sold them a bushel of DeKalb hybrid corn seed. In the fall, he went back to ask the farmers how they liked his seed. The farmers were generally enthusiastic. My father asked them about selling the seed to their neighbors, making a 15 percent profit for themselves.

The rest is history. Today, DeKalb's sales and distribution system includes some 7,300 farmer-dealers.

At Pioneer, a similar system had been developed. However, by 1933, Henry Wallace had been asked to serve as President Franklin D. Roosevelt's secretary of Agriculture.* Fortunately for Wallace, a few years before he assumed office in Washington, he met Roswell "Bob" Garst, a farmer and general-store owner from Des Moines. Garst soon began growing and selling seed for Pioneer, and eventually became the architect and driving force behind Pioneer's farmer-dealer system.

"I did not know anything about the genetics of corn and neither did any of the farmers," said Garst years later, "but I did know how to sell corn hybrids by demonstration. . . . I gave each farmer enough to plant about one acre." Describing how farmers eventually opted to use only hybrid seed, Garst explained, "It took about four years to get a farmer to buy enough of our hybrid seed to plant his entire corn acreage. The first year he would just buy a small amount, and he'd have good success with it. He would figure that this was an accident, so the next year he'd buy perhaps a bushel of the new

*The Wallaces, it seems, were always involved in national politics. The patriarch of the clan and founder of *Wallaces' Farmer,* Henry Wallace, also known as "Uncle Henry," was appointed to the Country Life Commission, along with Gifford Pinchot and three others, by President Theodore Roosevelt in 1908. Uncle Henry's son, Henry C. Wallace, continued as editor of *Wallaces' Farmer* and served as President Warren Harding's Secretary of Agriculture. His son, Henry A. Wallace, also an editor of *Wallaces' Farmer,* became Secretary of Agriculture in President Franklin D. Roosevelt's Administration and subsequently served as his Vice-President between 1941 and 1945. In 1948, Henry A. Wallace ran for President as the candidate of the left-wing Progressive Party. Wallace business associates, such as Roswell Garst, often made their own way into the public eye. In the 1950s, Garst, a rural innovator in his own right, met with Nikita Khrushchev in a much-publicized Iowa farm visit.

hybrid seed. It would do well again, but still the farmer thought it was just something that had happened twice in two years. So the next season he would get enough seed to plant half his entire corn acreage. That year he would really see the difference, so the following and fourth season, he would give us his entire order."

Since the days of Roswell Garst and Tom Roberts, the farmer-dealer system has been a huge success in selling hybrid corn seed. In the Midwest, for example, about 66 percent of all corn farmers buy their seed from farmer-dealers. Today, virtually every seed company utilizes the "farmer-dealer" system in some form or another, and the system is being used to sell fertilizer, pesticides, and farm equipment.

Although hybrid corn was first introduced to farmers in 1926, only about 1 percent of the acreage in the Corn Belt was planted to hybrid varieties by 1933. This changed rapidly, however, and by 1944 more than 88 percent of the Corn Belt was planted to hybrid corn. Yields increased dramatically; "corn power" had arrived.

In the race to become the General Motors of hybrid corn, DeKalb took the early lead over all comers, and held that position for nearly forty years. The DeKalb trademark, the winged ear of corn, first appeared in a 1936 *Prairie Farmer* magazine advertisement, invoking farmers to try the new corn hybrids with the message, "Let DeKalb Quality Hybrids Be Your Mortgage Lifter." And try it they did, by the thousands. By 1947, Richard Crabb wrote in his book *The Hybrid-Corn Makers,* "DeKalb has come to represent hybrid corn to more farmers than has any other name."

With hybrid corn, only those who knew the parent lines and breeding sequence knew how to make the high-yielding hybrids—called "closed pedigree" in the business—and this knowledge was legally protected as a trade secret. More importantly from a business standpoint, farmers could not save and reuse hybrid seed the following year and obtain the same yield, since "hybrid vigor" and yield would decline with continuing use of the seed. Farmers had to return to the seed companies to buy new seed each year. They willingly did so because the new hybrids brought higher yields which usually covered what they paid for the seed.

However, as agriculture had moved from wild corn, to domesticated corn, to open-pollinated corn, and finally to hybrid corn, something else was happening. As plant scientist Dr. James G. Horsfall from the University of Connecticut described it: "Corn, being open-pollinated, was a potpourri of genes. There were genes for yellow corn, white corn, red corn, flint corn, weaklings, dwarfs, giants—even genes for corn that grew along the ground like a vine." But the varieties carrying these genes were no longer in use. "By World War II," wrote Horsfall in a 1972 National Academy of Sciences study, "hybrid corn had essentially driven the old corn varieties from American fields." The genetic base of corn—and especially hybrid corn—was narrowing. By 1970, that narrowing would become a national issue not only for corn, but for other hybrid and nonhybrid crops as well. Even Henry Wallace once warned, "Neither corn nor man was meant to be completely uniform." Yet, in the heyday that followed the introduction of hybrid corn, the genetic uniformity of tomorrow was nowhere in the field of view. Hybrid power was the wave of the future.

HYBRIDIZATION SPREADS

Hybrid corn touched off a revolution in agricultural genetics; bringing with it a whole new approach to the science of yield that soon began spreading to livestock as well as other crop species. In fact, Henry Wallace predicted in 1937 that the principle of hybridization would spread from corn to chickens, from chickens to swine, and from swine to cattle. And as early as the 1920s, Wallace himself was involved in breeding hybrid chickens at the same time he was breeding hybrid corn. His son, Henry B. Wallace, later directed Pioneer's poultry breeding program, which first offered hybrid chickens commercially in 1942. DeKalb followed two years later with their poultry breeding program, marketing their "DeKalb Chix" in 1948. In addition to their work in hybrid poultry, Pioneer and DeKalb would also apply hybrid techniques to other livestock species; DeKalb with hogs and Pioneer with cattle.* But hybrid crops became the more lucrative area for both firms.

The late 1950s and early 1960s were the boom years for hybrid

*Pioneer bred poultry for both egg production and broiler production. Pioneer's "Hy-Line" hybrid layers were bred for reduced body weight, superior shell strength, and egg yield and size. The company's broilers were bred for more efficient feed-to-meat conversion, better "hatchability," and superior growth rates. By 1977, more than 1.2 million white-egg- and brown-egg-laying chickens were being pro-

sorghum, a popular new feed grain that was first hybridized in Texas in the 1940s. DeKalb initiated a sorghum research program in 1949, and marketed the world's first hybrid sorghum seed in 1956. Farmers adopted hybrid sorghum more quickly than they did hybrid corn, and by 1960, 70 percent of all sorghum acreage in the United States was planted to hybrid varieties. Pioneer introduced its sorghum lines in the early 1960s, and by 1970, the two firms were dominating both the hybrid corn and hybrid sorghum markets. By this time Pioneer had also begun a hybrid alfalfa program.

Wheat was an obvious candidate for hybridization too, but the nature of the self-pollinating wheat flower made controlled cross breeding in this crop considerably more difficult than in corn or sorghum. Nevertheless, DeKalb began a hybrid wheat program in 1961. Pioneer followed with their program in 1969. By 1972, DeKalb was spending $900,000 a year on its wheat program, but without much success in producing high-yield varieties. Ten years later, after sinking some $25 million and twenty years of research into the program, DeKalb decided to pull out of the hybrid wheat race. In February 1982, it sold the program to Monsanto, the nation's fourth largest chemical company.

Pioneer and DeKalb also moved into other areas of hybrid crop development. Pioneer began work on hybrid cotton in 1975 and DeKalb initiated a hybrid sunflower breeding program in 1979.*

duced annually, accounting for roughly $25 million in net sales. Pioneer sold layer poultry in seven midwestern and three western states. Broiler sales, however, had peaked in 1970 at $2 million, sliding to $845,000 by 1977. Pioneer also engaged in contract egg production with more than two hundred farmers in the early 1970s.

DeKalb was also involved in the hybrid poultry business. By the late 1970s, it held a 30 percent share of the domestic layer market, and its brown- and white-egg-layers were sold in fifty countries. DeKalb began a swine breeding research program in 1970, followed by the acquisition of the Lubbuck Swine Breeders in 1972. Five years later, DeKalb was marketing hybrid boars and gilts.

Between 1966 and 1976, Pioneer engaged in the breeding and sale of purebred and cross-bred cattle, focusing on five major breeds—Angus, Charolais, Red Angus, Hereford, and Polled Hereford—plus some cross-breeding with European breeds such as Simmental and Maine Anjou. Selling 1,000 to 1,500 bulls a year, as well as purebred heifers, Pioneer's breeding program focused on incorporating traits of commercial value such as high weaning weight, daily weight gain as a one-year-old, and desirable carcass quality. Pioneer sold its cattle through its seed agents and salesmen.

*Pioneer also became interested in breeding soybeans, but not from the standpoint of hybridization. In 1973, Pioneer saw the expanding use of soybeans in the United States—a dramatic fourfold increase since 1950—coming at the expense

By this time, these two American companies, as well as others, were selling hybrid seed all over the world.

THE HYBRID LEGACY

More than sixty years after President Warren G. Harding's Secretary of Agriculture, Henry C. Wallace, had told Tom Roberts and his friends about a revolutionary new kind of corn, hybridization had established itself as the premier technique of agricultural genetics, not only in crop breeding but in livestock genetics as well. In the wake of hybrid corn, a new commercial industry was born, hundreds of new seed companies formed, and American agriculture boomed. A significant scientific step had been taken which allowed for the privatization of some of the key and most important genetic ingredients in food production. Yet other things were changing, too.

While hybrid corn was being developed in the 1930s, a new legal system of intellectual property governing the ownership and sale of plants (and later, seed) was developing in the United States and around the world. Before long, these property rights—called patents—would be extended to plant and animal genes and biotechnology processes, and would help determine what kinds of agricultural products would be developed, and who would own them.

of corn acreage, and decided to jump into soybean research and nonhybrid seed improvement. (Part of the motivation for Pioneer's and other seed companies interest in non-hybrid seed improvement came on the heels of "seed patenting" legislation passed by Congress in 1970.) Soybean acreage was then running at about 80 percent of corn acreage. But soybean planting rates were roughly four times those of corn. R. Wayne Skidmore, then Pioneer chairman, explained to his stockholders that "the demand for both [soybean] oil and meal is expected to climb. Becoming a supplier of soybean seed could help stabilize our seed sales volume." By 1982, Pioneer was selling $22 million worth of soybean seed.

CHAPTER

4

DESIGNS ON NATURE

Nothing that Congress could do to help farming would be of greater value and permanence than to give to the plant breeder the same status as the mechanical and chemical inventor now have through the patent law.

—*Thomas A. Edison, 1930*

The lame-duck senator from Alabama was talking on the telephone as he stood behind his high-backed, padded chair in an office being emptied of personal effects. It was December 1980; Ronald Reagan and the Republicans had just captured the White House and the Senate in the November elections, and Alabama's junior senator, Democrat Donald Stewart, was one of the casualties. As Congress groaned through a pile of bills, Senator Stewart was tending to some last-minute business. He was talking on the telephone to an official from the Department of Agriculture.

"Corporate concentration doesn't happen in one big fell swoop," he explained, "it happens in bits and pieces, a little at a time; a little bit here in this sector, a little more over there in that sector. One day you wake up, and the whole damn thing is swallowed up by a few corporations." Yes, he said, answering the voice on the other end of the line, he knew that corporate concentration was taking place everywhere in the economy, but he wasn't dealing with the whole economy just now. He said he thought he might be able to do something about the problem in just one little part of the economy—the part dealing with seeds.

The senator from Alabama was spending one of his last few hours in the Senate this way because the seed industry and the Department of Agriculture wanted a "little bill" released from his agriculture subcommittee. As chairman, once he gave the word, the bill could be brought to the Senate floor and passed in the blink of an eye. No one would know the difference; bills were flying everywhere.

The "little" bill in question was an amendment to a 1970 "plant patenting" law, called the Plant Variety Protection Act (PVPA). The PVPA awards eighteen-year "certificates of protection" (essentially patents, although there are technical differences between U.S. Patent Office patents and PVPA certificates) to those who develop new varieties of seed-bred plants such as wheat, barley, and soybeans. The purpose of the law, as with most patent law, is to encourage innovation and investment in research—in this case plant breeding and the development of new plant varieties. By giving the "inventors"—the plant breeders—a limited monopoly on their new invention, the PVPA is said to spur innovation and the development of new varieties.

When the original PVPA was passed by Congress in 1970, a few canning interests, including the Campbell Soup Company, objected to the bill, and six "soup vegetables"—tomatoes, peppers, okra, celery, cucumbers, and carrots—were exempted. By 1980, however, the objectors had dropped their concerns, and the seed industry was pushing hard to finish the list of eligible crops. But in the seed industry, things had changed a great deal in the intervening decade. Senator Stewart, for one, was asking some hard questions about the whole situation.

"We're going to kill this bill," Stewart said into the phone with some devilish delight. "That's right," he repeated for the disbelieving USDA official on the other end of the line, "kill it." Stewart was concerned about the number of large corporations that were buying up small, independent seed companies: at least sixty such companies had by then been acquired by multinational corporations, which through their newly acquired subsidiaries were getting the lion's share of seed patents. Might there be a connection between the multinationals' newfound interest in the seed business and the market protection afforded by an eighteen-year patent?

There were questions, too, about rising seed prices for farmers—up 150 percent between 1972 and 1977—and the worldwide corporate marketing of new high-yielding crop varieties, some of which required more pesticides and fertilizers. There was also the matter of the International Union for the Protection of New Varieties of Plants (UPOV), an international organization headquartered in Geneva for the ostensible purpose of coordinating plant-patenting laws worldwide, but which some critics believed was a developing seed cartel. Other questions concerned the genetic vulnerability of new crops used in modern agriculture, and what selective inbreeding and hybridization might be doing to the nation's crop base. All these questions and more led to a mini-controversy in one little corner of the U.S. Congress—a controversy conducted, for the most part, out of the public eye.

But in that lame-duck session of Congress, the screws were turning. Lobbyists from the seed industry were calling the senator three times a day. Plant scientists back home at Auburn University were calling him to say that they were working on some new varieties of okra and bell peppers, and the right to patent them would sure be helpful. The canning people who put money into the university would look favorably upon this, the callers said.

Republican Senator Jesse Helms of North Carolina, flexing his new muscle as the incoming chairman of the Senate Agriculture Committee, was making some noises about the bill, and the outgoing chairman, Democratic Senator Herman Talmadge of Georgia, was hearing a lot about "this little bill." Under a suspension of the rules, the House of Representatives had hastily moved its version of the bill to final passage a few weeks earlier, so the senator from Alabama was the only remaining hurdle.

Finally, lame-duck politics prevailed. The senator from Alabama "came around," as they say, and H.R. 999, "a bill to amend the Plant Variety Protection Act of 1970," was soon on its way to President Jimmy Carter, who would sign it into law.

The story of the "little bill" amending the Plant Variety Protection Act of 1970 is but a small part of a much larger "bio-patenting" story; a story that has since been broadened by newer kinds of genetically engineered and biologically altered substances and the laws that protect them as private property.

The tale of how this body of patent law and politics came about, and how it continues to be changed to this very day, goes a long way in illustrating how commercial interests have staked out, protected, and perpetuated private ownership of some of the most crucial natural resources available to mankind: food-producing resources governed by genes. The story begins in Washington, D.C., more than fifty years ago.

IF FOR LIGHT BULBS, WHY NOT PLANTS?

In 1930, Herbert Hoover was President of the United States, and the U.S. Congress, under Republican control, was in its second session. On April 14, for a few fleeting moments, the debate on the Senate floor turned to a bill numbered S. 4015, a bill "to provide for plant patents." Senator Clarence Dill, a Democrat from the State of Washington and a member of the Senate Committee on Patents, rose to speak: "I want to say just a few words about this bill," said Dill. "The bill proposes to extend the right to secure a patent to those who invent or develop new plants by what we would call grafting." But, he continued, "the experience we have had with the monopolization of patents and the granting of patents for inventions raises grave doubt as to the wisdom of granting patents on new kinds of plants of a food-producing nature. On the other hand, the nurserymen and the various people engaged in the development of plant and food products are very anxious to have this bill passed."

After Dill had spoken, Democratic Senator Thaddeus Caraway of Arkansas, a member of the Senate Agriculture and Forestry Committee, joined him in a coloquy:

> MR. CARAWAY: What is this thing they are going to patent?
>
> MR. DILL: This is simply an amendment to the patent laws, and the senator will find by reading the bill that it provides for the securing of patents on new plants that are asexually reproduced, according to the language.
>
> MR. CARAWAY: I am curious to know. One is evolved gradually from the other?
>
> MR. DILL: Yes.
>
> MR. CARAWAY: At what stage do they fix their absolute right so that nobody else can further produce [it] or benefit?
>
> MR. DILL: That would be decided by the Patent Office. I suppose, when the Patent Office determined that they had pro-

> duced a plant sufficiently different. I have very grave doubts about the constitutionaltiy of the provision.
>
> MR. CARAWAY: The practicability of it is questionable. When are we going to lay our hand on nature and say, "You can go only this way and that way?" How are we going to control it? Are we going to say to everybody, "You can not take this plant and further improve on it"?
>
> MR. DILL: How can we say, "For seventeen years this plant is a product under control of the patentee"?
>
> MR. CARAWAY: Nobody may further improve it or touch it?
>
> MR. DILL: I will not say they would not improve it. They could not produce it without the consent of the man who developed it. . . . May I say that my reason for speaking was simply that the Senate might understand the remarkable kind of legislation it is. I do not want alone to take the responsibility for the passage of the bill, for it may be that my doubts are not justified; but I have felt that it is such a departure from anything we have ever done in the Senate that senators ought to realize what kind of legislation it is.

The bill Caraway and Dill were discussing on the Senate floor proposed to give the owners, discoverers, and developers of a certain class of plants—those that could be reproduced by vegetative grafting, budding, and cuttings—a patent or exclusive ownership right for a seventeen-year period. It was the first serious attempt by the Congress to extend the franchise of patenting to plants of any kind.* The rationale for doing so was that plant breeders and horticulturalists were just as much inventors as those who created toasters or light bulbs. Their "inventions," it was reasoned, were just as important to society as any other, and so should be protected and encouraged in the same manner. Additionally, horticulturalists could argue, as they did, that their services were part of agriculture and food production.

*In 1906, Congressman Allen of Maine introduced a bill authorizing the Commissioner of Patents to register and allow the exclusive use of new plants for twenty years under the Trade Mark Law, and in 1908 Congressman Clarke of Missouri proposed a bill to amend the patent laws for horticultural products, but neither of these bills was ever seriously considered.

Prior to 1930, patents for plants were not available from the U.S. Patent Office, principally because plants were believed to be part of nature and, therefore, not eligible for patenting.

A long-standing 1889 decision of the U.S. Patent Office, denying a patent on a fiber found in a pine-tree needle, was generally regarded as the legal precedent concerning the patenting of "products of nature." That decision found that allowing patents "upon the trees of the forests and plants of the earth . . . would be unreasonable and impossible." Yet in 1930, Congress thought otherwise, and found that the work of the plant breeder, "in aid of nature," was patentable.

Drafted as an amendment to the U.S. Patent Laws, the 1930 Act was originally introduced in Congress by Republican Senator John G. Townsend of Delaware and Republican Representative Fred S. Purnell of Indiana.

In some ways, the Townsend-Purnell Plant Patent Act of 1930 was a bill aimed at helping a few rural business interests of that day, especially, the orchard, flower, and nursery businesses. Senator Townsend himself owned 130,000 acres of orchards while shepherding the Plant Patent Bill through Congress. Operating in the shadow of the November 1929 stock market crash and a worsening economy, the Hoover Administration was also eager to be perceived as "helping agriculture." Hoover's Secretary of Commerce, R. P. Lamont, wrote Representative Albert H. Vestal, Republican chairman of the House Committee on Patents, to say that his department was "very much in favor of any bill that will put agriculture on the same basis as industry with respect to patents." Hoover's secretary of agriculture and his patent commission also spoke on behalf of the bill.

Writing in 1934, one observer, Robert Starr Allyn, would note, "The passage of the Plant Patent Act in 1930 was a timely gesture of sympathy to the farmer and plant breeder following closely on the heels of the decline of prices in 1929. . . ." Yet there were a few Congressmen who saw it differently. Representative Fiorello LaGuardia, Republican-Progressive of New York who later became the mayor of New York City, remarked sarcastically, jabbing the Hoover Administration and Congressional colleagues, that the plant patent bill "does not provide a patent or copyright for anyone who devises a real farm-relief program."

Despite the pointed objections to the bill of senators such as Dill, Caraway, and a few others, it was passed in the Senate on May 12, 1930. One day later, Congressman Vestal, chairman of the House Patent Committee, moved to take up the Senate-passed bill in the House. Under a unanimous consent agreement, with no opportunity for amendment or debate, the bill was passed on a voice vote. Ten days later, the bill was signed into law by President Hoover.

The brevity of the Congressional debate on the Plant Patent Act of 1930 reveals that few senators and congressmen who ventured opinions on the subject really knew what the bill was all about. The speed with which Congress acted on the measure prompted Robert Starr Allyn to write in *The First Plant Patents:* "The history of the Act shows that it was thrown together hastily and practically no thought given to many of the basic problems involved."

PAUL STARK, LUTHER BURBANK, & THOMAS EDISON

Three men whose names figured prominently in the fight to win passage of the 1930 Plant Patent Act were Paul Stark, Luther Burbank, and Thomas Edison. Paul Stark was a nurseryman-turned-lobbyist from Stark Brothers Nurseries of Louisiana, Missouri, who headed the plant patent lobby effort for the American Association of Nurserymen. Luther Burbank was the famous California plant breeder who had worked for Stark Brothers Nurseries between 1893 and the 1920s. And Thomas Edison was the famous inventor who befriended both Stark and Burbank and supported the fight to secure a plant patent law.

Paul Stark served as the point man for nursery and horticultural interests in Washington, generally acting as the liaison with congressmen and senators, gathering supporting materials, and generating mail and telegrams for use in the debate.*

Although Paul Stark was the recognized leader of the horticulturalists' effort to obtain a plant patent bill, it was the names of

*Stark and his network of nurserymen soon became adept lobbyists. When New York Democratic Senator Royal S. Copeland threatened to slow consideration of the bill on April 14, Stark and his network turned him around. "I have had protests from people in my state about the bill," said Copeland on the Senate floor, "and wish to study them before I consent to the passage of the bill." With Copeland's objection, the bill was passed over. During the next few days, Copeland's

Luther Burbank and Thomas Edison that really made the difference in moving the legislation.

Luther Burbank was one of America's most famous plant breeders. During his lifetime, which spanned the period 1848 to 1926, Burbank was responsible for introducing or improving some 800 varieties of trees, vegetables, fruits and flowers, several bearing the Burbank name. For example, the Russet Burbank potato, developed in 1872, is grown on more acres in the United States today than any other potato variety. Yet none of Burbank's "inventions" were ever patented during his lifetime, and his dismay over that fact left its mark on others around him, including his friend Thomas Edison and nurseryman Paul Stark.

Stark Brothers Nurseries, a well-established business in the 1920s, profited handsomely from a long-time association with Luther Burbank. In 1893, Paul Stark's father, Clarence, traveled to Burbank's plant-breeding laboratory in Santa Rosa, California, and estalished a remunerative working relationship between Burbank and Stark Nurseries. Many of Burbank's new plant varieties, developed in the years following that meeting were marketed by Stark Brothers Nurseries.*

Shortly before his death in 1926, Burbank wrote a letter to Paul

office was bombarded with telegrams from New York area horticulturalists and nurserymen urging him to support the Plant Patent Bill. By April 17, when the plant patent bill was brought up again, Copeland had changed his tune: "I have come to the belief that this is a very worthy measure," said Copeland, noting the "voluminous" correspondence he had received on the matter. After that, Copeland couldn't say enough about the good of Stark Brothers and the pending bill. "The Stark Delicious apple to my mind," said Copeland on the Senate floor, "is one of the most delicious apples I have ever tasted. It is well named. It required years of effort in Missouri before the apple was developed. They now sell it under a bond that no one who has the trees may sell or give away any of the grafts. I have 100 of them on my farm. I think it is such a remarkable product that I feel extremely thankful that somebody had the energy to work out the development of the plant. . . . For my part," he said, "I am very happy to join in support of the bill."

*In fact, Burbank deeded his research and horticultural estate to Stark Brothers. Dickson Terry, in *The Stark Story,* describes what Burbank left to Stark Nurseries:

> The horticultural legacy passed by Luther Burbank to Stark Brothers was indeed a great one. It included a great number of varieties, among them exciting new kinds of fruits and flowers which the great experimenter developed and had never marketed; 120 types of plums, 18 varieties of peaches, 28 varieties of apples, 500 hybrid roses, 30 cherries, 34 pears, 52 gladioli, and many others.

Stark outlining his views on the need for plant patents: "A man can patent a mouse trap or copyright a nasty song, but if he gives to the world a new fruit that will add millions to the value of earth's annual harvests he will be fortunate if he is rewarded by so much as having his name connected with the result. Though the surface of plant experimentation has thus far been only scratched and there is so much immeasurably important work waiting to be done in this line I would hesitate to advise a young man, no matter how gifted or devoted, to adopt plant breeding as a life work until America takes some action to protect his unquestioned rights to some benefit from his achievements."*

Stark used this Burbank statement in the Congressional debate on the patent bill, giving it to congressmen and senators he worked with. Burbank's widow, Elizabeth, submitted it to the House Committee on Patents in a telegram. The Burbank name and legacy were also used during the short House debate of May 5, 1930, particularly to quell the objections of New York's Democratic Congressman Fiorello LaGuardia. LaGuardia agreed that Burbank was "the outstanding American of his time," but he did not relent in his opposition to the bill.

During the 1930 debate, Thomas Edison, who had also dabbled in plant breeding, sent his own telegram to Congress urging support

Stark Brothers then, was greatly indebted to Burbank. Receiving these Burbank materials before the patent law was enacted, in fact, the nursery could then market and further develop them. Subsequently, under the Plant Patent Act, nine "Burbank patents"—five plum, two rose, one cherry, and one peach—were issued to Stark Brothers Nurseries. As of 1975, Stark still maintained the Burbank trade name and was marketing several Burbank varieties.

*In 1911, Burbank himself had expressed a somewhat different view on patenting:

> No patent can be obtained on any improvement of plants, and for one I am glad that it is so. The reward is in the joy of having done good work, and the impotent envy and jealousy of those who know nothing of the labor and sacrifices necessary, and who are by nature and cultivation, kickers rather than lifters.
>
> Happening however to be endowed with a fair business capacity I have so far never been stranded as have most others who have attempted similar work, even on an almost infinitely smaller scale.

But Burbank had obviously changed his mind about patenting by the 1920s. And it was this latter Burbank view—which favored patenting—that prevailed on Congress and Burbank's friends Edison and Stark.

of patenting rights for plant breeders. Congressman Purnell read Edison's telegram during the House floor debate and at the hearings held by the House Committee on Patents. Edison's telegram said, in part: "Nothing that Congress could do to help farming would be of greater value and permanence than to give to the plant breeder the same status as the mechanical and chemical inventors now have through the patent law . . ."

After the Plant Patent Act was signed into law, Edison told *The New York Times:* "Luther Burbank would have been a rich man if he had been protected by such a patent bill. As a rule the plant breeder is a poor man, with no opportunity for material reward. Now he has a grubstake." Yet, the real "grubstake" went eventually to the nursery and flower businesses that obtained the patents.

By 1940, it was clear that some of the largest nursery and flower businesses of that day—businesses that had worked the hardest for the Act, such as Stark Brothers Nurseries, Jackson & Perkins rose breeders of New York, and Hill Brothers nurseries and rose breeders of Richmond, Indiana—had gathered a good share of the patents issued under the new act.

Today, nearly 5,000 patents have been granted for plants such as roses, walnuts, apple trees, lilacs, African violets, mints, and sugar cane. Roughly 70 percent of these patents have been issued for roses and other flowering plants and shrubs; 20 percent for fruits such as apples, pears, avocados, and strawberries; and the rest for plants such as poplar trees, evergreens, pecan trees, lawn grasses, and hops.

Although the holders of these patents range from the Brooklyn Botanical Gardens to a subsidiary of the Superior Oil Company, in some individual categories of food plants, a few companies have been the largest recipients of patents. For example, Stark Brothers Nurseries alone accounts for nearly 30 percent of all apple patents; 15 percent of all plum patents; 11 percent of all pecan patents; and 10 percent of all peach patents. Armstrong Nurseries of Ontario, California holds nearly 10 percent of all peach, nectarine, and raspberry patents; Driscoll Strawberry Associates of Watsonville, California, holds 30 percent of all strawberry patents; and the U.S. Sugar Corporation holds seven of the ten patents that have been issued for sugar cane.

SEEDS LEFT OUT

While it may have opened the door to biological patenting, the 1930 Plant Patent Act did not extend patenting to hundreds of agriculturally important plant species such as wheat, barley, and tomatoes. Such seed-propagated plants were once proposed for inclusion in the Plant Patent Act, but the U.S. Patent Office, the Department of Agriculture, and the presiding congressional committees at the time opposed that action. Seed-propagated plants were regarded by biologists as inherently unstable, meaning that they were prone to natural variation and could change over time due to open-pollination. They could not always be reproduced exactly from one generation to the next. Moreover, patenting such naturally changing substances was then regarded as "useless," since a patent on such a changing property could not be enforced. In order to lay legal claim to a substance with a patent and be able to defend it in court from infringement, the substance must be distinct and stable. Seed-bred crops, in this sense, could not hold a patent.

But in the 1930s and 1940s hybrid corn was introduced. While hybrids are reproduced by seed, only the developer of the parent lines used in the cross-breeding knows the correct combination of parents needed to produce the seed. Hybrids, then, have a kind of "built in" legal protection; they do not need to be patented. Nevertheless, the commercial success of hybrid corn, and the emergence of a burgeoning seed industry that came with it, pointed up the business potential of improving other seed crops—particularly those that were difficult to hybridize. But from the seed industry's point of view, there was still one item lacking: a patenting law for seed-propagated plants.

JOHNSON COMMISSION TAKES AIM AT PLANT PATENTS

After the passage of the 1930 Plant Patent Act, the issue of further plant and seed patenting lay dormant until November 1966, when President Lyndon Johnson's Commission on Patent Law Reform reported its recommendations to Congress. One of the Commission's findings stunned both the seed and horticultural communities. While it acknowledged the valuable contribution of plant and seed breeders, Johnson's commission said it did not "consider the patent system the proper vehicle for the protection of such subject matter [plants or seeds], regardless of whether the plants reproduce sexually or asexually." The Commission later recommended repeal of the 1930 Plant Patent Act and urged further study to determine the most appropriate means of legal protection for the products of plant breeding. These recommendations, however, came at a time when the American seed industry was moving steadily toward patent protection for seed-propagated plants.*

During the 1950s, some American seedsmen conducting business abroad, had become involved with European seedsmen who were advocating "plant breeders' rights" and plant patent laws. One of the most influential American seedsmen to grasp the importance of the Europeans' push for seed patents was Allenby White, vice president for research and market development with the Minneapolis-based Northrup King Company.

Allenby White's "dream," according to one former USDA official who worked with him, was to achieve a high degree of cooperation and reciprocity in the international seed trade so that countries developing new varieties of plants would be eligible for each other's patents, facilitating the worldwide marketing of popular crop varieties. In 1961, the first step toward such international coordination was taken in Paris. Seed officials from twelve nations gathered at the "Convention of Paris" to discuss the seed trade and "breeders' rights" laws. That group would later be formalized into the International Union for the Protection of New Varieties of Plants (UPOV). It was White's hope then, and later that of the entire industry, to use the UPOV, and American involvement in UPOV, as the means for achieving a booming international seed trade. The first step along that path for the Americans, however, was the necessary adoption of a seed-patent law by the United States.

At the Paris convention of 1961, some Northrup King Company

*In the Johnson Commission's report, it was stated: "The Commission believes strongly that all inventions should meet the statutory provisions for novelty, utility and unobviousness and that the above subject matter (which included plants) cannot readily be examined for adherence to these criteria." The concern, it seemed, had to do with the changing biological nature of seed and plants, the arbitrary nature of granting patents on such materials, and the difficulty in enforcing such patents.

representatives, including Allenby White and Kenneth Christensen, and a few USDA officials, began planning a campaign to push for such a law. With White's help, and after much internal debate on strategy and what system of patent protection to pursue, the American Seed Trade Association (ASTA) settled on a bill that would bring seed-propagated plants under the 1930 Plant Patent Act.

When the Senate Judiciary Committee embarked on a series of lengthy hearings to address the Johnson Commission Report and proposed reform of the nation's patent system, ASTA seized the opportunity to push its seed patent bill. In February 1968, before Senator John McClellan's patent subcommittee, ASTA offered an amendment which simply added the words "or sexually" to the plant patent section of the 1930 Act.

Horticulturalists, protected by the Plant Patent Act for over 35 years, immediately viewed the seed industry's move as a threat to their interests. The Johnson Administration, led by the Departments of Commerce and Agriculture, also rejected the seed industry's attempt to move into the patent system. The Senate Judiciary Committee didn't like the idea either. And even the Farm Bureau opposed the seed industry's patent bill, lining up with horticulturalists and federal officials.

The Department of Agriculture, led by Secretary Orville Freeman, was particularly firm in opposing seed patents. In a 1968 letter to Senator McClellan, chairman of the Judiciary Committee, Freeman expanded on some of the same concerns raised against seed patenting in the 1930s: "Great difficulty is encountered in keeping a seed-propagated variety true to its original characteristics. Many varieties of crop plants exhibit a change in genetic composition from year to year, so that a variety, in a few years, would no longer even fit the description of the basis on which it was patented."

The seed industry, for its part, had the support of the National Cotton Council and the major cotton-seed companies. Those interests realized that money and profits were being lost to farmers who simply reproduced popular cotton varieties, selling the seed to their neighbors. This loss of revenue was unfair, the cotton breeders argued, especially since they shouldered the costs of developing the new varieties.

In addition to the cotton industry, a few other business interests also backed the pro-patent position of the seed industry. In a letter to the McClellan Committee, Milan D. Smith of the National Canners Association stated that "the vegetable canning industry relies on the seed industry for its most basic and vital ingredient, and therefore has an intimate concern for the economic welfare of that industry." But Smith also underscored his industry's need for new vegetable varieties that could withstand the rigors of machine harvesting: "Rapid advancements in mechanical harvesting and handling . . . have imposed certain restrictions in production on the grower and processor," he wrote, explaining that while "most machines are designed to make a single destructive harvest . . . the varieties available in several crops were created for multiple harvest by hand and do not yield heavily at any one time." What the canning industry needed, said Smith, was "a wide spectrum of new creations and greater effort devoted to vegetable improvement than the seed industry is willing or able to provide. Without legal protection for new varieties and [the] recovery of research and development costs, commercial breeders obviously cannot justify expansion to the point where they can meet our rapidly changing needs for new varieties. . . ." Remarks such as these foreshadowed developing ties between food processors and genetic suppliers, revealing even then how the genetic changes made by breeders in new crop varieties were very directly related to the needs of mechanical harvesting and modern food processing.

Yet despite the assistance of powerful allies such as the cotton and canning interests, the American seed industry did not have enough influence to overcome the objections of the Johnson Administration, horticulturalists, and scientific bodies such as the Crop Science Society of America. Seed industry officials would later concede that they had made a procedural mistake in attempting to add seed-patents to the Plant Patent Act and the general patent laws.

Dale Porter of Pioneer Hi-Bred International, summed up the seed industry's 1968 failed attempt this way: "Initially they were naive enough to think they could develop a system that would . . . merely expand the Plant Patent Act." Instead, said Porter, "they found immediate opposition to this from many sectors, and it proved politically unavailable to them. They then started looking for a different system that would be politically available."

That new system became known as "plant variety protection," a patent-like system that would be administered by the Department of Agriculture rather than the Department of Commerce. According to Porter, the seed industry regarded this alternative approach as "better than nothing."

A CHANGE IN POLITICAL FORTUNE

Since the seed industry was now willing to pursue a "patent-like" system administered by the USDA rather than the U.S. Patent Office, the legislative course for their bill in Congress also changed. The jurisdiction shifted from the Judiciary Committee to the Agriculture Committee, where a friendlier reception meant easier approval. Further, with the election of Richard Nixon in 1968, and the appointment of Clifford Hardin as Secretary of Agriculture, the political fortunes of the seed industry also changed dramatically. At the Department of Agriculture, the new assistant secretary for marketing was Richard Lyng, a friend of the seed industry.

Lyng came to the Nixon Administration from California, where he was once state director of agriculture, and also ran a family-owned pea- and bean-seed business in Modesto. Shortly before joining the Nixon Administration, Lyng sold his seed business to the Northrup King Company. Although he never played a publicly visible role in moving the seed bills in Congress, and tried to keep an arm's length from seed-policy matters at the USDA, Lyng did become involved in bureaucratic infighting when the Bureau of the Budget (now the Office of Management and Budget) threatened to veto the bill.

The first draft of the seed-patent bill, officially named the Plant Variety Protection Act (PVPA) was the product of a March 1969 meeting between seed-industry representatives and a small group of USDA officials.* In Congress, the bill was introduced in late 1969 by Democratic Representative Graham Purcell of Texas and Republican Senator Jack Miller of Iowa.

By June 1970, back-to-back hearings on the bill were held in the House and Senate agriculture committees, with little opposition in

*Representatives of the American Seed Trade Association (ASTA), the National Council of Commercial Plant Breeders, the Association of Seed Control Officials, the Association of Seed Certifying Agencies, and some state agricultural experiment station directors met with officials from the USDA research and marketing divisions to fashion the bill that would be introduced in Congress.

sight.* Many of the same organizations whose representatives testified in favor of the seed-patent law at the McClellan hearings in 1968 now urged adoption of a USDA-administered seed-patent law.

No consumer or general farm organization† appeared before either the House or Senate committees considering the Plant Variety Protection Act, and most of the objections raised about the bill were dismissed by the two committees, or regarded as minor and insignificant. However, the committees did move to accommodate the objections of the Campbell Soup Company.

Campbell's vice president for vegetable research, Eldrow Reeve, argued that the proposed bill was "not required" to provide protection for the development of new plant varieties, but would in fact "impede progress and be detrimental to the interests of the agricultural community and the consuming public." Reeve also noted his company's concern that the bill might create restrictions on the exchange of scientific information important in the development of new plant varieties. "As a result of activity in recent years leading up to the present proposed plant variety protection act," said Reeve, "there has been a perceptible reluctance among plant breeders to exchange genetical materials. We believe enactment of S.3070 would essentially eliminate exchange of valuable germplasm and severely curtail the development of new varieties."

*The Senate hearing, in fact, ran exactly 55 minutes. Democratic Senator B. Everett Jordan of North Carolina, chairing the hearing that day, was in a hurry to make another appointment and hustled fifteen witnesses through the record. After the hearing, Jordon remarked: "I do not think that I have ever held a hearing that was more for one side than this one has been—except one. That was when I happened to be on a committee with the power to accept money. Somebody wanted to give us $20 million for a building and I did not have a bit of trouble getting it through."

†Some farm groups did, however, submit written statements. Jack Felgenhauer, president of the Washington Association of Wheat Growers, a group whose farmer-members financially supported public wheat-breeding programs, expressed concern that commercially produced wheat varieties might not come up to the "end-use" standards needed to satisfy export markets. Felgenhauer recommended that "wheat be deleted from the Act" unless public breeders were protected and end-use qualities, such as those desired by the export market and the baking industry, assured. Marvin L. McLain, legislative director for the Farm Bureau, suggested that farmers be represented on the proposed Plant Variety Protection Board and that the proposed seventeen-year patent period be shortened. "We believe," wrote MacLain, "that farmer interests would be better served by a lesser period of years of plant variety protection."

In a bow to Campbell's and a few other food-processing interests, six "soup vegetables" were exempted from the proposed law (this exemption would bring the seed industry back to Congress ten years later to amend the law and reopen the issue of seed patenting).

When Congress finally moved to "consider" the Plant Variety Protection Act for final passage, it was under the worst of circumstances: a pre-holiday, end-of-Congress, lame-duck session in December 1970, with mountains of left-over business to attend to. The December 8 House floor debate reveals a scant three pages of proceedings on the bill in the *Congressional Record*. Less than one hour of debate actually transpired, during which seven members spoke.

One of the few members who did speak in opposition to the bill was Representative Robert Kastenmeier, Democrat of Wisconsin, then chairman of the House Judiciary Patent Subcommittee. Kastenmeier raised questions about the costs of patenting to farmers and consumers, as well as the problem of "inventor concealment."* But Kastenmeier was the lone dissenter on the substance of the bill, and he was overcome by the powerful Agriculture Committee chairman, Texas Democrat W. R. "Bob" Poage and others supporting the legislation, some of whom invoked the ravages of the 1970 Southern Corn Leaf Blight in their remarks.

*"I represent an agricultural district, too," explained Kastenmeier in the brief floor debate, "and I am concerned about the cost of this to the farmer, and ultimately, of course, to the consumer. It seems strange that we have gone all the way through our history up to 1970 without the need to resort to this sort of protection for some special interests.

"As a matter of fact," he explained, "plants have been developed over the years, have they not—the winter wheats and things we grow in the Dakotas—and without such recourse to protective laws, but rather through development through the State universities and the Department of Agriculture? This [Bill] will result, will it not, in a sort of hiding of development, as is often the case in the patent development protection?"

Later in that same debate, Kastenmeier and Poage sparred over the question of "concealment," with Poage charging that there was: "no way in the world for a seed producer to profit by his development if he hides it, because what he has to do is to make the development public and sell it on as wide a basis as he can." But Kastenmeier explained that he was refering to "the hiding of the development of the process." "An invention in process is often concealed by the inventor," said Kastenmeier. "Presently in our public institutions and otherwise, information with respect to the development of plants is commonly shared, and there is none of the concealment or hiding which would be implicit in an economic motive under this sort of legislation." This issue would reemerge in the 1980s when it was discovered that publicly funded plant scientists stopped freely exchanging plant material.

Final House approval of the bill occurred under a suspension of the rules, and the measure was passed on a voice vote.* The next day the bill went to the Senate for a confirming vote, after which it was sent to President Richard Nixon.

NIXON'S REACTION

Although the seed patent bill had smooth sailing in Congress, it met some resistance in the White House review process, which was then coordinated by the Bureau of the Budget (BOB). As with all bills sent to the President, a "legislative referral memorandum" is prepared, which generally lays out the pros and cons of each bill, recommending either approval or veto. In this case, nine federal agencies and executive-level offices were involved in the review. Five had no objection or no recommendation on enactment; two, the USDA and the U.S. Civil Service Commission, recommended approval; and two, BOB and the President's special assistant for consumer affairs, Virginia Knauer, recommended a veto.

"In short," wrote Knauer in her memo to BOB, "the granting of protection appears to be unnecessary, and of considerable scope. It [the bill] seems likely to create some degree of market power in producers, and will require a sizeable government apparatus to administer. . . . Any public benefit from stimulus to innovation seems dubious," said Knauer. "The most likely results are increased prices to consumers, increased taxes to taxpayers, and increased revenues to plant breeders." Knauer recommended a pocket veto.

The Justice Department had no objection to the bill, but Richard Kleindienst, then deputy attorney general, noted in his letter "the policy question [of] whether the major food and fiber crops which would be covered by this legislation should be subject to private restraints."

"Furthermore," wrote Kleindienst, "we have no indications that [the] development of new varieties of sexually reproduced plants in this country has been retarded by the lack of private protection. It would seem that the contrary is true," he continued, "[as] evidenced

*The only recorded roll-call vote ever taken on any of the plant patent bills—the 1930 Plant Patent Act, the 1970 Plant Variety Protection Act (PVPA), or the 1980 amendments to the PVPA—was the 1970 vote on the *rule* under which the PVPA would be considered. No recorded roll-call votes were ever taken on the substance of any of the plant patent bills or amendments offered in Congress.

in part by the award of the Nobel Peace Prize on December 1, 1970, to Dr. Norman E. Borlaug of this country for his research into and development of new grains. Dr. Borlaug's technological breakthrough is now in the public domain, but could have received patent protection for seventeen years under S.3070."

Meanwhile, a sparring match between BOB and USDA officials was underway. At one meeting, the assistant secretary of agriculture for marketing, Richard Lyng, argued strongly for the bill, according to one BOB examiner present at the meeting. Lyng insisted that the bill had to become law, and that he knew what he was talking about because he had come "out of the seed industry." However, after Lyng and the other USDA representatives had left that meeting, Arnold Webber, an assistant director at BOB who had public policy reservations about the bill, told his staff to "write up a veto memorandum, I've heard enough."

The memorandum recommending a veto was prepared by BOB on December 23, 1970, two days before the President had to act on the bill, and a veto message had also been prepared.*

*The veto message prepared by BOB read as follows:

> I have today withheld my approval of S. 3070 which would have established within the Department of Agriculture a Plant Variety Protection Office to administer a program of protection for breeders, developers and discoverers of novel varieties of sexually reproduced plants. This protection would have been organized along the line of the present patent laws.
>
> Commercial protection in the nature of a monopoly can only be extended at the expense of limiting competition in the market place and its benefits of the lower possible prices and increased output. Therefore, before any patent-like protection is extended there should be established a clear case of need in order that the public interest may be served.
>
> No persuasive evidence has been brought forward to show that the commercial protection which would be afforded by this bill would significantly increase the present high of innovation in this field.
>
> On the other hand, the monopoly rights which this bill would make available to plant breeders, developers and discoverers would result in increased prices of seed to farmers and related disadvantages to consumers of agricultural products. In addition, this bill would have a deleterious effect on the free exchange of information between private and public plant breeders which had been vital to the impressive level of innovation that has characterized the development of new varieties of food and textile-producing plants.
>
> Finally, the effective enforcement of this legislation is uncertain because the varieties of plants to be protected may be subject to significant natural changes.

But the veto message was stopped at the eleventh hour by BOB director George Shultz, later President Reagan's secretary of state. A hand-written note scribbled on a white BOB route slip dated December 24, 1970, from "the director," instructed Wilf Rommel, then assistant director for legislative reference, to "change memo to recommend approval. Include 1-page summary—the attached summary of arguments might help." There is no other explanation in BOB's file that accounts for the last-minute reversal, and many of the arguments for disapproval remained in the final memorandum that went to President Nixon.* Some retired USDA officials who were involved in the USDA/BOB review of the patent bill say that the seed industry had a contact in the White House who helped persuade BOB and the President of the merits of the Act. On December 24, 1970, President Nixon signed the Plant Variety Protection Act into law—the seed industry had its Christmas present.

SEED PATENTS FOR THE "SOUP VEGETABLES"

The loose ends left hanging by the exclusion of the six "soup vegetables" from the PVPA in 1970 brought the seed industry back to Congress in 1979 with a bill to include those vegetables. This

> In view of the lack of clear evidence of need; the potential disadvantages to the general public and to the primary user of new varieties of plants and seeds; and the uncertainties with respect to administration, I do not believe this bill warrants my approval.
>
> The White House
> December, 1970

*A comparison of the last paragraphs of memorandums prepared on December 23 and 24, reveals that only a few sentences were changed, while the negative tone remained. The paragraph from the memorandum recommending a veto read: "On balance, we believe the extension of a monopoly right to this type of food and textile producing plant varieties has not been justified by a clearly demonstrated need. In addition, there are other practical and administrative problems which would benefit from further study. Accordingly, we recommend disapproval of the bill and have prepared the attached draft memorandum of disapproval for your consideration."

On the following day, the new last paragraph read: "In summary, protection of food and textile producing plant varieties of this type has not, in our view, been justified by a clearly demonstrated need. In addition, there are other practical and administrative problems which would benefit from further study. Accordingly, we think that more extensive consideration of this proposal by the congressional committees concerned would have been desirable. However, we do not believe that these concerns warrant your disapproval of the bill."

time, the industry had the cooperation of the Carter Administration. The Departments of Agriculture, Commerce, and State—concerned with both the domestic and international aspects of the seed trade—coordinated work on the amendments, which included two basic provisions: addition of the six soup vegetables, and lengthening of the patent period from seventeen to eighteen years. The latter provision would bring the American seed law into conformity with European laws and the UPOV convention.

In Congress, Representative Kika de la Garza, Democrat of Texas, and Senator Frank Church, Democrat of Idaho, both introduced bills to amend the PVPA on behalf of the Carter Administration in January 1979. Representative de la Garza, besides being chairman of the House Agriculture subcommittee having jurisdiction over the bill, represented a portion of Texas' Rio Grande Valley that was then, and is today, a leading producer of bell peppers and carrots, two of the vegetables excluded from the PVPA. Idaho is a major vegetable seed-producing state, and Senator Church was then up for reelection.

In sponsoring their amendments, neither Church nor de la Garza could have imagined the controversy that would swirl around this legislation during the eighteen months that followed. They believed, as did seed-industry and Carter Administration officials, that the new bill was merely a "minor housekeeping bill," comprising "technical" amendments to a ten-year-old law.

The first signs of controversy appeared in July 1979, when three "outsiders" came to testify before de la Garza's subcommittee. These three—Cary Fowler of the National Sharecroppers Fund of North Carolina, Gary Nabhan, an agricultural botanist from Tucson, and Dan McCurry, a director of the Consumers Federation of America—questioned the impact of plant patenting on small farmers and consumers, how seed patents might affect the genetic diversity of agricultural crops, and whether such patents might encourage a trend toward "corporate takeovers" of family-owned seed companies. This trio—each of whom opposed amending the law until further study was undertaken by Congress—also raised a range of other questions.

OPPOSITION & NEW ISSUES

Cary Fowler is a soft-spoken activist in his thirties who owns a small farm in North Carolina, and has been involved in agricultural issues for a number of years. Fowler came to the National Sharecroppers Fund in the late 1970s after working with Frances Moore Lappé and Joe Collins on *Food First,* a book about Third World hunger amid agricultural plenty. In that book, Fowler wrote sections dealing with the Green Revolution and its impact on Third World countries, and in the course of his research, became intently concerned with the problems of crop uniformity, the displacement of native crop varieties by new hybrids, and the resulting problem of genetic vulnerability.

After *Food First* became a national best-seller in 1977 and 1978, Fowler had the opportunity to lecture in the United States and other countries, and would sometimes venture into the topic of seeds and genetic vulnerability by way of discussing the Green Revolution. Few people were moved by Fowler's concern over seeds, but one who was was a Canadian named Pat Roy Mooney. Mooney would later help to "internationalize" the seed issue with his paperback book *Seeds of the Earth.*

In the late 1970s, Fowler published a pamphlet for the National Sharecroppers Fund entitled *Graham Center Seed Directory,** which was aimed at helping small farmers and gardeners identify "alternative" seed companies and nurseries that carried hard-to-find traditional plant varieties—those not sold by the large commercial seed companies and mail-order houses. In his directory, Fowler also wrote a short essay entitled "Reaping What We Sow: Seeds and the Crisis in Agriculture" in which he raised the issue of plant patenting. To his surprise, the Graham Center Directory and his essay "went over big," and second and third printings followed. The mailing list that Fowler accumulated by way of requests for his Graham Center Directory, provided him with a roster of people and organizations who were concerned about the "seed issue."

When the Church and de la Garza seed bills were introduced in Congress, Fowler was drawn into the fray. He then turned his mailing list loose on Congress, went to the press, wrote articles about the issue, and traveled around the country alerting other groups to the "seed patenting" issue. Fowler rallied scientists and church interests, and wrote to secretary of agriculture, Bob Bergland, urging him to consider the impact of rising seed costs on small

*The Frank Porter Graham Demonstration Farm and Training Center is affiliated with the National Sharecroppers Fund.

farmers. Shortly, Fowler became a one-man lobby opposing the seed-patent amendments.

One of Fowler's central concerns for all crops was genetic uniformity, a concern supported by a 1972 National Academy of Sciences (NAS) study, entitled *Genetic Vulnerability of Major Crops*, in which the Academy found that most major crops were genetically uniform and genetically vulnerable. The NAS noted that programs such as seed certification "adversely influenced genetic diversity," since they emphasized uniformity and purity in seed stocks. Moreover, said the NAS, the PVPA might narrow the range of genetic material used in plant-breeding programs and thereby increase crop vulnerability to disease and infestation. Using the example of wheat, and noting that wheat breeders generally could increase genetic diversity, the NAS explained that because the PVPA stressed novelty and uniformity, there was an implied need for genetic homogeneity to qualify for patent protection. "Indeed," said the Academy, the new law "may discourage wheat breeders from pursuing the genetic diversity concept of plant pest biological control . . . [and] if so, it poses a danger."

In addition to the NAS study, reputable scientists such as biologist Garrison Wilkes expressed apprehension about "improved varieties" displacing local and or primitive crop varieties and driving them to extinction along with the genetic diversity they held needed to sustain all agriculture. Other plant breeders, some from prestigious research institutions such as the International Maize and Wheat Improvement Center (CIMMYT) in Mexico, also took a stand against plant patenting because it often worked to restrict the free exchange of genetic materials among scientists.

In Congress, meanwhile, the reverberations from the first criticisms of the PVPA began to have an impact. Following de la Garza's April 1979 hearing, newspaper stories appeared throughout the country focusing on the objections to the seed-patenting amendments. And at about the same time, according to Fred Wahl, a former aide to Senator Frank Church, "the damnedest thing happened."

BIRCHERS GO AFTER FRANK CHURCH

American Opinion magazine, allegedly connected to the John Birch Society, began to use the seed-patent controversy in an attempt to link Church, then chairman of the Senate Foreign Relations Committee, to a Communist-inspired "seed conspiracy" plot. Church was portrayed as the tool of the "socialist-run" UPOV, and his bill as a measure that would dictate what seeds could be grown. There was even a scenario, according to Wahl, that included the spectre of "federal police arresting Grandma" for growing illegal vegetables. That hypothetical ploy played well in Idaho, especially among conservatives who wanted Church out of the Senate. With reelection in the balance, Church was forced to defend himself.

As a point of clarification of his seed bill, Church issued a press release explaining what the bill would and would not do. It was "ludicrous" to think that his bill would make it illegal for home gardeners to grow certain kinds of seed, said Church. "My bill is not intended to outlaw certain seeds, banish backyard gardens, nor force genetic uniformity," he stated in the release. "Least of all is it designed to hand over big profits to a few seed companies. . . . What must be remembered is that plant-variety protection has worked to protect the new varieties developed by small companies against exploitation by larger seed companies."

Church, a liberal Democrat who took a special interest in the rising power of big business, was certainly no friend to multinational corporations. His landmark Senate hearings in the 1970s probing the power and operation of the multinationals, set him apart from many of his Senate colleagues as someone who was not afraid to take on the very powerful. However, in the coming years, more than one hundred American seed companies would be bought by larger companies, and even as Church ran for reelection, Sandoz, the major Swiss pharmaceutical corporation, had acquired two vegetable seed companies in his own backyard—Idaho's Rogers Brothers Seed Company and Gallatin Valley Seed Company. Perhaps if Frank Church had survived his reelection bid, he might have launched his own Senate investigation into the emergence of the multinational seed business.

At the time, however, the "seed issue" was a bothersome one for Church; both he and his campaign staff had to spend a good deal of time explaining what his seed patent bill was all about. Eventually, the issue was put to rest in Idaho when Church noted that his Senate campaign opponent, Republican Representative Steve Symms, was a co-sponsor of the House version of the seed bill, and that Idaho's other senator, Republican Jim McClure, had supported the original Plant Variety Protection Act in 1970.

SEED BILLS IN CONGRESS

In Washington, co-sponsors of the seed-patent bill were receiving voluminous amounts of mail about it. In the House, a second hearing was scheduled for April 1980, in the hopes that the controversy would subside. Instead, gardening, environmental, farm, and church interests joined the fray, raising still other issues.

On the Senate side, the seed bill had to travel through the agriculture subcommittee chaired by Senator Donald Stewart, Democrat of Alabama, who was not as eager to move it as were his House counterparts. Stewart suggested at one hearing that the larger corporations entering the seed business might be "less innovative" than the smaller companies, and that this might hurt consumers. He also had other concerns, and decided to study the matter further.

Meanwhile, the American Seed Trade Association (ASTA) was completely frustrated by the bill's slow progress. Before the 1980 October election recess, ASTA asked Representative de la Garza to request that the bill be considered under a suspension of the House rules—a procedure under which a bill is voted on without amendment and with only limited debate. But a bill considered under suspension requires a two-thirds majority for passage, and at that time, ASTA determined it did not have the votes, and asked de la Garza to withdraw the bill from consideration. ASTA then waited until the lame-duck session, when de la Garza, using the same procedure, pushed the bill through on a voice vote.

ASTA lobbyist Harold Loden had also been urging Democratic Senator Herman Talmadge of Georgia to pry the bill loose from Senator Stewart's subcommittee. Under pressure from interests at home, as well as from his Senate colleagues, Stewart agreed to adopt the House version of the bill. On December 8, the Senate passed it without debate or a recorded vote and sent it to the President.

Jimmy Carter, by then a lame duck himself, received some advice from federal agencies, such as the Council on Environmental Quality, to veto the bill, but instead he allowed it to become law. Thus, in late December 1980—in almost an exact repetition of the circumstances surrounding the passage of the original PVPA ten years earlier—the PVPA amendments became law. And once again, despite a brief controversy, few people really knew much about seed patenting.

Today, more than a thousand seed patents have been issued by USDA's Office of Plant Variety Protection. But seeds are only a part of the agricultural resource base that is now eligible for patenting and private ownership. What subsequently opened up the entire field of agriculture—and all of biology for that matter—to commercial patenting and private ownership, was the U.S. Supreme Court's June 1980 decision in the case of *Diamond v. Chakrabarty,* the first case involving the patenting of a genetically engineered microbe.

THE SUPREME COURT OPENS THE FLOODGATES

In 1972, a General Electric research scientist named Ananda Chakrabarty filed a United States patent application for a genetically engineered bacterium. His action immediately raised legal questions about whether a living microorganism could be patented under the existing statutes. Eight years later, in June 1980, the U.S. Supreme Court ruled, in a landmark five-to-four decision, that Mr. Chakrabarty's bug was "patentable subject matter." "The patentee has produced a new bacterium with markedly different characteristics from any found in nature, and one having the potential for significant utility," wrote the court in its decision. "His discovery is not nature's handiwork, but his own; accordingly, it is patentable subject matter." In the court's opinion, "the relevant distinction was not between living and inanimate things, but between products of nature, whether living or not, and human-made inventions."

When the Supreme Court ruled on Chakrabarty's oil-eating microbe it looked simply at the criterion for patentable subject matter, and found that Chakrabarty's bug was no longer a natural product: it had been changed by man, there was no twin for it in nature, and it was novel and useful. Therefore, it was patentable.

The court's decision brought some sharp criticism, such as that from MIT molecular biologist Jonathan King. "I don't know what we've learned in two hundred years of biology, if we can have a Supreme Court that says there is no difference between the living and the non-living," said King. "We may be made of carbon, hydrogen, and nitrogen, but typewriters are too, and they don't reproduce themselves, or talk, or give interviews." For others, including those seeking patents, the close decision raised more questions than it answered, and the debate continues to this very day.

Despite the fifty years of history in patenting plants such as fruit trees, the more recent history of patenting seed-propagated plants such as wheat and tomatoes, and now the Supreme Court's ruling that man-made microbes can be patented, there are some important questions about the use of patents in agriculture and food production that have never been adequately debated by either Congress or the general public. Among these questions are the following: whether food-producing resources should be patented at all, and if so, whether they should be treated differently than other kinds of patents; whether patents on agricultural supplies such as pesticides and livestock antibiotics always contribute to innovations of benefit to society; and whether such patents facilitate economic concentration and economic power in the food and farm system. We will return to these and other questions in a later chapter.

CHAPTER 5

THE NEW BREED

He likes to be on the cutting edge.
—*A Santa Fe acquaintance of David Padwa*

When David Padwa lived in Santa Fe in the early 1980s, he could be seen driving around town in a white, late-model Volvo station wagon with a license plate bearing the acronym of the world's most important molecule—"DNA," the stuff of inheritance. DNA is also the stuff of David Padwa's latest business venture; the stuff inside seeds—"genetic envelopes" he calls them. Padwa's new company sells seeds to farmers throughout the world; seeds that he says will be improved through biotechnology and genetic engineering.

The Agrigenetics Corporation is the name of the ten-year-old company that the fifty-one-year-old Padwa founded and now runs from a modern, red-brick headquarters building in northeast Boulder, Colorado. Both the man and the company are from a unique mold, and their respective histories tell an interesting story about changes now taking place in agricultural science and commerce.

David Padwa is a bright, effusive businessman and intellectual of many descriptions and, some might say, many lives. Those who know him say he just might be a modern-day shaman or prophet—the kind of person who is good for advancing ideas and bringing new ways of doing things into society and business. Others aren't so sure. Some see him as a fast-talking wheeler-dealer, and wonder just what he's up to. A few friends who know him personally say he is a very generous, very religious person, yet in public he is highly competitive and aggressive, capable of business elbowing with the best of them. All agree, however, that he is highly intelligent and, at times, close to the socially grating edge of genius.

Whatever he is, David Padwa is shaking up the American seed industry and turning heads in corporate boardrooms. Creating what may prove to be the prototype agricultural genetics company of the future, he has come like a whirlwind into an industry used to moving at a tortoise's pace. He has brought change, controversy, and a new kind of commerce to that industry. And like the metamorphosis he is effecting in the seed business, Padwa himself is a man of many transformations.

David Padwa was born to immigrant parents in New York City in 1932. His mother was Lithuanian, his father Austrian. At an early age, he began moving through life with an intellectual precociousness that would become his trademark. After leaving the Bronx High School of Science at the age of fifteen, he enrolled in the University of Chicago in 1947. Three years later, he received his bachelor's degree. By 1957, he added a law degree from Columbia University, and then began working toward a Ph.D. in law and government. In 1960, just short of his Ph.D., he paused to found a small company—"with my life savings, a couple of thousand dollars"—called Basic Systems, which sold training programs and educational materials for computer operators. In 1965, Padwa sold Basic Systems to Xerox for $6 million in stock, stayed on at Xerox for one year, then served as a visiting lecturer at Harvard University's School of Public Administration for a year. Then, at age thirty-six, he dropped out. As he put it: "I climbed mountains in Asia and Europe, sailed the Atlantic in a small sailboat and generally got a lot of fantasies out of my system."

One of the people Padwa met while at Harvard was Dr. Richard Alpert, a psychologist and friend of Timothy Leary who subsequently became interested in eastern religions and later took the name Ram Das. In 1967, Padwa became interested in Buddhism, and he and Alpert made a trip to the East* where they traveled

*In his book, *Be Here Now,* Alpert writes of meeting "David" and their departure to the east: "Then, along came a very lovely guy . . . an interesting guy,

through India, Afghanistan, and Nepal, and visited Buddhist leaders such as the Dalai Lama.

SANTE FE DAYS

In addition to his travels abroad, Padwa spent a good deal of time in the American West and Southwest in the early 1970s, where he moved within a liberal community of friends, artists and activists from California, New York, New Mexico, and Arizona. In Santa Fe, he bought a tract of land on the east side of the city, where he had a very attractive adobe-styled home built, designed by architect Bill Lumpkin. At the time, Padwa also owned an "architectural gem of a home," according to one friend who stayed there, on New York's East Side off Park Avenue, designed by Stanford White. But it was Padwa's house in Santa Fe that would become a gathering spot for his growing assortment of friends.

The people who paraded through Padwa's life during the late 1960s and early 1970s included a potpourri of Harvard intellectuals such as Alpert; East Coast poets such as Allen Ginsberg; West Coast artists and environmentalists like Stewart Brand; and "new age" eastern holy men.* He also became involved with the Black Mesa Defense Fund, a public-interest organization in the Southwest that was then challenging coal-industry plans for the strip mining of sacred Hopi and Navajo lands. At that time, Padwa also served on the board of the Lama Foundation, a New Mexico religious community in the mountains near San Cristobal. In 1972, Padwa and Jack Loeffler, a classical musician who had become concerned with

who had gone to the University of Chicago in his early teens and had taught seminars in Chinese economics, had started a company called Basic Systems, which had been sold to Xerox, and now he had retired. He was about thirty five and he had retired and taken his five million dollars or whatever he made, and was now becoming a Buddhist. He wanted to make a journey to the east to look for holy men and he invited me to go along. He had a Land Rover imported into Teheran and this was my way out."

*Padwa is described today by friends as a "devout Buddhist," and at his Santa Fe home he had a Buddhist temple built for Tibetan Buddhists who would occasionally travel through the United States. In his attempts to learn more about Buddhism, Padwa once tried to gain entry to the Himalayan nation of Bhutan, only to be denied a visa. Later, after learning that the King of Bhutan had a fondness for archery, Padwa sent him an expensive archery set from the United States, still trying to obtain an entry visa.

Indian rights and organized the Black Mesa Defense Fund, brought some Hopi and Navajo elders to Sweden for the first United Nations Environmental Conference, held in Stockholm.

BACK TO WORK

By 1974, Padwa began to think about rejoining the business world. At the prodding of some old New York friends, such as F. Palmer Webber of the Wall Street brokerage firm of Speer, Leeds & Kellogg, and Gilbert de Botton, a college chum who had become director of Zurich's Rothschild Bank, Padwa began to think about business and the future, and where good investments might likely be made. Agriculture and food first caught his eye, particularly given the predictions of a world food crisis during the 1980s. Looking for the right business niche to satisfy his own interests, as well as those of a small group of investors, Padwa formed Agricultural Equities, a holding company designed to build a diverse investment portfolio of small, agriculturally related businesses. In surveying the prospects for this new investment fund, Padwa looked at farming and ranching operations, the irrigation equipment business, feedlots, alfalfa pelleting, and farm-implement companies, but none of those really appealed to him. Then he discovered the seed business.

Padwa's initial interest in the seed industry grew partly out of his involvement with the Black Mesa Defense Fund. With the understanding that massive strip mining was soon to begin throughout the southwestern and western United States, Padwa knew that western grasses would have to be planted on the land following mining to help restore it, and that the seed for such grasses would be in great demand. In fact, the federal strip-mining law then wending its way through Congress indicated that the mining industry would have to abide by fairly stringent reclamation and revegetation requirements. At that time, however, commercial supplies of seed for native grass species were practically nonexistent. There was also a demand for native grasses along highway medians and rights-of-way, as well as for public and private range lands. To fill this void and make a few dollars while doing good, Padwa began narrowing his investment plans and set out to begin a native seed business.

Padwa soon discovered that one of the problems of producing and harvesting large quantities of native seed was the lack of adaptable machinery to do the job adequately and cost effectively on a commercial scale—native grass seed was, for the most part, hand-harvested. Padwa confronted this problem by bringing together a few engineers and scientists to devise new kinds of harvesting equipment for the hard-to-handle native seed. According to one former associate, Padwa "totally inhaled the entire technology" of combines and native-seed harvesting equipment. The machinery engineers couldn't believe Padwa's grasp of the jargon, his working knowledge of the equipment, and his acumen in a rather arcane field of engineering. The result, under Padwa's prodding, was something of a breakthrough in native-seed harvesting; the group produced designs for new combines capable of harvesting thousands rather than hundreds of pounds of seed.

By 1975, Padwa's native seed business was called Grassland Resources, Incorporated. Initially, he set up shop in an old, thick-walled building that was once a meat-storage locker in the railroad yards of Santa Fe, the cool basement of which was ideal for seed storage.

THE TURNING POINT

By all accounts, Grassland Resources was a business very much in the Santa Fe mold, a fairly "laid back" operation. Then, in 1975, Padwa met Bob Appleman, the sixty-five-year-old owner of the Arkansas Valley Seed Company, a small, $3-million-a-year farm-seed operation located in Rocky Ford, Colorado. Appleman's seed company was involved in range grasses, turf grasses, and small grains such as wheat and barley.* Through Appleman, Padwa began to see real possibilities in the seed business, far beyond those of his tiny native seed company. He also began to see the important commercial connection between the newly emerging science of biotechnology and the seed industry.

While not a scientist by profession, Padwa is the product of an excellent and early scientific education, and a person who has always managed to stay on the fringes of microbiology and genetics. For example, when he taught at Harvard in 1966, Padwa had the good fortune to sublet the Cambridge apartment of James Watson—of

*Some of Arkansas Valley's largest grass seed accounts then and now were large strip and open-pit mining operations and reclamation sites belonging to Utah International, Kennecott, ARCO, and others.

the famous Watson and Crick team that discovered the double-helix structure of DNA. Padwa's brother, Albert, holds a chair in biochemistry at Emory University in Atlanta, and a former in-law, Dr. Andrew Lewis, a virologist at the National Institutes of Health, was one of the first scientists in the mid-1970s to hybridize viruses, and was also involved in the Asilomar Conference, convened by scientists to address early fears of the possible escape of genetically engineered organisms. All of these associations—plus "being a reader of *Nature* and *Science* for more than twenty years"—figured into Padwa's consciousness, he says, as he looked more deeply into agricultural science.

Padwa had also immersed himself in scientific literature pertaining to agriculture and crop-genetics research, and he knew that any genetic changes made in crops in the laboratory had eventually to be engineered into—and marketed to farmers—in the form of seeds. With his native-seed business, and in the process of prospecting for other investments, Padwa was also learning about the arcane roles of "cytoplasmic male sterility" and the "mitochondrial genome" in the hybridization of crops, and recalls that this was "very exciting from a genetic standpoint."

The strategy for a new kind of agricultural genetics company soon became clear to Padwa: acquire good specialized seed companies, build a modern biotechnology research capability, and then use the business revenues from acquired seed companies to help finance the research. The only problem with this plan was that Padwa was an outsider in the seed business. But Appleman, a respected seedsman, helped solve that problem, first by selling Padwa his own seed company, and then by becoming part of Padwa's organization, subsequently opening doors for his new boss throughout the industry. After studying the seed industry carefully, Padwa began to put an aggressive acquisition strategy into motion. By the time the dust had settled, Padwa had assembled twelve companies in six years. Agrigenetics was now the seventh largest seed company in the United States.*

*In 1977, Padwa acquired the Colorado Seed Company of Monte Vista, Colorado, which specialized in barley seed. In 1978, he added the R. C. Young Company of Lubbock, Texas, specializing in sorghum seed, and Seed Research Associates of Sun City, Kansas, specializing in wheat seed. In 1979, he acquired a small Plymouth, Indiana, soybean seed company, V. R. Seeds. In 1980, three

Today, the combined annual sales of the Agrigenetics companies are in the $80-to-$100 million range, and they sell more than fifty species of crop seeds. Profits in 1981 reached $4 million, but have since declined, owing to a combination of the government's PIK program and a heavy debt load. Approximately half of Agrigenetics' present income derives from hybrid corn seed, about 15 percent from vegetable seed, 13 percent from sorghum seed, and 10 percent from soybean seed.

In the seed industry, David Padwa and Agrigenetics are regarded with a combination of fascination and stand-offishness, and sometimes outright disdain. Some observers simply consign the activities of Agrigenetics to the far-out, high-risk future of genetic engineering. Others simply regard most of Agrigenetics' moves as "hype." Yet some observers who might otherwise specualte about Padwa's motives note that Agrigenetics does have an impressive collection of seedsmen. Jim Carnes of International Seeds, an Oregon seed company, points out that Agrigenetics "has some very capable people, some who have twenty-five to fifty years of experience in the seed business." And Padwa himself has learned a few things about the business, too.

In the course of putting together the various companies that now comprise Agrigenetics, Padwa studied more than two hundred seed companies, and says he "looked closely at six companies for every one acquired." Today, Padwa probably knows as much about the American seed industry as anyone in the country.

more companies were brought into Padwa's organization: Sun Seeds of Bloomington, Minnesota, specializing in peas, beans, and sweet corn; the Taylor-Evans Seed Company of Tulia, Texas, specializing in hybrid corn and hybrid sorghum seed; and Agricultural Laboratories of Columbus, Ohio, a company specializing in the preparation of bacterially treated seed for legumes. In 1981 Padwa acquired the Keystone Seed Company of Hollister, California, a company specializing in vegetable seed; the Jacques Seed Company of Prescott, Wisconsin, specializing in hybrid-corn seed; the family-owned McCurdy Seed Company of Freemont, Iowa, specializing in corn and sorghum seed; and the Growers Seed Association of Lubbock, Texas, a small seed cooperative that Padwa says he "rescued" from the precipice of bankruptcy. In 1982, Padwa made a play for the then-failing hybrid wheat program at DeKalb AgResearch, Incorporated, but Monsanto, the large St. Louis chemical corporation, beat him out at the eleventh hour. But Padwa didn't go away empty-handed; he hired DeKalb's chief wheat breeder, Dr. James Wilson.

THE UNION CARBIDE DEAL

Of all the transactions David Padwa made while building his seed-company empire during the late 1970s and early 1980s, the one he made with Union Carbide in 1980 was perhaps the most important. In that $80 million transaction, Padwa not only acquired the Jacques and Keystone seed companies, but also came to know, and later hire, some key Union Carbide scientists and administrative personnel who are now at the center of Agrigenetics' business and its future.

One of these, R. N. "Sam" Dryden, a thirty-five-year-old Kentuckian with an easygoing, down-home manner, is president of Agrigenetics. Dryden was the man at Carbide charged with selling off the company's seed and plant research holdings, which at the time also included a biotechnology program. He remembers well why Union Carbide decided to pull out of biotechnology.

Union Carbide originally became involved in the seed business in the early 1970s when it purchased the Keystone Seed Company, the New Seed Company, and the AmChem Chemical Corporation, which owned the Jacques Seed Company. At that time, Carbide had organized its agricultural products division into two subdivisions—agricultural chemicals and plant sciences—with the three seed companies and a small plant biotechnology research effort comprising the latter subdivision. Initially, Carbide officials were planning on some business "synergy" between their biotechnology and seed divisions, according to Dryden, and with that view, more research money was pumped into biotechnology and plant research, which then put Carbide in front of most other corporations in this novel area of research.

But few people at Carbide fully realized how much research money was really needed to make a meaningful commitment to plant biotechnology. By 1976, with a mounting cash-flow problem, company officials decided they couldn't finance both their traditional chemical research and a new research push in plant biology, and elected to sell the plant sciences division, including the three seed companies. According to Dryden, "Carbide's decision came from the top. They decided, 'we're going to do what we do best, stay with chemicals and plastics; stay with our roots'. You know, back to batteries, the 'Eveready-made-us' kind of thing. . . ."

Dryden, then in Carbide's strategic planning division, was given the task of mounting an effort to sell the company's plant sciences division and its seed companies. During the course of his research into which corporations might be in the market to buy Carbide's plant sciences division, Dryden talked to corporate officials at Shell, Sandoz, Upjohn, and Ciba-Geigy among others, and he learned a great deal about who was doing what in the seed business and plant biotechnology.

In early 1980, Dryden had narrowed his list of prospective buyers to three or four large companies, and finally obtained a letter of intent from Limagrain, a large French seed company. He flew to Paris to close the deal, but at the last minute, Limagrain officials tried to change the price, much to Dryden's dismay. That same evening, Dryden was scheduled to have dinner with David Padwa.

At that time, Padwa and Agrigenetics were in the running to acquire one of Carbide's seed companies—Keystone, which sold vegetable seed and was then valued by Carbide at about $10 million. Agrigenetics, however, wasn't on Dryden's list for Jacques Seed Company—the larger midwestern hybrid corn seed company—because he thought the $70 million price tag was out of Agrigenetics' range. And Dryden's first preference was to sell the Carbide companies as a block.

During his dinner with Padwa, Dryden discussed his frustrations about not being able to close the Limagrain transaction. Padwa asked what the price tag would be for both companies, and Dryden replied it would be in the $70 to $80 million range. Padwa then asked Dryden to give him two weeks to see if he could raise the money. With a little help from Carbide and the New York investment brokerage firm of Morgan Stanley, Padwa came up with the money sixty days later and made the deal. In late 1980, both seed companies were sold to Agrigenetics. In the transaction, all of the research material and information in Carbide's plant sciences division—the "germplasm, technology, notebooks, etc." according to Dryden—went to Agrigenetics. But that wasn't all.

A few months later in New York, Dryden and Padwa met again for dinner, and Padwa offered Dryden the presidency of Agrigenetics and Dryden accepted. Today, from his perch at the new company, Dryden looks back on Union Carbide's decision to divest its "biology side" as a wrong move. "In hindsight," he says, "they should have divested their chemical side." But Dryden does say that Carbide's initial instincts in the plant sciences were good, and calls its

1972–73 move into biotechnology as one of the first of its kind and "farsighted." Corporate tradition and conservatism prevailed, however, as Carbide retrenched to protect the traditional chemical side of its business.

What had been a seeming albatross to Union Carbide, however, was a mother lode in seed and science for Agrigenetics. One Union Carbide scientist who came to Agrigenetics after a brief interlude at the Upjohn Corporation was Bob Lawrence, a forty-year-old biochemist who established and directed the plant-cell and tissue-culture program at Carbide's Agricultural Sciences Laboratory between 1976 and 1980. Today, Lawrence is the Director of Agrigenetics' Applied Genetics Laboratory in Boulder, Colorado. While at Carbide, he had begun to work on certain cell- and tissue-culture techniques that permit various crop species to be rapidly hybridized, as well as other techniques that permit the mass cloning and production of certain hybrid vegetable seeds. When he came to Agrigenetics in 1981, Lawrence brought his know-how—and his pending U.S. patent applications for these biotechnology techniques—with him. In 1982 and 1983, Lawrence and Agrigenetics were awarded patents for these techniques, which have since become keys to Agrigenetics' strategy for developing hybrid wheat, cotton, and vegetables. And for the lawyer in David Padwa, patents are an important consideration.

"In case you don't know it," he told an audience at the Los Alamos National Laboratories in March 1982, "copyright and patent are forms of intellectual property. Intellectual property is not a dirty word, you understand, and trade secrets—which are usually practical collections of petty lore, constituting a kind of 'witchcraft'—are also a form of intellectual property.

"In the case of *Diamond v. Chakrabarty* [June 1980]," Padwa continued, "the Supreme Court told us we could patent new life forms and have intellectual property in a gene. This may sound far-reaching but it isn't really, and eighteen years of protection is, of course, merely the twinkling of an eye in the evolutionary scheme of biology."

Padwa told another group of businessmen and scientists attending a 1981 Battelle Memorial Institute conference, "I remind everybody here that it may be of marginal commercial utility to develop something that isn't proprietary. You may get a Nobel Prize, a Kettering

Award and a whole bunch of other honors, but if you're trying to make a return for your shareholders, you've got to have a proprietary consideration, and you've got to have a way of protecting your product."*

ADVISOR, ENVIRONMENTALIST, SEMANTICIST

Although not a scientist, David Padwa has served on the National Research Council's Special Review Committee on Agriculture, and on an advisory panel of the Congressional Office of Technology Assessment for its 1984 study *Commercial Biotechnology: An International Analysis.* He has been a visiting lecturer at Harvard's Graduate School of Public Administration and a guest speaker for the Invited Colloquia Series at the Los Alamos National Laboratories. Padwa also helped found the Industrial Biotechnology Association, a new Washington-based trade association, and he has given speeches on biotechnology all over the country. In 1983, Colorado Governor Dick Lamb appointed Padwa to serve as a Commissioner of the Colorado Advanced Technology Institute.

Padwa claims to be an environmentalist and plant conservationist, pointing to his involvement in the first U.N. Environmental Conference, where, he says, he and others such as Huey Johnson—former Director of Natural Resources under California governor Jerry Brown—were among the first to raise the issue of genetic-resource conservation.

*Today, Agrigenetics is the third leading holder of patent certificates issued under the Plant Variety Protection Act, including twenty-seven in vegetables, fifteen in wheat, ten in soybeans, and seven in cotton. Besides this, Agrigenetics' subsidiaries hold two patents and nineteen trademarks issued by the U.S. Patent Office. One of the company's more recent patents—issued in April 1982—is a biotechnology patent, described by the company as "a major technological breakthrough." The patent covers a new laboratory method that uses both tissue-culture and genetic methods, and permits one or both of the parent plant lines used in making a hybrid cross to be genetically varied rather than inbred for uniformity, as is the standard practice for making hybrid crops. The technique has the potential for reducing the number of parent plant generations needed to develop a hybrid seed, and has been described as having "broad commercial ramifications" because it can be applied to many crop species. Some analysts say that with this technique Agrigenetics has the potential to condense the hybridization time for tomatoes, broccoli, and cabbage from twelve years to eighteen months. And since Agrigenetics holds the patent, companies interested in using the new process will have to come to Agrigenetics for a license.

Seed-industry critics and patent-rights foes, such as Cary Fowler of the National Sharecroppers Fund in North Carolina and Canadian Pat Mooney, author of *Seeds of the Earth*, are in Padwa's view responsible for "mobilizing political action in the seed trade," and spurring the formation of SEEDPAC, the seed industry's political action committee.

"You liberals put in Nixon and Reagan," he says, recalling that most liberals supported McGovern rather than Humphrey in 1968, which he regards as a colossal mistake. The left-wing liberals, he says, "can't stand the guy in the middle, but that's where reality is." Padwa believes that right-wing conservatives can be pushed by liberal "hawks" to the point that one day, "they'll be responsible for putting a guy like [James] Watt in the White House." Padwa was an early supporter of liberal Democratic Senator Alan Cranston's 1984 bid or the Democratic Presidential nomination, and has often spoken disdainfully of "Cap Weinberger's war machine."

Genetic engineering critics like Jeremy Rifkin—whom Padwa debated once on ABC's *Nightline*—have also irked the Agrigenetics chairman. Their concerns, he believes, are somewhat misplaced. "Even now the President is asking for 17,000 new warheads, but it is the molecular biologist who makes us nervous," Padwa said in a 1982 Los Alamos speech. He also takes a dim view of the "self-appointed Ayatollahs"—meaning some scientists and university faculty involved in agricultural science policy, and is often warning his listeners of the lesson of T. D. Lysenko, a Russian scientist who persuaded political leaders of his neo-Lamarkian view of genetics, and closed off other scientific findings to the contrary during the late 1940s.

What goads Padwa most is what he calls the "abysmal degree of scientific illiteracy on the part of the lay public," coupled with the media's tendency to focus on the dramatic. "While many professionals may still be scrupulous in their use of [scientific] terms" such as genetic engineering, gene splicing, and biotechnology, says Padwa, "the lay media is not." He believes that "the vague and shifting use of scientific sounding language is a persistent source of confusion and wasted effort, and at times the cause of legislative irresponsibility." In 1982 Padwa wrote, "the terms genetic engineering and gene splicing are now like a kind of crabgrass and presently colonize a wide spectrum of biological events. . . . At this rate the phrase 'genetic engineering' will shortly be applied to the normal act of procreation, and like Molière's bourgeois who was delighted to learn that he was speaking 'prose,' our lay contemporaries may thrill to the notion that we are all 'recombinants.'

"Biotechnology," insists Padwa, is both as old as the "neolithic beer makers of Mesopotamia" and as near as the doctor or dentist, all of whom, in his view, are biotechnologists.

A LITTLE HELP FROM HIS FRIENDS

When David Padwa needed money in 1975 to begin financing his seed company acquisitions, he went first to his friends. One—Gilbert de Botton—a college classmate who had become managing director of Zurich's Rothschild Bank and later director of Rothschild, Incorporated, particularly helped Padwa raise money through the people he knew at the bank. All in all, about 200 investors helped provide backing for Padwa's initial round of seed company acquisitions.*

In the course of his social and business travels during the 1970s, Padwa had also befriended Robert O. Anderson of the Atlantic-Richfield Corporation, who maintains a large ranch in Roswell, New Mexico, and was also one of the founders of the Aspen Institute, a corporate think-tank once headquartered in Aspen, Colorado. When Padwa was groping for new ventures in the early 1970s, he would occasionally visit the Institute, where he met other influential corporate leaders, such as Canadian oil magnate Maurice Strong, who once served as Canada's minister of the environment, and who played a major role in the 1972 U.N. Environmental Conference. (Strong also had a hand in "alternative" ventures such as the New Alchemy Institute.) Some of these acquaintances or their friends also helped Padwa with financing. Among some of those

*Among some of those who invested in Padwa's venture were: Max Palevsky, a computer whiz who had sold a company to Xerox when Padwa did; Drummond Hadley of the Busch family who Padwa knew from his travels in the Southwest; B. J. Pevehouse, president of Adobe Oil; James Niven, president of Pioneer Ventures Corp; and the Marsh family of Amarillo, Texas, acquainted during Padwa's Santa Fe days. European contributors contacted through the Rothschild Bank included the Flick family of Germany; the Bonniers of Sweden; and the Rothschilds of England and France.

who have put hundreds of thousands—and in some cases a million dollars or more—into Padwa's company or one of its limited partnerships are Henry Ford; General Motors heir Stewart Mott; William Weyerhaeuser of the Weyerhaeuser Corporation; and Michel Fribourg, chairman of the Continental Grain Corporation.

In 1980, Padwa became a member of the board of Biotechnology Investments, Limited, a London-based venture-capital firm established by Lord Victor Rothschild* of N. M. Rothschild & Sons.

In its first six months of operation, Biotechnology Investments raised $46 million from investors eager to bet on biotechnology. Four American-based companies—Agrigenetics, Applied Molecular Genetics, Applied Biosystems and Repligen—were among the first companies in which the new Rothschild account invested.

Yet Padwa discovered that his plans for Agrigenetics would require more money than he could raise from his well-connected friends. And that's when he met S. Leslie Misrock, a patent attorney with the New York law firm of Pennie & Edmonds. Misrock told Padwa that the $5 to $10 million he planned to raise among his immediate circle of friends was "much too small." Misrock taught Padwa about Section 1235 of the Internal Revenue Code, which allows income from patents to be treated as capital gains. Coupled with financing opportunities available through "research and development limited partnerships," which had been created by Congress in 1981—wealthy investors could obtain federal income tax deductions—and win royalties in the event that the partnership hit big with some new product or invention—by putting their money into private research ventures created by companies like Agrigenetics. Moreover, any income accruing to investors from such partnerships would be treated as long-term capital gains, taxed at a maximum rate of 20 percent.

By late 1981, with Misrock's help, the brokerage house of Op-

*Lord Rothschild, who holds a Ph.D. in biology from the California Institute of Technology, is a scientist of some renown and wide experience. Before retiring from the Royal Dutch-Shell Company in 1970, he moved the company into single-cell protein research in the late 1960s, as well as other research, including fermentation technologies and the study of the insecticidal properties of certain bacteria. In 1970, he headed a "think-tank" for the British Government that focused on Britain's research base and national goals, culminating in what was called the Rothschild Report.

penheimer & Company had sold $40 million worth of shares in Agrigenetics Research Associates to interested investors. By January 1982, a second offering raised an additional $15 million. Compared to Genentech's highly publicized $35 million stock offering of 1980, this $55 million in private funds was very creative indeed. Agrigenetics was a trailblazer once again; the first biotechnology company to take advantage of a novel scheme for financing research and development. Others would soon follow.

In securing the financing for Agrigenetics, Padwa has generally avoided major corporations. However, in June 1982, the Kellogg Company—the Battle Creek, Michigan, breakfast-cereal giant—invested $10 million in Agrigenetics, and now owns about 5.6 percent of it. William LaMothe, Kellogg's chairman, said that the investment would give his company access to "a field becoming increasingly important to Kellogg's long-term future." Kellogg and Agrigenetics are conducting joint research on increasing the protein content in cereal grains, mold resistance in corn, and new corn lines capable of producing higher yields of corn starch.

Besides Kellogg's investments, Hoffman-La Roche, a major Swiss pharmaceutical corporation, has put more than $25 million into Agrigenetics, and today owns about 15 percent of the company, as well as conducting collaborative research with it. Hoffman-La Roche is quite interested in plant biotechnology, because it derives about 50 percent of its pharmaceutical products from plants. This pharmaceutical giant may also help Agrigenetics develop an "artificial seed" technology for encapsulating mass-produced, pre-seed embryos, expected to come with biotechnology. Joint research is now being conducted by La Roche and Agrigenetics on the chemistry and biochemistry of the plant proteins called lectins; on antiviral agents in plants; and on the effects of interferon on virus diseases in plants.

Throughout all of his empire building and various financial transactions, David Padwa has not forgotten his friends. Many of them—both old and new—serve as principals in either Agrigenetics or one of its partnerships. Gilbert de Botton, F. Palmer Webber, James Niven, S. Leslie Misrock, and Bob Appleman all hold seats on the board of directors, or are otherwise involved in one of the company's limited partnerships. The Rothschilds, who in Padwa's words are "truly backers," own about 12 percent of Agrigenetics.

Those involved with Padwa in the business often give glowing reports of the man and the company. Bob Appleman, for one, a past president of the American Seed Trade Association, speaks very highly of Padwa. "David Padwa has done everything he said he would do," says Appleman, referring to Padwa's purchase of Arkansas Valley Seed Company and the terms of the agreement. "I've found him to be a very honorable person. He's a go-getter, and a very sharp person. He thinks deep, speaks well, and he's on his way to going places." Appleman believes that Agrigenetics is "a year ahead of everybody else in terms of research," and that in the years ahead, "you're going to be hearing much, much more about Agrigenetics."

Some people in the investment community tend to agree with Bob Appleman about Agrigenetics' future, and give the company high marks as a going concern. E. F. Hutton's Zsolt Harsanyi, a biotechnology analyst, says that "Agrigenetics is top of the lot," and adds that he expects it to "grow into a company with a half billion in sales in less than a decade."

"UNDER GENETIC CONTROL"

With Agrigenetics, David Padwa is attempting to build one of the prototype plant genetics businesses of the future. So far, Padwa and his scientists have made plans for developing new crop varieties, unraveling the riddle of nitrogen fixation in cereal crops, and developing hybrid wheat and cotton strains. Agrigenetics' scientists claim they have developed a new kind of harder, hybrid tomato that costs less to produce and is easier to process. They say they have also developed a high-protein wheat strain yielding flour that absorbs water 30 percent faster than ordinary flour, which is ideally suited to the baking industry. For Padwa, all of this begins with the gene.

"The annual primary genetic input to the American agricultural system," he explained to an audience of business executives in 1981, "has a current value close to $5 billion. This consists largely of the DNA encompassed by seeds, vegetative cuttings, and to some extent in [livestock] semen and ova.

"From this genetic beginning," he told his listeners, "our agriculturalists produce and ship an annual biomass with a value of about $150 billion at the farm gate. The entire process whereby $5 billion is turned into $150 billion is, of course, under genetic control," he continued. "Manipulations in this area become the science of yield.

"Traditionally, throughout the food industry," Padwa explained, "crop biomass is still treated as fungible stuff, where any lot is nearly interchangeable with any other. . . . In the future, crop biomass will not merely be a fungible raw material on which various operations are performed, but will have particular genetically engineered characteristics valuable to an end user." Moreover, according to Padwa, "such genetic improvements will flow through an 'identity preserved pipeline' from scientist, to seedsman, to farmer, to food processor." Achieving these "end-quality" changes in food products genetically "promises to be substantially more cost effective than subjecting food raw materials to elaborate chemical and industrial treatment," says Padwa.

Another and major question, however, is whether such end-quality characteristics—usually sought to facilitate food processing—will come without eroding present nutritional qualities.

OF FARMERS, COSTS, & CHEMICALS

Although there are a great number of farmers throughout the world, and even many in the United States, who save and re-use their crop seed year after year, David Padwa says that this practice is "obsolescent." In the United States, he says, "farmers are happy to give seedsmen 25 percent of the added value they receive in genetically improved seed.

"The name of the game in farming all over the world," says Padwa, "is minimization of risk, rather than maximization of opportunity; and every farmer knows that if he is using expensive labor, putting on expensive chemicals and expensive herbicide on expensive land, and paying expensive interest and expensive tractor time—he's not saving money or minimizing risk by buying cheap seed."

Padwa believes that the work of Agrigenetics will reduce the cost of farming by genetically incorporating into crops such as soybeans the trait for disease and pest resistance, which today must be controlled with chemicals. Agrigenetics' President Sam Dryden sees an agricultural future without chemical pesticides. "In two decades," he says, "we won't be spraying crap on plants anymore." Padwa has much the same view. "In time," he says, "the entire insecticide industry may be totally displaced by plant genetics."

But whether the chemical industry will see it that way is quite another question. In fact, Dryden, a former employee of Union Carbide, knows full well what a vested interest in the chemical industry can mean. "David hasn't worked in a Union Carbide," says Dryden of Padwa. "He doesn't understand the momentum behind plant, equipment and chemistry—and *careers* in chemistry. People there want to think, naturally, 'what I did with my life is important.' Now [with biotechnology] it's not important. People can't admit their life's work is bad, or gone, or obsolete. The Shell's, etc., . . . These guys have made $300 to $700 million investments to build huge chemical plants. They have a vested interest in chemicals." On the other hand, Dryden adds, "the Du Pont's, ICI's, BASF's, etc, are smart enough to say, 'what technology's next.' . . . Monsanto and Du Pont will say, 'if it obsoletes our business, let's move into it.' " At Agrigenetics, Dryden says, "there is no vested interest. We're a young company."*

BIOTECHNOLOGY & THE NEW BREED

David Padwa is a man with great faith in innovative science, small business, and the creative powers of entrepreneurs like himself. On the other hand, he holds some disdain for lumbering corporate giants, which he charges suffer from the classic big-business affliction of "analysis/paralysis." Says he, "I remember when some of the largest business equipment companies turned their nose up at Chester Carlson, the inventor of the xerographic process."

When asked by a *Forbes* reporter in 1982 what he might have done differently in his own business, Padwa replied: "I would have started at it earlier and hit it harder. I wouldn't have been so tentative the first couple of years. Instead of buying a company with a couple of million dollars, I would have started out looking for something

*Even the work at Agrigenetics is not completely free of pesticides. The company is conducting research on a seed-coating process it has patented that may, according to company literature, involve the use of certain combinations of "bacterial innoculants, plant starter fertilizers, pesticides, herbicide safeners, fungicides and growth regulating hormones."

for $10 million to $20 million. I think the main mistake at the beginning was not thinking big."

As a businessman and an intellectual, Padwa clearly enjoys the risk-taking and money-gambling aspects of American capitalism, particularly when mixed with science and technology. "The capitalization of technological possibility," he says, "is one of the most creative aspects of our economic system. It is at the heart and soul of our restless, innovative society." But it is science coupled with commerce that Padwa sees as the shaping force in society. "As Joseph Schumpeter was telling us fifty years ago at Harvard's Economics Department," he told a group of financial analysts in Boston in 1982, "science is the revolutionary force in this world, not the working class. Science, particularly in a business form. . . .

"Biology is now *the* major technological frontier—not solid state physics or integrated circuits," he tells an audience of financial analysts, advising them to broaden their "conventional definition" of high-technology stocks to include biotechnology.

Yet David Padwa, for all his flair and seeming acumen, is not the only "new breed" entrepreneur to enter the high-stakes race of agricultural biotechnology. Others like him are found throughout the new industry. They are, as a group, risk takers, uncomfortable with the strictures of large corporate bureaucracies or the plodding course of establishment science. They are people like Zachary S. Wochok, a plant physiologist who left a tenured position at the University of Alabama ("people thought I was crazy") and spent time in key posts at both the Monsanto Company and the Weyerhaeuser Corporation, but who is now president of the California-based biotech company Plant Genetics.

Or they are people like Ray Valentine, a co-founder of Calgene Incorporated, who tired of writing grant applications to the National Science Foundation and longed for more active application of his scientific knowledge. They are—people like Wochok and Valentine—impatient with convention, anxious to test their mettle.

"Monsanto and Weyerhaeuser," says Wochok, reflecting on his former employers, "are two very good, well-managed companies, and they are overwhelmed with great talent. You have to wait in line to do your thing." Waiting in line, however, is not Wochok's style, but his 5 years of experience heading up Weyerhaeuser's tissue culture program and several years with Monsanto working on the business side of that company's biotechnology program prepared him well for his new role as entrepreneur. Of his new company, Wochok says, "If you want a pension plan, this is not the place to be." Of the biotechnology business generally, Wochok says "the whole thing is a dice roll."

The "new breed" of business and science entrepreneurs now cropping up in biotechnology companies are not far removed from the experience of big corporations, and many, like Wochok, have worked for them, or like Padwa, have sold previous businesses to them. While entrepreneurs like Padwa and Wochok may serve modern society for a time, advancing a new idea into commercialization—and making money in the process—their companies are not usually survivors, especially when it comes to conducting expensive, long-term science and establishing a viable marketing system. Major corporations are the more enduring beasts.

LUBRIZOL BUYS AGRIGENETICS

In late September 1984, the Agrigenetics Corporation, facing a mounting debt load and a sour stock market,* was sold to the Lubrizol Corporation in a $110 million transaction. Lubrizol, an $800-million-a-year chemical maker headquartered in Wickliffe, Ohio, had already made other investments in biotechnology companies as well as the seed industry,† and was especially interested in plant genetics for developing vegetable oils and industrial lubricants. In

*Despite Agrigenetics' immediate financial problems, the company has become a formidable organization. It employs some 950 people in eighteen states; operates thirty field offices which manage 138,000 research plots at more than 2,000 sites around the country; has a network of 13,000 seed dealers, most of whom are farmers; and is spending about $20 million annually on research. The company's research facilities include over 100,000 square feet of laboratory and greenhouse space, among them, the Agrigenetics Research Park—the company's core genetics laboratory in Madison, Wisconsin (home of the University of Wisconsin, one of the top three agricultural schools in the country); and laboratories for applied genetics at its headquarters in Boulder, Colorado; tissue-culture work in Hollister, California; cereal chemistry in Scott City, Kansas; microbiology in Columbus, Ohio; hybrid cotton in Lubbock, Texas; and plant pathology in Farmington, Minnesota.

†In 1979, Lubrizol was one of the first major corporations to invest in biotechnology, purchasing a 25 percent share of Genentech. In 1982, the chemical maker acquired its first seed company, Sigco Research, Incorporated, a hybrid

fact, one of Lubrizol's seed company subsidiaries—Sigco Research—has developed a new sunflower variety whose seeds produce an oil that can be used in cooking at higher temperatures than vegetable oils now available. Lubrizol will plant 50,000 acres of the new variety in 1985, and has set up another subsidiary to enter the vegetable oil business. Now, with Agrigenetics under its belt, Lubrizol is in an even better position.

"The acquisition of Agrigenetics," says Lubrizol chairman Lester E. Coleman, "is like a ten-year jump for us. . . . We see this whole area [i.e., plant genetics] as a source of speciality chemicals [while] at the same time having a base of self-supporting business. . . . When all is said and done, we'll be number one in the oil seed area. I don't think anyone will come close to us."

But even before the ink had dried on the Agrigenetics deal, there was some speculation that Lubrizol was being eyed by even larger corporations as a possible takeover target. Royal Dutch-Shell was one of the companies rumored to have an interest in Lubrizol. But some observers doubted that Shell would acquire Lubrizol unless it was threatened by a hostile bid from another corporation. For it turns out that Lubrizol and Shell have some long-term joint ventures ongoing in the construction of new chemical plants. Shell is also Lubrizol's biggest customer for industrial lubricants.

However, what happened to the seed companies that became the Agrigenetics corporation, what soon befell Agrigenetics itself, and what might still happen to Lubrizol, is only a microcosm of the economic change that is sweeping through the seed industry and the nascent biotechnology industry worldwide—change that is based on genes and the economic wealth they will generate. Today, a dramatic consolidation of economic power is occurring throughout food-producing systems worldwide, and the Agrigenetics story is only one small part of a much larger story.

sunflower producer located in Breckenridge, Minnesota. Since then, Lubrizol has taken a 28 percent equity position in Sungene Technologies Corporation of Palo Alto, California, a biotechnology company also working on sunflower, and has acquired the Lynnville Seed Company of Lynnville, Iowa, a company specializing in soybeans. With the Agrigenetics companies, Lubrizol now owns 16 American companies in seed or seed-related businesses.

CHAPTER

6

FROM THE GROUND UP

I think it's immoral. . . . They've got control of the pump and now they've got control of the food when they have the seed.

—*A California farmer reacting to the Atlantic Richfield Corporation's acquisition of a local seed company in 1980*

When Marc Moret gazes out the window of his seventeenth-floor office suite in Basel, Switzerland, he can see the Rhine River and the French border in the distance. Also on the landscape before him is the sprawling chemical works of Sandoz, Limited, the $3-billion-a-year Swiss multinational corporation which the fifty-eight-year-old Moret now directs. What Moret cannot see from his office window, however, are the various American companies that Sandoz has acquired since the late 1960s. In fact, since 1974, Moret's company has spent more than $300 million acquiring American seed and agrichemical companies, the biggest of which is Minneapolis' Northrup King & Company, one of America's oldest seed companies.

Jesse Northrup and Charles Braslan came from the east coast of America to the growing Mississippi River town of Minneapolis in 1884. There the two men met and built a small mail-order seed business they hoped would grow with the country. The name of their company was Northrup, Braslan & Company. Later, a third partner joined the firm, and it was renamed Northrup, Braslan and

Goodwin. By 1890, the new company was prospering nicely, buying and selling seed in Minnesota's farm country. However, after the financial panic of 1893 and a warehouse fire, the company went bankrupt. Undaunted, Jesse Northrup reorganized with a new partner, Preston King, and in 1922 the company moved into hybrid corn breeding, and later into vegetables, alfalfa, sorghum, cereals, sunflowers, grasses, and cotton. By the 1960s Northrup King had become one of the world's largest "full-line" seed companies.

In 1968, the company became a publicly held corporation traded on the New York Stock Exchange. In the following eight years the company's revenues quadrupled, its stock price rose to respectable levels, and it became by all outward appearances one of the world's most profitable seed companies. Yet by 1974 and 1975—even though the company was by then posting profits of $14.2 million on $153 million of revenues—there were signs of trouble. In those years, the company was plagued by bad weather, poor planning in its seed fields, and corporate mismanagement, particularly in its hybrid corn division, and its market position began to erode dramatically. In 1975, the company's stock earnings and profits had tumbled by 50 percent, and by August 1976, its stock had fallen to $7.87 a share, off $20 a share from its peak mark. Recovery, at least in the eyes of market analysts, was a long way off. Then came the lifesaver.

In Sepember 1976, Sandoz offered to pay nearly $200 million in cash for Northrup King. Sandoz's offer of $19.40 a share for the company's 1.01 million shares of outstanding stock was more than $110 million above the company's book value and $80 million more than the value of its stock. "We were stunned," said Northrup King's president, James B. Massie, "We simply could not believe it." The *Minneapolis Tribune* called the Sandoz bid "a breathtaking offer."

Sandoz, which had already acquired the Rogers Brothers Seed Company of Idaho Falls, Idaho in 1974, was apparently very eager to acquire Northrup King. Over a three-month period, Sandoz had indicated it might be willing to pay fifteen to sixteen dollars a share for the Minneapolis-based seed company, but on September 15, 1976, three senior officers of Northrup King turned down a Sandoz offer. Five days later, on September 20, after Sandoz had raised its bid to $19.40 a share, the Northrup King board unanimously accepted the takeover offer.

At the time of the Sandoz acquisition, 10 percent of Northrup King's stock was held by directors and employees of the company, and as much as 50 percent of its stock was said to be closely held. When the company was finally sold, eighteen "corporate insiders" grossed more than $7 million on the transaction.*

Money, no doubt, explains why Northrup King's management and its board of directors were willing to sell their company. But "why," asked *Minneapolis Tribune* reporter Dick Youngblood, "was Sandoz willing to pay such a premium price?" Youngblood, who had been covering the ups and downs of Northrup King's business, and who had interviewed company officials about the Sandoz transaction, offered the following explanation for Sandoz's keen interest in the deal:

> The attraction for Sandoz is varied, but the key element is the long-term bullish outlook for the food industry in a world where rising population is crowding production capacity.
>
> Moreover, there is striking similarity between the seed industry and the pharmaceutical and specialty chemical fields.
>
> Both Sandoz and Northrup King are in basic industries, both understand that large sums must be spent on research that may not pay off for some time.
>
> Finally, Sandoz's agricultural chemical business faces, worldwide, the kind of environmental restriction now emerging in the United States. Thus, the opportunity to move into another basic agricultural industry with such familiar patterns is viewed as very attractive.

The early 1970s were a precarious time for world food supplies, a time of tight markets and rising prices brought on by a combination of events—severe winter weather in the Soviet Union during 1972–73 causing a significant shortfall there followed by huge Soviet purchases of American grain; meager harvests and famine

*Chief among the inside gainers were Preston King ($5,282,930), a member of the founding King family as well as a director and manager of the company's farm-seed division; D. Kenneth Christensen ($535,857), board chairman; and Richard Utter ($412,250), Allenby White ($288,672), Ralph Kelly ($233,382), and Kenneth Erikson ($126,100)—all officials with the company.

throughout Asia and Africa; falling world grain reserves and declining emergency food assistance; an American soybean embargo in 1974 to protect domestic supplies and consumer prices; and the Arab oil embargo of 1973–74 which contributed to rising oil prices and subsequently the rising cost of food production and distribution throughout the world. All of these events fomented the much-written-about "world food crisis" of 1973–74. And it was this food crisis that also helped to pique the interest of some multinational corporations in the food business and the basic industries essential to the production of food, such as the seed industry.

For many U.S. seed companies, this was also a peak earnings period, culminating ten years of generally good profits. Goldman Sachs & Company, the New York investment house, found that between 1965 and 1974 the three leading U.S. corn seed companies—Pioneer, DeKalb, and Northrup King—had achieved annual sales growth of 14 to 15 percent for corn seed, with pretax profits averaging between 20 and 25 percent a year. L. William Teweles, a seed-industry consultant well-known for helping lure prospective corporate buyers to the United States, explained in 1975, "this is one hell of a profitable business." And by this time, several industrialized nations, including the United States, had also enacted seed patenting laws.

Sandoz wasn't the only major corporation that bought up seed companies in the mid- and late-1970s. Celanese, Ciba-Geigy, International Multifoods, ITT, the Kay Corporation, KWS AG, Occidental Petroleum, Olin, Royal Dutch-Shell, Pfizer, Purex, Southwide, Stauffer Chemical, and Upjohn also acquired one or more American seed companies during the 1970s. Similar acquisitions were also occurring in Australia, Canada, Japan, and Western Europe.*

Many of the corporations investing in the American seed industry knew exactly what they were looking for. Typically, the hybrid-corn-seed companies were at the top of their list, especially the ones with experienced plant breeders and a record of market performance. Eight of the nation's ten leading hybrid-corn-seed compa-

*See, for example, Pat Roy Mooney, "The Law of the Seed: Another Development and Plant Genetic Resources," *Development Dialogue,* 1983:1–2, the Dag Hammarskold Foundation, Uppsala, Sweden. Mooney identifies 762 "corporate changes" in the seed industry in eighteen countries that have occurred since the adoption of seed-patenting laws in those countries, or since 1970.

nies—which together account for more than 80 percent of all hybrid corn seed sold in the United States—are now owned by larger corporations. Of the top ten, only Pioneer Hi-Bred—a firm that holds nearly 40 percent of the market by itself—has remained independent of mergers or outside corporate influence.*

Foreign corporations buying up American seed companies knew what they were looking for too. When Claeys-Luck, the second largest seed company in France, purchased the California-based Neuman Seed Company in August 1980, the French firm was eyeing American and larger world markets. Claeys-Luck holds exclusive sales and distribution rights to certain French crop varieties which could be adapted to, and marketed in, the United States. In addition, with facilities and outlets around the world, Claeys-Luck will be able to market Neuman Seed varieties more widely as well.

Even predominantly mail-order, home-gardener type seed companies attracted the eye of corporate investors. In 1978, for ex-

*The eight firms today ranked behind Pioneer are listed below in order of the approximate standing in the hybrid-corn-seed market, with their new corporate owners, and the year in which the merger or acquisition occurred:

#2-DeKalb-Pfizer Genetics	Pfizer acquired Trojan in 1973 and then merged Trojan with DeKalb in 1982.
#3-Cargill	Cargill acquired PAG in 1971 and ACCO in 1980.
#4-Ciba-Geigy	Ciba-Geigy acquired Funk's in 1974.
#5-Golden Harvest	Golden Harvest was formed by a merger of 6 smaller companies in 1973.
#6-Sandoz	Sandoz acquired Rogers Brothers in 1975 and Northrup King in 1976
#7-Upjohn	Upjohn acquired Asgrow in 1968 and bought O'Gold's from Central Soya in 1983.
#8-Lubrizol	Agrigenetics acquired Jacques in 1981, and Lubrizol acquired Agrigenetics in 1984.
#9-Stauffer Chemical	Stauffer Chemical acquired Blaney Farms and Prairie Valley in 1978 and RBA in 1980.

ample, the second largest mail-order seed house in North America, the Joseph Harris Seed Company of Rochester, New York, was acquired by the Celanese Corporation; the largest North American mail-order seed house—Burpee Seed Company—was acquired by General Foods in 1970 and subsequently sold to ITT in 1978. Both Burpee and Harris were gathered up by corporate investors because they were, according to one industry source, "the only mail-order houses doing any significant amount of [plant] breeding."

The specific reasons for corporate investments in the American seed industry during the 1970s varied with each corporation. Some agribusiness corporations buying and selling wheat or corn in world markets wanted to improve their control over the seed-to-buyer movement of a particular agricultural commodity. Energy corporations were looking for ways to invest surplus capital, making diversification moves, or banking on a possible future in "energy" crops. Some food companies wanted to obtain a captive supply of the seeds and plants they used for food processing and manufacturing. Chemical and pharmaceutical firms were eyeing "synergistic" research and marketing strategies for seeds, agricultural chemicals, and fertilizers. Foreign corporations saw an opportunity to buy American firms at bargain prices with stronger currencies while penetrating American markets. Other corporations were and still are hedging, investing in a wide assortment of agricultural and food industry enterprises, seed companies among them.

By the late 1970s, however, there was an additional and perhaps more powerful reason for moving into the seed industry—the emergence of biotechnology. According to A. G. Laos of the Stauffer Chemical Company, all corporations now involved in plant genetics "realize that even if they have a breakthrough [in biotechnology] they are not going to be able to do much with it—if they don't have a good hold on the seed industry." In this regard, some of the big chemical and pharmaceutical companies were just catching up to David Padwa.

Corporations investing in plant biotechnology needed the seed companies' marketing networks. During the early 1980s, for example, corporations such as Atlantic Richfield, Lubrizol, Monsanto, Rohm & Haas, Pfizer, Sandoz, Stauffer, and Upjohn had purchased seed companies or had otherwise strengthened their positions in the seed industry through contracts or partnerships. Yet some people had already begun to raise questions about corporate inroads into the seed industry.

SEED MONOPOLY

During 1980 Congressional hearings on the seed-patent bill, several organizations, including the National Farmers Union, the Consumers Federation of America, the Environmental Policy Center, and Ralph Nader's Congress Watch all raised the issue of corporate control in the American seed industry.

Testifying before a House Agriculture subcommittee, Woodrow Wilson, then director of legislative research for the National Farmers Union, observed that "Domination of the seed industry by large, multinational conglomerates may leave farmers with few alternative suppliers of one of their most basic inputs. Just four companies—DeKalb, Pioneer, Sandoz, and Ciba-Geigy—account for approximately two-thirds of all seed sales in the United States for corn, America's biggest crop," said Wilson, and "these four companies also sell 59 percent of the sorghum seed in the United States."

Wilson also complained of rising seed prices and lack of competition. He pointed out that along with the acquisitions of seed companies, seed prices had risen rapidly since 1970, with an 150 percent increase in the total cost of seeds and plants to American farmers in the five years between 1972 and 1977, and an 164 percent rise in the average cost of seeds and plants per farm. Raising the spectre of monopolization by crop and by region, Wilson noted that "Major seed companies often specialize in certain varieties and concentrate their marketing strategies on a regional basis, reducing effective competition. The growing potential for price manipulation," concluded Wilson, "may cost American farmers hundreds of millions of dollars each year."

Some businessmen with small seed companies have also expressed concern about the emerging influence of major corporations. George Pickering, owner of the Pickering Seed Company of Lewisville, Indiana, says he has watched big energy and petrochemical concerns buying up seed companies. Pickering explains that one seed business he was involved with in the 1960s—Midwestern Associated Growers, since acquired by Royal Dutch-Shell's Kansas-based company,

North American Plant Breeders—didn't have enough financial strength to compete. "It's a shame for all these seed companies to be controlled by a few large corporations," says Pickering. He adds, however, that he is not against big business, but thinks it is "important to keep as much diversity in the ownership of the industry as possible."

Others in the seed industry do not share Pickering's views. They believe that the industry is inherently competitive, and that domination by a few companies is implausible. "You just don't dominate the seed business," says Bob Hartmeier, a California production plant manager for Upjohn's Asgrow Seed Company. He points out that "There are a lot of people in the seed business, and it's easy to get into. . . . You just contract a grower, sell your seed, and branch out from there." Jeff Schrum, director of Asgrow's research station in San Juan Batista, California, agrees with Hartmeier: "I don't see how a few conglomerates could control seed sales. . . . The germplasm is spread all over, and there is a diversity of markets."

However, in the seed market that matters most—hybrid corn—it isn't so easy to set up a new business. Established companies now have an advantage. One Harvard Business School case study examining the position of Pioneer Hi-Bred International concluded: "Heavy research costs coupled with the lengthy development and testing period of a successful hybrid produced significant barriers to entry for potential new competitors and prevented rapid shifts in a company's market share." The only shifts in such a market now—and most other crops will invariably be hybridized—seems to be those brought about by larger *outside* corporations buying up established companies.

ESTATE TAX BLUES

Many seedsmen in the United States who sold their family-owned businesses during the 1970s felt they were forced into doing so because of the threat of confiscatory federal tax laws. In fact, the American Seed Trade Association (ASTA) went so far as to suggest that "the basic reason for the merger and sale of seed companies" during the 1970s was the U.S. tax code. ASTA argued that a combination of estate and capital gains taxes levied on retiring owners put "extreme pressure" upon family businesses to sell out. During the 1970s and 1980s, a number of the owners of "mom and pop" seed companies that were formed in the 1940s and 1950s were faced with retirement, and some found that there was no one left in the family who wanted to take over the business.

"You need to sell the business, or part of it, to satisfy the government upon death," says William Park of Park Seeds who has not sold his business but is nearing retirement himself. "And of course," he continues, "most would rather sell to a publicly owned corporation and get readily tradeable stock rather than a note from another seed company, or cash."

One owner of a midwestern seed company is particularly bitter at having to sell his business in 1978 for fear of federal inheritance taxes. "If I could have seen a way to pass this company on without the Federal government taking most of it in taxes," he said, "I would not have sold out to a larger corporation." Recounting a lifetime of building his business from scratch, he says: "I went without a lot of things that other people had. . . . I could have spent my money early. . . . I worked my whole life building this business, only to have the government waiting to take it all away. . . . it was a hard decision for me to sell this business; it was doing quite well, and I had no financial problems. But when I found out what the government would take, I had no choice."

In fact, this particular seedsman, who now manages his former company for a large chemical corporation, blames the federal government for the demise of small businesses generally. "Government doesn't like small business," he says, "they would rather deal with one big outfit than 20 smaller ones. The tax laws," he says flatly, "have been anti-small business." On the other hand, he says, publicly held corporations have lots of small shareholders, and don't face the estate tax problems that small businesses do.

TAKEOVERS & COMPETITION

Agrigenetics' David Padwa says that the "takeover" issue in the American seed industry is exaggerated, and that the only things really being "taken over" are warehouses. The only reason such warehouses are incorporated as businesses, says Padwa, is because they are real property, which is ideally handled or traded in corporate form. According to Padwa, half of the "seed companies" that claim membership in the American Seed Trade Association are "nominal" seed companies which primarily handle seed and do not develop their own breeding lines or do research—a key distinction between the "seed trade" and the "seed industry," although both terms are often used interchangeably.

Yet in 1984 there were fewer seed companies of any kind than in 1970, and certainly by 1990 or 2000 there will be fewer still.

In a 1981 staff document prepared by the Federal Trade Commission (FTC), it is noted that economic concentration and anti-competitive activity in the seed industry did not appear to be much of a problem. The FTC paper cited a "wide variety of seeds available from a large number of companies," as well as numerous varieties developed by USDA agricultural experiment stations. However, the FTC paper did point out that in the hybrid corn market, prices and economic concentration were "substantially higher" than in other sectors. The FTC paper also noted that farmers are generally required to return to seed companies to buy new hybrids, and that competing breeding activity at the USDA's agricultural experiment stations for top-line corn varieties might not be much of a competitive factor. The study also noted that 80 percent of the hybrid corn seed market "is shared by fewer than 10 large firms." For the seed industry as a whole the FTC paper concluded: "From an antitrust perspective, the seed industry, at least for the foreseeable future, is not likely to be the scene of substantial anticompetitive practices."* However, with the trend toward crop hybridization and the use of biotechnology, that could change rather quickly. And then there is the question of the impact of seed patenting on the industry's structure.

In a 1983 University of Wisconsin study, entitled "An Economic Evaluation of the Plant Variety Protection Act," prepared for the U.S. Department of Agriculture, agricultural economists L. J. Butler and B. W. Marion note: "There is little doubt that the PVPA does

*However, the FTC paper did add one important *caveat* in its study: "This is not to imply that there are no reasons for concern about the growing concentration of new seed research efforts in the hands of pharmaceutical, and especially pesticide manufacturers. Such acquisitions may have serious implications in terms of the genetic diversity of our crops and the possibility of their increased dependence on agricultural chemicals for survival."

provide a mechanism for increasing market power. This is part of the intent of any patent system. If, in the future, the majority of seeds sold in some species are privately protected varieties, then it is likely that some increase in concentration will take place; competitive market forces may be hampered to some extent. When, and to what extent this will occur, is impossible to predict."

THE ACQUISITION MAN

In the United States, the man to see if you want to buy or sell a seed company, or make a deal in agricultural biotechnology, is L. William Teweles of Milwaukee, Wisconsin. Teweles was himself a seedsman for twenty-six years, and, with his brother Robert, owned the L. Teweles Seed Company. That company was sold to Kent Feeds in 1972, and that's when Bill Teweles decided to go into the business of selling other people's seed companies to multinational corporations.

In plain language, Bill Teweles is a broker. He works to match up American seed companies or biotechnology firms with prospective corporate partners, foreign or domestic. By Teweles' own count, he and his consulting firm have initiated or assisted in at least eighty seed and plant science company acquisitions since 1973. Today, Teweles' client list includes some of the biggest names in industry, ranging from Anheuser-Busch, Exxon, and General Mills in the United States, to Royal Dutch-Shell, Mitsubishi, and Nestlé abroad.

Part of Teweles' business is to convince interested corporations that the international seed trade and agricultural biotechnology are well worth looking into. In a 1978 prospectus Teweles wrote, "for multinational corporations, seed lends itself to world-wide commercialization. . . . the global seed trade is one of the fastest-growing, most profitable industries in the food chain."

"In our opinion," says Teweles' associate James Kent, "the restructuring of the seed industry is in its infancy . . . Many believe that seed- and plant-science-related acquisitions, divestitures, diversification, vertical and horizontal integration, and the pooling of resources have just begun."

Teweles keeps regular tabs on American seed companies eligible for acquisition. He writes and often phones the proprietors of seed companies to find out their business status and to determine which ones might be interested in selling. "He calls a few times a year, at least," remembers Dennis Stamp, owner of Wilson Hybrids, Inc., a family-owned seed business in Harlan, Iowa. In the early 1980s, Stamp received mail from corporations such as Elf Aquitaine, the giant French petrochemical corporation, naming Teweles as the company's exclusive representative in the event the Iowa seedsman became interested in selling. In 1984, Stamp sold his seed company to LaFarge Coppee, a major French corporation and Teweles' client.

Beyond his role as an industry proponent and matchmaker, Teweles is also a provider of specialized information on the seed and biotechnology industries—expensive and very private information. In the late 1970s, he put together a mammoth document called *The Global Seed Study.* At its price of $25,000 a copy, only the largest companies interested in the world seed market could afford one. In December 1983, Teweles released a new 760-page study entitled *The New Plant Genetics;* it sold for $30,000 a copy. This study focused on how genetic engineering would impact the market prospects for twenty-eight important crop species, stating that "The new plant genetics is expected to add $5.6 billion to the annual value of crops before the year 2000," and that after that, the crop value from new plant genetics improvements would skyrocket to $20 billion annually.* Teweles increasingly finds himself researching prospective biotechnology partnerships and acquisitions as much as seed company buyouts.

At the conclusion of a slide show Teweles routinely gives when speaking about the U.S. seed industry, he says, "I find it very hard to believe that seed companies and genetics firms can exist in the future without one another." By the year 2000, says Teweles, "there will be about twelve U.S. seed companies operating on a worldwide basis, twelve in Western Europe, and five multinationals operating

*According to the Teweles study 85 percent of the estimated commercial impact of genetically improved seed is expected in ten major crops—wheat, barley, alfalfa, sugarbeets, corn, sorghum, cotton, rice, soybeans, and tomatos. These ten crops represent 80 percent of annual seed sales in eleven countries—United States, Australia, Denmark, Spain, Canada, France, Holland, United Kingdom, Japan, West Germany, and Italy. And these eleven countries account for 90 percent of annual retail seed consumption in the developed free world.

in Japan and the Far East."* And between now and then, Bill Teweles and his consulting firm plan to be helping that prediction come about.

AND LIVESTOCK GENETICS TOO

Just as the new biotechnologies are transforming the American seed industry and the entire science of plant breeding, new bioengineering techniques are also revolutionizing the American livestock industry and the science of animal breeding. Today, the animal embryo is the focus of attention for genetic alterations in cattle, swine, poultry, and sheep, just as the seed is for genetic alterations in crops.

Through techniques known as "superovulation" and "embryo transfer," beef and dairy herds can be greatly expanded, almost without limitation. Livestock embryos can now be frozen, stored and later transferred to surrogate mothers, greatly enhancing their use as well as the geographic distribution of superior genetic material.† Scientists are also perfecting ways to split developing em-

*In contrast to Bill Teweles' view of the seed industry is that of the American Seed Trade Association, whose former representative Harold Loden denied that multinationals were making inroads. "The U.S. seed industry," said Loden in 1979, "is composed of several hundred companies from small to large and the premise that the responsibility for plant variety development is largely in the hands of multinational companies is unfounded. . . ." While spokesmen for the seed industry typically say there are as many as 600 seed companies, Teweles finds there are about eighty "fully integrated" American seed companies with annual revenues of $10 to $50 million, and which engage in all of the four principal seed trade functions: research, production, conditioning, and marketing. And it is these highly integrated companies that have become the targets for acquisition.

†Through the use of follicle-stimulating hormones, cows can be made to release ten to twelve eggs at a time instead of one—i.e., superovulation. In a few cases, scientists have succeeded in stimulating the release of as many as one hundred eggs in a donor cow, not all of which have survived. After a superovulating cow is artifically inseminated, the resulting multiple embryos in her womb can be flushed out while still microscopic, and then "stored" in liquid nitrogen or transferred into the wombs of surrogate mothers. With these techniques, a high-quality cow can be made more or less the equivalent of a stud bull, spreading her superior genetic materials over many offspring that she conceived, but did not herself bear. Already, one superior donor cow has produced 110 calves using these techniques, which is the reproductive equivalent of about 150 years in normal cow/calf production time. In 1982 some 35,000 calves were born as a result of embryo transfer techniques. Today, there are approximately 140 companies in the United States performing embryo transfers, doing about $25 million worth of business annually.

bryos again and again (before they differentiate, up to the 60-to-80 cell stage), in effect making many sets of identical twins and implanting these embryos into surrogate mothers. It is also possible to take two or more embryos from popular breeds and fuse them together to mix their traits, and then transfer the resulting embryo to a recipient cow. Microscopic seven-day-old embryos can now be selected on the basis of sex, and scientists foresee the day when they will be able to microinject genetic traits for growth and other desired characteristics into such embryos.

"The future in the embryo business should almost be like an assembly line," says Charles Srebnik, Chairman of the Board at Genetic Engineering, Incorporated, a Denver-based livestock genetics company. "If we can create an animal that would have greater milking capabilities or greater capabilities of converting feed to beef, or an animal that will have twins consistently, that really is where the embryo business is going."

And just as the changes in plant biotechnology and crop genetics have drawn a number of major corporations into those fields, a similar pattern of corporate investment and research has come to the field of livestock genetics.

"Obviously," says Robert Zimbelman, manager of experimental agricultural sciences for the Upjohn Company, "both the feed industry and animal drug industry have a stake in the genetic makeup of the animals to be fed or treated and how this might change in the future." Alteration of the genetic basis of food-producing animals, Zimbelman explains, could influence their response to required nutrients, their efficiency in the use of nutrients, or their susceptibility to diseases. Zimbelman suggests that drug and feed companies not now in the animal genetics business need to think about doing so for two reasons: as a new business venture for supplying animal products to the world, and as a "defensive mechanism" for survival of their existing businesses.

And a number of corporations operating in the animal products market in the United States have a lot to be defensive about. Many do a wide-ranging "animal health" business, selling everything from specialized feed additives to species-specific growth hormones. Pfizer, American Cyanamid, Cargill, DeKalb AgResearch, Diamond Shamrock, Eli Lilly, International Minerals & Chemicals, W. R. Grace, International Multifoods, Merck & Company, Monsanto, the

SmithKline Corporation, and Upjohn are among those corporations already involved in the conventional livestock products market. At least twenty such corporations in the United States are now pursuing their own biotechnology research programs in the animal health field, and many also have investments in, or research contracts with, biotechnology companies.

In May 1984, for example, H. J. Heinz Co. signed a three-year biotechnology research contract with Biotechnica International, Incorporated of Boston to pursue among other things, "products of interest to the animal feeding industry." The SmithKline Corporation is backing biotechnology work at the Cetus Corporation, and American Cyanamid is supporting research at Molecular Genetics. Both biotechnology companies are working on livestock vaccines, with Molecular Genetics planning to produce as many of thirty different animal vaccines over the next six years. Applied Molecular Genetics, Inc. (AMGen) of Thousand Oaks, California—partly backed by Abbott Laboratories—is working on a hormone for chickens that would speed the growth of broilers by 15 percent and would bring the birds to market size a few days earlier than is now possible.*

FOOD-CHAIN GENETICS

While some corporations are investing in biotechnology from the standpoint of developing either new crop varieties *or* livestock breeds, others are investing in bioengineering for both crop *and* livestock applications, as well as the myriad of products such as feeds, fertilizers, and pesticides used to sustain crops and animals in agricultural production. Such corporations are eyeing the genetics of the entire food chain as a whole, seeing clearly the genetic thread that runs through a broad spectrum of agricultural and food-

*Worldwide, the potential market for animal vaccines and various kinds of livestock growth hormones is huge, conservatively valued at between $700 million and $1 billion. "The potential for making money from veterinary vaccines is enormous," says Vincent Scialli, manager of Genentech's animal sciences group. "There are over 20 billion farm animals in the world that are susceptible to diseases that vaccines should be able to prevent." In 1981, more than 500 million conventional doses of hoof-and-mouth vaccine were used in South America. And there are at least two dozen animal diseases for which no effective or economical vaccines currently exist.

producing products. They are typically acquiring seed or animal genetics companies, buying equity in one or more biotechnology companies, conducting in-house research in crop or animal science, and/or entering into university agricultural research contracts. (See Appendix.)

As early as 1972, for example, Monsanto, the St. Louis agrichemical giant, took a gamble on an unknown company named Genentech, acquiring a small share of equity. By 1980, that investment had ballooned twenty times in value and Monsanto became a believer in biotechnology. In addition to Genentech, Monsanto has also invested in three other genetics firms, has ongoing research projects at several universities, and has built a new Molecular Biology Center at its corporate headquarters in St. Louis.

In 1980 Monsanto put together $100 million in investments to form a nutrition unit focussing on new sources of animal feeds using genetic engineering techniques, and in March 1981, with Genentech, announced the development of a genetically engineered "bovine growth hormone" that could be used in dairy cows and beef cattle. Monsanto studies on animal growth hormones found that such substances could "enhance meat production" and that dairy cows using the hormone could produce more milk without increasing feed requirements. Genentech and Monsanto have a joint development agreement to produce various kinds of growth hormones for food animals.

In the plant area, Monsanto has a substantial in-house biotechnology research program, which has been recently buttressed by several outside acquisitions. In 1982, for example, Monsanto acquired DeKalb's hybrid wheat program, and by 1983, the company had acquired Jacob Hartz Seed Company, an Arkansas company specializing in soybean seed.

"We feel very strongly," says Monsanto's director of plant sciences James E. Windish, "that in the 1990s, some combination of wheat and soybean breakthroughs will generate $300 million to $500 million in annual revenue for Monsanto." Monsanto's management sees huge potential in biotechnology. "I believe—and many hard-headed scientists agree with me," says Monsanto's Howard Schneiderman, director of R & D, "that with the new biotechnology almost anything that can be thought of can ultimately be achieved—new organisms, new limbs and organs, new treatments for disease, new ways of controlling pests, crops which produce their own pesticides, disease-free domestic animals, whole new industries that will sell products that even today cannot be imagined, let alone made." Monsanto, of course, is not alone in seeing biotechnology's broad-based potential.

W. R. Grace, the nation's fifth largest chemical company, has dabbled in the seed and livestock genetics industries since the 1960s. As early as 1965, Grace acquired the Rudy Patrick Seed Company of Kansas City, and two years later purchased Pfister Associated Growers, another seed company in Illinois. Grace soon tired of these companies' low profitability, and in the early 1970s sold them to Shell/Olin and Cargill, respectively. But Grace had also acquired American Breeders Services (ABS) in 1967, a Wisconsin livestock breeding company. ABS is the nation's largest dairy breeder, and is also involved in breeding beef cattle. Meanwhile, W. R. Grace has become one of the nation's largest suppliers of beef feed supplements, especially to feedlot operations in the Central Plains states. Grace also sells premium feeds to hog producers in Iowa, Nebraska, and Wisconsin.

In December 1981, Grace became a 16 percent owner of AGRI, Inc., a Berkeley, California, bioengineering company working to develop genetically engineered animal vaccines, and the company has at least three contracts with universities to develop genetically engineered vaccines. Says Robert J. Klunge, head of Grace's technical research group: "[Agricultural] vaccines available today are far from perfect, but we are in a good position to improve them."

But Grace's biggest deal in agricultural biotechnology came in June 1984, when it formed a joint venture—named Agracetus—with the Cetus Corporation's agricultural operations in Madison, Wisconsin. Prior to this transaction, Cetus Madison was one of the largest and best known agricultural biotechnology efforts in the country. Grace's 51 percent share of the venture is valued at more than $60 million. "This collaboration in biotechnology for agribusiness," said Grace executive vice president Lloyd L. Jaquier, "should enable both of our companies to stay at the forefront of this developing field by drawing on the expertise of the parent companies. Together, we expect to build a niche position in this new specialized, agricultural business." But Grace was also investing for other reasons. Because "the heyday of hydrocarbon chemistry in such fields as selective herbicides and insecticides" might be coming to an end, said Jaquier, Grace wanted to be ready to meet the future.

Agracetus plans to genetically improve corn, cotton, soybeans, wheat, and rice; to develop new agricultural microbes; and to produce new livestock vaccines and novel growth hormones. Grace expects to see genetically modified crops on the market in five to ten years, and new animal health products in two to four years.

All of this investment, of course, has to do with food and population growth. "If we talk about the world's population growing by fifty percent over the next twenty years, and the world's arable land increasing by only four to five percent," says Atlantic Richfield vice president Jim Caldwell, "that means we're going to have to have repeats of the productivity or yield increases that we've had over the past 20 years. And that's going to come from applications of the new biotechnology."

Atlantic Richfield owns two seed companies and has established a $10 million in-house agricultural genetics subsidiary called the Plant Cell Research Institute. "Over the next twenty years," says Caldwell, "there's going to be a lot of technology brought to bear on agriculture. . . . This is a revolution in agriculture. . . ."

J. Eugene Fox, director of ARCO's Institute adds, "ARCO isn't interested in immediate returns on its investment here. We're probing the long-range possibilities." Although ARCO is working in biotechnology to be poised for a non-renewable energy future when the oil runs out, they are also investing for other reasons. "Their interest," says Fox, "is food."

In November 1980, when Archie Dessert sold his Imperial Valley vegetable seed company to ARCO, he was looking at the financial resources that the oil giant could put into research and development. Dessert noted that his company had never ventured into DNA research because the technology was too expensive. "We always had R and D," said Dessert, "but it has been difficult to expand it to the degree . . . it needs to be expanded. . . ." When he sold his company to ARCO, Dessert believed that the larger corporation would move his company into more basic DNA research to improve seed varieties, develop new vegetable hybrids, and produce varieties with better disease and insect resistance. "Research and development is the name of the game in the seed business," said Dessert, "and ARCO already gambles millions every time they drill a well."

Similarly, when Northrup King sold out to Sandoz in 1976, one of the immediate benefits to the seed company was a much-needed infusion of cash for research and development—a much-appreciated fact by Northrup King management. Says NK president Doug Lohman: "European companies in general, and Sandoz in particular, are long-term planners compared with American businesses. As an American, I'm used to the short-term view. I've lived it all my life. Even now, six years after Sandoz purchased Northrup King, I am still amazed at Sandoz's willingness to plop down $3 million or $5 million today when the payout may be ten or more years from now."

Some corporations like ARCO and Sandoz have very deep pockets when it comes to financing what they perceive as important future markets. But increasingly, ARCO and Sandoz will find themselves in the company of other corporate giants also moving into the business of agricultural genetics.

At his office in Dublin, California, ARCO's Gene Fox points out that "all the large oil companies have some involvement in plant biotechnology," and he ticks off their names—Shell, Chevron, SOHIO, Phillips, Exxon, and Standard Oil of Indiana. Asked if one should be worried about such companies controlling the genetic base of food production, Fox replies, "twenty or thirty large corporations will dominate agricultural biotechnology, but it will still be very competitive."

Major corporations with investments in the seed, livestock genetics, and/or biotechnology industries, are probably in the best position to win the agrigenetic sweepstakes for several reasons: they have enough "patient money" to invest in research, they have commercial agricultural products which are already sold to farmers through existing distribution networks, they have solid bases of corporate income and established rates of return for investors, and they have the ability to wait out the results of long-term research projects. Moreover, such corporations have the regulatory experience, the legal talent, and the political clout to see to it that their interests are protected and advanced.

The bottom line in all of this means that the emergence of the hundreds of new biotechnology companies will probably be short-lived, with only a few "survivors" making it as independent companies.* This does not mean, however, that the biotechnologies will die, or that new genetic products will not be forthcoming; only that the science, developing product lines, and decision making behind those products will be in different hands. As for the seed industry, further consolidation and increased corporate ownership will undoubtedly continue, driven largely by the advances in the new genetic sciences and the prospects for booming world food growth.

Although corporate investments in, and acquisitions of, smaller seed, livestock genetics, and biotechnology companies may be necessary for the successful development and commercialization of new agricultural products, it nevertheless raises disturbing questions about control and decision making, direction of research, what kinds of products will be marketed, and how and when such products will be marketed. Particularly troubling from a market domination standpoint, is that a few corporations are now moving into a broad range of agricultural product research and development that can be integrated by way of the genetic command. For it is this long-armed, "broad-spectrum" influence in the food-making process from seed to table that is worrisome; an unmistakable concentration of power organized around the gene that should not be taken lightly.

*Since 1980, about twenty biotechnology companies have "gone public" on Wall Street attempting to raise money with varying degrees of success. The most fortunate of these have been the earliest such offerings, most notably Genentech and Cetus. Cetus managed to bank about $100 million from its initial offering, which it is still using as working capital. Other biotech companies haven't been so fortunate, particularly given a finicky market. But even among companies that have been successful with public stock offerings, there is also some reliance on corporate financing and/or corporate research contracts. In some cases, corporations which invested in genetics firms during the late 1970s have tightened their grip on the industry, expanding their equity shares in more than one company, and in a few cases, completely acquiring those firms that stumbled.

Biotechnology companies now faced with long product lead times, heavy and continuous capital requirements, and no commercialization or marketing experience, will have little choice but to turn to corporate backers, merge among themselves, or go bankrupt. Some firms plan to be absorbed by their larger corporate partners anyway. "Clearly," says market analyst Scott King, "it's the basic strategy of many of these concerns. How else to explain why a company that plans to remain independent would sell half of its equity to one or more large corporations?" *Chemical Week* agrees: "Many biotechnology companies may well fall into permanent junior status as research arms of major corporations, either as subsidiaries or suppliers."

CHAPTER

7

BIGGER COWS, FEWER FARMERS

> We're talking about a technology
> with a potential to transform agriculture.
> —*Thomas B. Rice*
> *Director of Plant Genetics Research*
> *Pfizer, Inc.*

Astronauts orbiting the earth in the space shuttle *Columbia* could see tiny green circles clustered on the American land surface below, and anyone peering out of the window of a passenger jet flying over the nation's heartland today can see them even more clearly—thousands of them, covering millions of acres of land from Idaho to Texas; green circles of land with crops growing on them. What makes them green, circular, and visible even from outer space is something called "center-pivot irrigation."

Tethered to a center-pole well, and propelled by a gas, diesel, or electric motor, a center-pivot irrigation system is essentially a revolving pipe on wheels. As it rotates, the center-pivot arm sprinkles water on the crops and land it traverses. Like the hand of a clock sweeping a clockface, it swings slowly over as much as 160 acres of land with each full rotation. In recent years, center-pivot systems—or "circles" as they are called in farm country—have proliferated across American farmland. In Nebraska alone the number of center-pivot systems rose from 2,500 to some 24,500 between 1973 and 1983. Millions of acres of land, much of it previously rangeland, are now irrigated by center-pivot systems and planted to crops such as corn, potatoes, and sugarbeets.

Center-pivot irrigation is not cheap; one fully-installed 160-acre system can cost as much as $70,000. In some states, with the help of non-farm investors, federal tax laws, and major corporate development projects, hundreds of pivot systems have been installed over thousands of acres, often dramatically changing local farming patterns. In 1975, for example, the Center for Rural Affairs in Walthill, Nebraska found that "new corn" areas had mushroomed across the entire state, and that regions once limited to grazing livestock were "suddenly sporting grain elevators and feedlots."

Center-pivot irrigation is one of the most visible parts of the nation's changing farm scene; one of the many technological advances that has altered the appearance of the agricultural landscape and helped to make the American farmer the productive envy of the world. Yet modern farm technologies such as center-pivot irrigation do not operate in a vacuum, but rather tend to be interactive. For example, not long after center-pivot irrigation started in the United States, new hybrid corn varieties began to be developed to work better under the "high-stress" conditions (i.e., many plants, high humidity, more pests) of irrigation.

"Hybrid selection is a vital component of irrigated corn management," says DeKalb AgResearch in a brochure entitled *Irrigated Corn Management*. "Maximum yields under irrigation require high plant populations, and this in turn necessitates hybrids that will respond to more intense management . . . hybrid[s] that can take the crowding and still maintain stalk quality and ear development."*

Pointing to its own research, DeKalb claims that "Top irrigated yields demand at least 26,000 plants per acre," and tells farmers that "with the right hybrid, water and fertility program," still higher yields may be possible, even at planting rates above 29,000 plants

*DeKalb sells both hybrid corn seed and irrigation equipment. Since its 1975 acquisition of the Lindsay Manufacturing Company—whose motto is "Lindsay Makes It Rain"—DeKalb has been selling center-pivot and lateral movement irrigation systems. "Many scientists feel the genetic potential of present day hybrids may run as high as 600 bushels per acre," says DeKalb. "But while genetic potential has been increasing, rainfall hasn't. Irrigation offers today's corn grower the opportunity to set and consistently achieve yields and profits far beyond the wildest dreams of the nonirrigating corn producer."

per acre. Yet more plants per acre in a high-humidity irrigation environment means a greater likelihood of crop disease, blight, and insect damage. Plant breeders, to their credit, have tried to breed into the new hybrids the characteristics needed to resist the "stress-related" factors associated with irrigation, but such crops still show a great if not increasing need for fertilizer, insecticide, and herbicide. Conveniently for irrigators, fertilizer and pesticides can be applied directly to crops through sprinkler irrigation systems.

As this example illustrates, chemicals, machinery, and genetics are the ingredients of agriculture that have evolved an ever-closer working partnership in farm country during the past few decades. This partnership is a high-tech, capital-intensive arrangement that will soon be augmented in new ways by genetic engineering and biotechnology, and one that may soon bring dramatic change to the practice of farming. The key question for farmers everywhere, however, is: will agricultural biotechnology continue to serve the interplay between chemicals, machinery, and capital, or will it instead help to cut costs, reduce capital outlays, and diversify the agricultural landscape?

A SHIFT IN POWER

The American farmer once served as the sole craftsman of agriculture, assembling, juggling, and mixing the ingredients of farming—seed, soil, his own labor, and the right combination of crops and livestock. Much of what he needed for farming was taken from his own land; grain was saved for seed, animal manure was spread for fertilizer, and crops were used for livestock feed. Mixing these home-grown ingredients with his own hard work, the whimsical elements of nature, and a bit of intuition, the farmer hoped for a good harvest.

Today, the farmer is still a mixer and juggler of farm ingredients, but the new technologies of agriculture, rather than being of his own making, are increasingly manufactured away from the farm and sold to him by businesses and corporations that specialize in producing them. In fact, about 70 percent of all the farm ingredients now used in agriculture come from the "nonfarm sector" of the economy.

American farmers spend nearly two-thirds of their cash receipts each year to purchase farm supplies, and selling them the goods and services they need to grow crops and raise livestock is big business, worth billions each year. In 1981, for example, farmers spent $18 billion for purchased feed, $9 billion for fertilizer, $3 billion for pesticides, $4 billion for seed, and $9 billion for farm machinery.* Worldwide the markets are much more lucrative.

The manufactured ingredients of agriculture have contributed dramatically to increasing American farm productivity. Corn yields, for example, have doubled in the last twenty years. Indeed, today's average farmer produces enough agricultural bounty to feed seventy-eight people. Yet what is now called the productive power of the American farmer is not really his power at all, but rather those who supply him. The power of productivity has moved off the farm, and in a sense to the city—to the university and the corporation—to the centers of high science.

FARM CONSOLIDATION

Since the late 1930s, the number of farms in the United States has dropped from a peak of 6.8 million to a total of about 2 million today.† Farmers now comprise less than 3 percent of the American population—down from 25 percent in the 1930s. At the same time, farms have continued to become larger. The average American farm has grown from about 200 acres in 1950 to 450 acres in 1983.‡

*Seed companies, chemical firms, and pharmaceutical corporations spend a great deal of money wooing farmers through advertising. Five American seed companies—Pioneer, DeKalb-Pfizer, Northrup King (Sandoz), Funk's (Ciba-Geigy), and Jacques (Agrigenetics)—each spend more than $1 million annually on print advertising. In the area of agricultural chemicals, animal drugs,and antibiotics, Eli Lilly, American Cyanamid, and Ciba-Geigy each spend more than $4 million on print advertising; and Dow, Du Pont, and Stauffer spend $2 to $3 million annually. Even more is spent for TV and radio time.

†Another, more dramatic illustration of this change is the decline in the number of counties in which agriculture accounts for 20 percent of the local income. Thirty years ago there were 2,000 such counties; today there are less than 700.

‡These trends are expected to continue in the future. The USDA has warned, for example, that by the year 2000, 1 percent of all farms will account for 50 percent of all food production; 4 percent of the farms will hold 60 percent of all farmland; and the number of young persons entering farming will decline by 40 percent. Given present trends, in other words, the nation will lose 700,000 farms by the year 2000—and that is probably a conservative estimate.

Farms and farmers "go under" for a variety of reasons: drought, pestilence, poor prices, speculation, urban sprawl, taxes, indebtedness, lack of heirs, and high interest rates among them. Usually, though, a combination of circumstances prevails upon the farmer over one or more years to contribute to bankruptcy or the farmer's decision simply to get out of the business. The important distinction here is that there are voluntary reasons for going out of business, such as the retirement of older farmers, and involuntary reasons, such as low crop prices. In the latter category of involuntary reasons are also federal loan and price support programs, federal and state tax policies, credit policies, agricultural research policies, and technology. While no single factor is typically responsible for delivering the final economic blow to a struggling farmer, together they can have a decided effect on his bottom line.

Technology has been important to the farmer because it has enabled him to save labor costs and expand the amount of land he can farm. Bigger tractors and more potent pesticides, for example, have been both labor-saving and land-expanding developments. However, the interplay of farm technologies has often created a need for, or prompted the growth of, other technologies. High-yielding hybrids, thicker planting patterns, and heavier fertilization, for example, have generated the need for larger, more powerful combines. Some technologies have contributed to agricultural specialization; insecticides, fungicides, and herbicides have facilitated one-crop farming, and in some cases planting the same "cash crop" year after year. Thus, with the introduction in the mid-1960s of special insecticides and fungicides aimed at controlling cutworms and rootworms, more farmers began repeatedly planting corn on the same land.

The net effect of modern farm technologies working together has been to increase farm size and agricultural productivity, thus decreasing the number of farms needed to do the job. While modern agricultural technologies have helped farmers and ranchers increase their productivity, they have also made farming and ranching more expensive,* and some might say riskier. In fact, the entire progression of modern, industrialized agriculture has been to move farming

*Irrigated land in the United States, for example, has tripled since 1940, and now encompasses about 60 million acres. The capital investment in the irrigation plant alone is staggering and a number of other inputs and services are needed to keep it going.

from a reliance on a few simple, mostly home-grown farm supplies to an absolute dependence on an extensive range of complex agricultural technology that has its origin away from the farm.

THE GENETIC CORE

Genes are at the hub of all agricultural production; they are the fundamental, unseen ingredient in every farmer's purchase of seed, orchard planting, stock, or new livestock breeds. And practically every other agricultural product used on a farm or ranch—from fertilizer and feed to pesticides and irrigation equipment—is, to some extent, dependent upon the genes of crops and livestock.

Consider seeds. They carry a certain genetic potential for crop yield, but achieving that yield clearly depends upon such external factors as water, climate, and soil nutrients. Seedlings, of course, must be protected, fed, and nurtured in many ways to realize their full genetic promise. Although most hybrid corn varieties have been developed for improved yield, a corollary requirement to obtain that yield is increased use of fertilizer. Thomas Urban, president of Pioneer Hi-Bred International, notes that his company is "steadily producing corn (varieties) with a high yield potential, so we will have to use more fertilizer." Consequently, when a corn farmer buys hybrid seed, he is not only buying the seed, but indirectly the genetic requirements for a whole range of other supporting inputs, from fertilizer and fungicides to herbicides and irrigation equipment.

Thus, in many ways, at least some of the costs of operating a farm or ranch are determined by the genetic code in crops and livestock. If the highest yielding crop varieties are those that respond primarily to fertilizer, irrigation, and certain plant growth chemicals, they will carry *genetically* certain fixed costs in their use. Similarly, a particular prize-winning livestock breed that presupposes the use of medicated feeds, growth hormones, and antibiotics is a breed that carries certain *genetically based* accessory costs.

Indeed, the genetic information contained in crop varieties and livestock breeds can influence, at least to some extent, the kind of machinery used on a farm; how much energy is consumed in planting, harvesting, or milking; how much feed is needed in a poultry, dairy, or cattle operation; and how much money is needed to equip and run the modern farm or ranch.

Thus far, genetic technologies in agriculture have not by themselves been so powerful, and have more often been used to "fit into" or accommodate existing practices and technologies. But this could change rapidly with the fine hand of gene splicing, for what is now approaching from the benches of biotechnology is a change in the *mix* of genetic and biochemical ingredients in agriculture; a change that could determine what farming will cost, and in the end, who will do the farming.

THE GENETIC VARIABLE IN AMERICAN AGRICULTURE

The history of the genetic variable in American agriculture is one of subtle rather than dominating influence, yet there are signs that crop and livestock genetics—working in concert with other technologies, federal farm programs, and tax incentives—have helped to change the nation's farming system, making it more specialized, less diversified, and increasingly the province of off-farm agribusiness.

For example, prior to 1960, and as late as 1964, cattle feeding in the United States was dominated by midwestern family farmers with small feedlots located in the heart of the Corn Belt. Ten years later most cattle were being fed in large, absentee-owned feedlots in the West and Southern Plains, particularly in the Texas panhandle. By 1978, the Southern Plains region was outproducing the Corn Belt by a two-to-one margin, or by more than 6 million head of cattle a year. This dramatic shift in the location and control of a major agricultural industry would not have been possible without the development of hybrid sorghum.

The development of hybrid sorghum in the mid-fifties, says the USDA, was possibly "the most important biological innovation . . . [and] . . . in conjunction with new irrigation technology,*

*Accompanying the introduction of hybrid sorghums in the 1950s was the advance of light weight, aluminum irrigation technology. Much of the sorghum expansion that occurred in the Southern Plains during the 1960s was irrigated expansion, for sorghum grown under irrigation typically yields three to four times more grain than that grown under dry-land conditions. Consequently, hybrid sorghum varieties were developed that could use more fertilizer and tolerate higher planting densities for use with irrigation. By 1969, more than one-fourth of all sorghum grown in the United States was irrigated, much of it in the Southern Plains.

enabled the Southern Plains states to become a major feed grain production area."

Sorghum, a crop believed to have been first domesticated in Africa, is a plant with an upright growth habit and a distinctive elongated grain head of several inches, often reddish brown, found at the top of the plant. Also called "milo" and "Sudan grass," sorghum has been grown in the United States since the 1800s. In 1905, short sorghum varieties were first introduced, and by the 1940s sorghum had been adapted for combine harvesting. However, it wasn't until the introduction of commercially developed hybrid varieties that sorghum production began to take off.* For example, sorghum production more than doubled in Texas between 1956 and 1966, rising from 124 million to 312 million bushels. By 1970, the Southern Plains states, and predominantly Texas, began taking an increasing share of all cattle fed and marketed, and midwestern agriculture became a little less diversified in the process.

In the nation's vegetable industry too, the genetic variable has helped to bring about profound change. Prior to World War II, vegetable production was scattered throughout the East and Midwest with farmers growing vegetables on small farms, often as a secondary crop. There was great variation in how farmers grew vegetables then, as well as the kinds of vegetables grown and how they were marketed. The vegetables used by the food processing industry of that era were not uniform in quality or grade, and often came in unpredictable and uneven supply streams, making "efficient" use of factories and processing equipment very difficult.

Following World War II, technological changes in food processing and freezing created a demand for vegetable production and vegetable varieties that could meet the new factory specifications. Vegetable varieties grown for food processing became quite distinct from those used in the "fresh market" sector.

*Federal farm programs also played an indirect role in the rise of sorghum in the Southern Plains. Acreage restrictions on wheat and cotton farming had the effect of moving more farmers into sorghum, sometimes the only other crop that could be profitably grown in areas such as the Texas High Plains. With a growing national demand for meat during the 1960s and 1970s, and a developing feedgrain surplus in the Southwest fueled by the highly productive sorghums, a new feedlot economy began to emerge in some parts of the Southern Plains. Tax shelters, however, also played a very formative role in shaping the new style of livestock feeding, luring wealthy, absentee investors into the business.

During the 1950s and 1960s, machinery and equipment engineers, working with plant breeders, began making advances with mechanical harvesters for peas, green beans, potatoes and tomatoes. "Development of mechanical harvesting equipment and vegetable varieties suitable for mechanical harvesting often were undertaken simultaneously," explains the USDA. "The mechanization of tomato harvesting, for example, was the result of a joint effort of engineers and plant breeders to simultaneously design a suitable tomato and a machine to harvest it."* With the subsequent adoption of the tomato harvester in California between 1969 and 1974, the average size of tomato farms in that state increased by 25 percent.

In fact, by 1977, California alone had captured more than 50 percent of the nation's processed vegetable production (due in part to federally-subsidized irrigation projects and cheap water) while the entire North Atlantic region accounted for a shrinking 5 percent share. And with that East Coast-West Coast shift in the processed vegetable industry, plant breeders and seed companies—working increasingly to "fit" new vegetable varieties to the demands of mechanical harvesting, factory processing, and long-distance shipping—would continue to play an important role in the production of vegetables.

In the poultry, swine, dairy, and cattle industries, too, breeding

*The close working relationship between vegetable processors and plant breeders is not new. As early as 1905, for example, one of the largest pea canners of that day, John H. Empson & Daughter Incorporated, hired Luther Burbank to develop a specific kind of pea plant. Empson wanted a pea variety that would produce abundant and uniformly small peas that would ripen at the same time to facilitate harvesting and processing. Empson knew that consumers would pay a premium for small peas, usually considered more flavorful. He also knew he could cut production costs if he had a pea plant that would grow uniformly as well as produce uniformly small peas. Pea vines that grew at the same rate in the field could be cut with a mower rather than harvested by hand, and uniformly small peas would not have to be run through a screen sieve for sorting in the factory. Further, if all of the peas were the right small size, they could be given the top grade on the label.

Under contract to Empson, Burbank began growing two pea crops a year at his Santa Rosa, California research station, selecting the seed of those plants that exhibited the desired characteristics. After three years of work, and six generations of selections for yield, vine size, pea size, uniform ripening and flavor, Burbank settled on a variety he believed filled the bill, a variety he named the *Empson*. By 1912, John H. Empson & Daughter was growing hundreds of acres of the new peas to supply its canning operations, and the pea was being sold with Burbank's name on the label.

for specific characteristics has become increasingly important. "A number of basic advances were made in poultry breeding, nutrition, and disease control during the late forties and fifties," explain USDA researchers Don A. Reimund, J. Rod Martin, and Charles V. Moore. "Foremost among these advances was the development of new fast-growing strains of chickens bred specifically for meat production." Businessmen like Frank Perdue of Perdue Farms, who hired geneticists to breed such birds, were among the first in the industry to capitalize on the genetic variable in developing their poultry businesses.

Recently in the hog industry, breeders have begun to produce animals that can withstand the rigors and stress of "confinement"—that is, hogs that can be raised in large-scale "hog factories" instead of out of doors.*

Traditionally, hogs bred and raised by family farmers were housed in makeshift, dirt-floored hog pens and other kinds of outbuildings. Large confinement facilities, however, are often constructed with steel or concrete floors, and some hogs do not respond well to this kind of environment. In some confinement operations, 15 to 20 percent of hogs develop feet and leg problems. So today, swine breeders are producing hogs that "fit" these operations. "Breeding stock companies compete on the basis of genetic engineering," say Marty Strange and Chuck Hassebrook, "emphasizing their hogs as bred for skeletal development to withstand the hard floors the factory hog must endure." Adds Kelly Klober, a small-scale Missouri farmer who dislikes this trend, but points to other sources of pressure, "Today, breed associations, show rings, and extension offices push (the) message: Breed for concrete and confinement. . . ." However, breeding hogs with the genetic traits for confinement has led to an increase in back fat and a reduction in red meat per animal.

Sometimes, a genetic change in a particular crop or livestock

*In the 1950s, antibiotics were introduced in the hog industry that could be mixed with feeds. When these antibiotics were fed continuously to hogs at low levels, disease could be prevented. With disease in check, large numbers of hogs could then be brought together in confinement facilities. In confinement, however, certain kinds of stress and disease became more prevalent (confined hogs have, for example, exhibited lowered sex drive and other behavioral problems, such as biting off the tails of their confined fellows), leading to the use of more antibiotics, vaccines, genetic screening and/or special breeding programs to overcome such problems.

strain, or even the introduction of a new strain, can have a subtle but pronounced effect on a given farming economy, and on who winds up prospering. Consider, for example, the impact that one new strawberry variety named "Aiko" had in California.

"Aiko" was developed by Dr. Royce Bringhurst of the University of California at Davis and was first introduced for commercial production in 1975. Grown predominantly in the state's central coast region, the Aiko strawberry today accounts for about 16 percent of all California strawberry acreage. The Aiko has been popular among growers because it is a high-yielding variety, is easy to harvest, and is tolerant of some viruses. It is also popular among shippers because its berries are "firm" and can withstand the rigors of long-distance travel to eastern markets.

According to Dr. Bringhurst, the Aiko was first observed as a seedling in 1965, but was not released as a new variety until 1975. Steve Huffstutlar, a USDA marketing specialist, says that Bringhurst originally kept the new variety in his lab because he was not satisfied with the plant's characteristics. The Aiko berry was not tasty, it was not red inside, it did not "color-up" evenly, and it was more hollow at its core than that of other strawberry varieties. However, Huffstutlar explains that Bringhurst was pressured by strawberry growers from the Watsonville area of California to release the Aiko for the long-distance market. Bringhurst says that the Aiko was released because of its "continuous fruiting quality" and the fact that it could be harvested over a nine-month period.

As a crop in the field, under favorable soil and moisture conditions, the Aiko is a strawberry plant capable of yielding as much as sixty tons per acre per year. However, in locations where such favorable conditions do not exist, it will yield only twenty to thirty tons per acre per year. In terms of its physical structure, the Aiko is a somewhat temperamental and weak plant, with few leaves and a root system that is poor at foraging out nutrients. The Aiko, in other words, is a high-achieving pedigree, good at yield only when planted in the right environment with the right support system.

According to Steve Huffstutlar, the Watsonville growers were especially well-positioned to use the Aiko variety; their farms were located in the right microclimate, with the right sandy soils. These farmers also had drip-irrigation technology at their disposal. Aiko, in other words, fit right into the Watsonville area, and a number of farmers there made out quite well. But Huffstutlar charges that while the Aiko strawberry may have made some well-positioned farmers "very rich," it also "created a class of impoverished strawberry growers." For when farmers with smaller operations tried to use the high-yielding Aiko, they were often unable to obtain the Aiko plantlets from commercial nurseries. The plantlets were limited in supply, and the nurseries favored the larger growers. Moreover, those small-scale farmers who did obtain Aiko seedlings were often located on less suitable land. And finally, few of the small farmers were able to obtain financing for drip irrigation systems. As a result, Huffstutlar points out, "A big percentage of these farmers went broke; many of them were put in a plant materials bind *and* a technology bind. . . . They had a lack of financing and a lack of technology. Some have persisted, but it's taken them ten years to catch up." In sum, he says, "The Aiko strawberry penalized small farmers, and made others rich."

Asked about Huffstutlar's charge that some farmers made out better with the Aiko than did others, Bringhurst says that "people who are ready to exploit something new do it, others wish they had. . . . The same thing happens in every other crop."

But now with biotechnology, there is even a greater question about which farmers will be ready to exploit new genetic breakthroughs, as well as how many might be placed at a competitive disadvantage or otherwise made technologically obsolescent.

GET SOPHISTICATED OR GET OUT

Some corporate officials and biotechnologists, when asked about how their new genetic products will affect agriculture, insist that these new technologies, like those of the past, will be "scale-neutral"—meaning all farmers will benefit equally. But others, like the Cetus Corporation's Michael Goldberg, Director of Agricultural Business Development, explain that the new biotechnology products "are going to sell at high prices," and that could affect the shape of the nation's farm system, as well as who might farm, in the future. "It [biotechnology] will certainly not slow the pace of increasing farm consolidation," he says, "and may accelerate it."

But biotechnology also has a potential for diversifying agriculture and reducing the farmer's costs, a development that might help all farmers stay in business. "Imagine a strain of wheat that grows well in the dry lands of western Kansas—without heavy irrigation," says

Monsanto's Executive Vice President, Nicholas Reding, addressing a January 1984 meeting of Kansas' Board of Agriculture. "Or a corn plant that fixes its own nitrogen. Or soybean plants that have even higher protein, or don't need to be processed before animal consumption. Or cattle that convert protein to meat with an efficiency we only dream about today. The potential is awesome.

"These new technologies, like those of the past, will give American farmers the edge they need to remain the most productive in the world . . . [and they] will substantially reduce the cost of producing food and fiber. . . ."

Sam Dryden, president of the Agrigenetics Corporation, points to certain kinds of natural molecules that some plants use to repel insects. Engineering such traits into agricultural crops, he argues, "could be several orders of magnitude more cost effective than traditional chemical approaches to formulating, manufacturing, and applying biocides."* And such a prospect might be helpful to small-scale farmers as well as larger ones.

However, Cetus's Michael Goldberg explains that it's the larger farm operation that can usually take advantage of new products. "The guy who can afford to innovate is the guy with a little extra money," he says, "the guy with a little extra acreage. He's the one that benefits from increased yield or lower production costs. Almost by definition, the rich get richer. That *may be* the trend with biotechnology; to affect it [the farm system] in one direction."

Norman Goldfarb, president of Calgene in Davis, California, sees the same pattern a little bit differently. With agricultural biotechnology, he says, "there will be an increasing selection for smart farmers; the ones who are technology oriented." And he adds, "smarter farmers will be more environmentally conscious."

American agriculture, it is said, is the last of the nation's important economic activities to follow the path of industrialization. By

*Yet using this kind of approach might reduce yield as it reduces cost. "The added value in this case," explains Dryden, "may be one of cost reduction [eliminating the need for insecticide application], not of increased harvest index [yield]." And Dryden recognizes the "simple-minded focus on yield" as a formidable barrier in adapting such approaches. "In any case," he says, "we should be helped if we refined our use of the word 'yield' to mean return on investment rather than as a measure of gross biomass per acre."

bringing the factory to the field, and applying the principles of specialization and mass production, "efficiency" has been brought to agriculture.

But industrializing the agricultural landscape has meant putting aside the notion of the diversified family farm, the farm that had a little bit of everything—chickens, hogs, cattle, and several kinds of crops. In many ways the diversified family farm made good, solid economic sense. It was the perfect hedge against the boom-and-bust agriculture that today haunts every specialized farmer. It was also perhaps the most reliable agricultural safety net; the frugal alternative to "putting-all-your-eggs-in-one-basket" agriculture. Yet this kind of agricultural diversity was deemed "inefficient" by all the experts. "Get big or get out" was the message trumpeted to farmers from the 1950s through the 1970s. Secretary of Agriculture Earl Butz urged farmers in 1973–74 to plant "fence-row-to-fence-row"; the USDA and agribusiness counseled "efficiency through expansion"; and the modern ingredients of farm production—hybrid seed, fertilizer, pesticides, and mechanization—made it all possible.

But today, coming on top of the "get-big-or-get-out" philosophy of the past, is the "get-sophisticated-or-get-out" school of high science. And increasingly, the approach of this agricultural philosophy is a high-tech approach; an approach founded on genetics and chemistry, management and mechanization; an approach that has little room for the traditional view of family-farm agriculture.

"I submit to you," says Pioneer Hi-Bred International's President Thomas Urban addressing a January 1984 gathering of the National Governors' Association, "that our food system has quietly become the largest high-tech business in America. Any farmer or [other] participant in the food chain from input companies to consumers knows this. . . . What technology, for example, is higher than the manipulation of genes in living organisms?

"To keep current with food production technology," Urban told his audience, "we must dismiss our traditional perceptions of agriculture." He had earlier objected to the use of terms such as "parity," "family farm," and "corporate farming" as being replete with rhetorical baggage and useless as metaphors in describing today's agriculture system. "Like it or not," Urban continued, "food production has been transformed from a local operation into an international business dependent on high technology." He then urged the Governors' Association to work for farm policies which would allow "the 700,000 or so large-scale farmers, and the attendant sectors of the food system, to add to their productivity."

From the standpoint of major agribusiness suppliers as well as government policy makers faced with rising price-support costs, there is lots of "dead wood" in the nation's farm system, consisting of thousands of small, "inefficient" producers who are not, for the most part, consumers of high-tech agricultural products, and whose farming output does not measure up to that of a "modern farm." Although there are 2.4 million farms operating in the United States today, the largest 700,000 account for nearly 85 percent of all farm production and use well over half of the nation's farmland. Roughly 7 percent of all farms—about 190,000 farms nationwide—account for 56 percent of all gross agricultural sales. Bigger yet are the 81,000 farms with gross sales of $200,000 or more, and the 6,000 "superfarms" with gross sales of over $1 million each. These are the high-tech farms—the ones likely to use the most sophisticated line of agricultural supplies.

From the perspective of companies with high-tech agricultural supplies to sell, it is not so important that the market of potential farm customers be large in numbers as it is that the farmers in that market have the right high-tech attitude. Agribusiness suppliers, government officials and, indeed, many farmers themselves believe that fewer "modern" farmers means a "more efficient" agricultural system.

"Politicians in Washington seem enamored of the concept of the family farmer," says Cetus's Michael Goldberg. "Many [farmers] are not efficient contributors to our farm economy. There isn't any reason in the world why we should subsidize their existence. Dairy subsidies are ludicrous. A dairy farmer will produce as much as he can because of the subsidies. It becomes a self-fulfilling system. What is needed is more efficiency. Subsidies take the economics right out of the system." And besides, he adds, "milk is non-strategic. We'd be better off putting subsidies into our strategic mineral reserve."

In the dairy industry, however, breeders and biotechnologists now talk about a day when "supercows" will replace today's conventional cow. A June 1982 *Business Week* story on the livestock genetics revolution featured a color picture of a farmer standing next to, and dwarfed by, a giant dairy cow. The caption beneath the picture noted that "a dairy cow as big as an elephant and capable of producing up to 45,000 lbs. of milk a year may become more than science fiction. . . ."

"In the future," explains Carnation Company's Dr. Autar Karihaloo, director of embryo transfer and research, "a cow, whether it is a beef or dairy cow, will be very efficient in terms of utilizing the feed and increasing the production. . . . Say, for example, an animal could go to slaughter at four months of age as against two years. By the same token, a cow producing 30,000 pounds of milk could easily produce 60,000 with far less feed." Add to that the availability of purer and more potent genetically produced hormones, vaccines, and feed additives, and the prospect for enormous changes in livestock productivity, herd size, and farm operations—not only for the dairy industry, but other agricultural industries too—lie directly ahead.

A genetically manufactured bovine growth hormone enabling cows to produce 40 percent more milk on less feed could result in a reduction of the nation's dairy herd by one-third, according to some estimates. That would mean a dramatic reduction in farmers and farm numbers. A similar development in the beef or pork industries would not only affect cattle ranchers and hog farmers, but also farmers who produce feed grains for those industries. Such changes, in turn, would lead to substantial pressure to change federal farm policy.

But when it comes to federal farm programs, some of today's agricultural biotechnologists throw up their hands in disgust. Calgene's Norman Goldfarb, for one, is quite upset. "Government involvement in agriculture," he says, "is a total disaster. Look at tobacco. That market is so screwed up it's not in anybody's interest to do anything commercial [with biotechnology] in tobacco."

Cetus's Michael Goldberg believes that we should insulate the farm economy from what government programs can do to it. "You don't need PIK [the federal government's payment-in-kind program to give farmers incentive to lower production] if you drive the inefficient producers out of business. Today, we have a production reserve of excess farmers." What about the notion that it is important to keep the food-producing resource base widely held? "There is not a good hard economic reason for having farm production widely dispersed," he says. "Americans fear concentration in any area," he explains, "but the evolution of economic systems is toward increasing concentration.

"A change from 500,000 farmers to 10,000 farmers may seem like enormous economic concentration on the face of it, but could they act collusively?"

Yet the real question is not whether farmers are capable of acting collusively (indeed, history has demonstrated that they can't) but how corporate suppliers and government policy makers will go about the business of guiding and directing the application of agricultural technology in the years ahead. The chief ingredient of agricultural power—the gene—is now in their hands, not farmers'.

MADE-TO-ORDER AGRICULTURE

In many of the agricultural industries where the genetic variable and other technologies have contributed to "modernization," there has been rampant consolidation of farming operations, and a corollary increase in the involvement of nonfarm corporations.

Since 1954, for example, more than 40 percent of the nation's broiler farmers have gone out of business, and most of those that remain now produce for large poultry processors.* These companies are "vertically integrated"; that is, they produce and supply the chicks and feed to farmers, buy back the finished birds, and process them for market. "Already in South Dakota," says farm extension economist Mark Edelman, "you can't sell or produce poultry products on any sizeable scale unless you're under contract to a processor. There just isn't any open market available."

In the hog industry, a similar pattern of consolidation and vertical integration has occurred. In 1954 there were 2.4 million hog farmers in the United States; today there are about 600,000. In 1974, less than 3 percent of the hogs marketed in the United States came from operations producing more than a thousand a year. By 1979, some 40 percent of the nation's hogs were produced in confinement operations of this size. Today, some corporate hog producers such as Tysons Foods are capable of turning out as many as half a million processed hogs annually.

*ConAgra, Gold Kist, Holly Farms, Perdue Farms, Tysons Foods, Continental Grain, and Central Soya are among the nation's largest broiler producers.

What is happening throughout America today is that the nation's farm system, and what is produced in that system, is being driven from the processing end of the food chain—by the specifications and orders set by major corporations, supermarket chains, and in some cases, even the federal government. In this process of "top-down" agriculture, the genetic variable—now made more certain and more predictable with biotechnology—will play an increasingly important role.

Today, for example, the largest 422 feedlots in the nation handle an average of 30,000 head of cattle per year; some handle as many as 90,000 a year. Feedlot operators with these kinds of facilities want a steady flow of cattle that are uniform and predictable in their feeding and weight-gaining habits—what is called "uniform feedyard performance" in the industry. Meat packers, too, have an influence on what kind of cattle feedlot operators buy. "The preferred carcass size in today's era of boxed beef," says University of Missouri extension agent Vic Jacobs, "is set primarily by the size of the box."

Feedlot operators are now being encouraged to buy only feeder cattle that are "genetically packaged"—that is, cattle that are uniform in size, weight, sex, age and originate from a common genetic source. This means that feedlot operators will be looking for "genetically packaged" lots of cattle of one hundred head or more. And according to W. H. Shirley, president of Focused Feeding, Incorporated, "before long, the genetic package will set the market prices" for feeder cattle. "Those animals that cannot be genetically packaged," he says, "will be discounted heavily in the marketplace."

Written contracts between farmers and processors is another way that particular crop varieties and livestock breeds can be specified. Farmers engaged in all kinds of agriculture sign contracts with grain millers, food processors, and meat packers.* Many of today's farm contracts in crop agriculture, for example, specify planting date,

*In the United States today, nearly one-third of all agricultural production is determined through a combination of contract farming and vertically-integrated companies. "Contracts and vertical integration have increased in importance in agricultural commodity markets," says the Congressional Office of Technology Assessment (OTA). "In 1980 total farm output either contracted or integrated was 30 percent, which represented a 40 percent increase from 1970. Evidence suggests that these coordinating mechanisms are accelerating. . . .

irrigation schedules, fertilizer and pesticide use, harvest date and price at delivery, and a few specify crop variety.

Adolph Coors Company, makers of Coors Beer, has contracts with some 2,500 farmers who grow a Coors-owned barley variety named Moravian III. This variety is grown over several hundred thousand acres in Colorado, Wyoming, and Montana. The Dixie Portland Company, a Kansas flour miller, has made contracts with Kansas wheat farmers to grow a specific variety of wheat it owns called "Plainsman Five." Del Monte and Green Giant specify certain vegetable varieties in contracts they have with Wisconsin vegetable growers. The Quaker Oats Company publishes a recommended list of white corn hybrids for farmers who grow corn for Quaker under contract in Iowa, Kansas, Missouri, the Ohio Valley, and the Southeast. And there are others.

Vertically integrated food and agribusiness corporations already own or control much of what they need in the food production process, but some are heavily engaged in supplying "genetic inputs" and related supplies to farmers.

Cargill, the huge international grain trader, provides an example of what might be called a "full service" farm supplier. In addition to buying and selling grain worldwide, Cargill also sells animal feeds, hybrid seed, fertilizers, and breeding stock. In Arkansas and Canada, Cargill hogs and feeds are sold to farmers under contract with the company. Cargill is also involved in soybean crushing, flour milling, corn milling, oilseed processing, poultry breeding, feedlots, and commodity trading and merchandising. Nor is Cargill alone.

The Dreyfus Corporation, the French equivalent of Cargill, has in recent years been eyeing the American hog industry. DeKalb AgResearch, the nation's second largest hybrid corn-seed producer, is also involved in poultry genetics, swine breeding, commodity futures, and, as discussed earlier, center-pivot irrigation manufacturing.

In addition to this corporate activity in "genetic inputs" and

"There are many benefits from contracting and vertical integration," says the OTA, "such as control of quality and timing of production, reduction in price risk, and insurance of a market outlet; but there are also drawbacks. There is concern that such coordinating mechanisms require farmers to relinquish some or all of their independent decision-making responsibility. Also it pits large nonfarm firms with substantial economic power against much smaller family farms. Still others argue that a few large agribusiness firms may use these coordinating devices to dominate agricultural production and secure unnecessarily high food prices from the American public."

related supply areas, the federal government also has its hand in the genetic composition of food, although somewhat indirectly, through USDA's program of "marketing orders." Covering some thirty-three different kinds of fruits, nuts, and vegetables, marketing orders were first established under the 1937 Agricultural Marketing Agreement Act to facilitate the orderly production, supply, and marketing of fruits and vegetables. Marketing orders were created to regulate the flow of crops to market, to insure, for example, that fruits and vegetables wouldn't all be picked at once creating a glut on the market during one part of a season and shortages later.

However, in order to regulate the supply of fruits and vegetables, marketing orders have become very specific, establishing requirements for size, grade, and volume of production. Of the forty-seven marketing orders in effect today, forty-four include some specific type of requirement related to size, grade, or maturity of crop. These specifications directly affect the kinds of crop varieties grown. And in at least three cases—Valencia and Navel oranges and Bartlett pears—the orders themselves *are* variety specific.

In the export market, United States government grading standards already help to insure a measure of uniformity and consistency in the nation's exported grain crop. And when foreign governments come into the American market to buy grain, they normally stipulate the kind and quality they want.* These specifications can and do work their way back to plant breeders and seed companies that develop new varieties of corn, wheat, soybeans and other crops. In the future, it is quite likely that the genetic factor will figure more prominently in meeting the demands of foreign customers.

FEWER FARMS, MORE TECHNOLOGY, GREATER VULNERABILITY

Modern agriculture, it appears, is headed for further economic consolidation, and genetic ingredients will play a more powerful

*In the grain trade, the orders are often very large, coming from buyers such as the Soviet Union's Exportkhleb or the Food Agency of Japan. Moreover, such orders can be fairly specific. For example, when the Soviet Union and Brazil buy hard red winter Kansas wheat, they want at least 40 percent of the kernels to be hard, dark red in color, and "vitreous," or glassy, in appearance. In fact, about 40 percent of the foreign countries that buy American wheat out of Gulf of Mexico ports use similar standards, which also include specifications for the test weight, moisture content, and protein level of the wheat.

role than ever before. The genetic changes coming to agriculture will have ramifications for who survives in farming. Increasing productivity in crop and livestock agriculture will mean fewer, larger farms.* The businesses supplying farmers, as well as those that handle wheat or process vegetables, are also consolidating around the economic advantages of the gene.

In many ways, this food-system consolidation is a continuation of past trends; trends which have been driven in part by new technological applications. Such technologies, we are told, produce greater efficiencies and higher productivity. Historically, we have already seen the results of the mechanical and energy revolutions in agriculture, and the mass decline in farm numbers they wrought. But now, farmers are becoming the tenders of genes they do not control, and something more than declining farm numbers is at stake.

Although the "efficiency" of fewer farmers may well prove unsurpassed on one level, there are other considerations too often forgotten by the economist and government official—such as food-system security, agricultural sustainability, and control of what is produced—considerations that do not always yield clear, quantifiable data, but are nonetheless as valuable as, and indeed the basis for, any economic return. In the demise of this nation's farm culture and its people—both now and in the past—we have never squarely faced those kinds of considerations.

However, beyond the decline in farm numbers and the social dislocation that will surely accelerate with bioengineering in agriculture, there are other structural changes, too. Much of the "order power" and specification-making in food production is now shifting more rapidly to the processing and supply ends of the food chain—where genetic instructions can determine the agricultural performance of a seed or the processing durability of a tomato. From this vantage point of supply and production—at food's genetic point of origin—corporations that contract, buy, or sell, and government agencies that set specific standards for marketing and shipping, will increasingly determine how food is finally shaped. In this "top-

*In fact, some very large "farms" in California have positioned themselves in either the seed industry or in biotechnology research. In 1974, the Tejon Ranch acquired the Waterman Loomis Company, an alfalfa seed operation. More recently, J. G. Boswell, one of America's largest farms, bought out Phytogen, a biotechnology company working on alfalfa, cotton, and sunflowers, all crops that Boswell grows.

down," gene-directed process, productivity and efficiency may rise, but a vulnerability of the kind inherent in too few producers, too few suppliers, and too few processors will certainly increase as well. And should such systems collapse, or be unable to react quickly, it will be farmers and consumers who will likely bear the first weight of "economic adjustment," not the architects of government policy or corporate strategy.

CHAPTER

8

NUTRITION: LAST IN LINE

Campbell has spent decades developing
nutritious and sturdy tomato varieties. . . .
We specially bred the tomato
for mechanical harvesting. . . .
—*A. M. Williams, President*
Campbell U.S. Division, 1981

"Rio Verde is one hybrid cabbage that can take the long trip to market," says the magazine advertisement from the Northrup King Company in *Seedmen's Digest.* Northrup King, the Minneapolis-based seed company, is pitching a new hybrid cabbage named Rio Verde to potential buyers in the food-processing industry, and it is touting the new variety's shipping and processing qualities. The text of the advertisement—entitled, "Rio Verde, One for the Road"—is superimposed over a color photograph shot from ground level over the top of a waiting pile of green cabbages on a loading platform. Roaring by in the background is a large, yellow tractor trailer making its way to market over a two-lane rural highway.

"Over the road or riding the rails, in mesh bags or cases," the ad continues, "this shipping hybrid comes through beautifully. It's got a firm, uniform head that's well protected by good, strong wrapper leaves. . . . Its flattened globe head is six to seven inches in diameter and five to six inches deep. And its blue-green color has proven appeal—Rio Verde is one of the most popular hybrids in use today. Make Rio Verdė your one for the road. Talk with your NK distributor."

Most consumers never see advertisements like this because they are found in out-of-the-way trade journals such as *Seedmen's Digest* or *Vegetable Grower*. Through the pages of these magazines, seed companies sell their latest botanical creations to the people who will grow, process, and ship them; people whose businesses depend on how the seed becomes a crop and how that crop, in turn, "performs" in field and factory.

But a consumer accidentally discovering Northrup King's Rio Verde advertisement might rightly think that the company is actually developing cabbages for trucks and packing crates rather than homely dinner tables. And in fact, seed companies spend little time worrying about consumers. Consumers, they say, buy vegetables mostly on the basis of appearance, texture, consistency in cooking, or apparent freshness. Yet there are other considerations, such as nutrition. Cabbages, for example, happen to be a very good source of vitamin C, ranking just behind oranges and just ahead of grapefruits. They are also a source of minerals such as calcium and phosphorus. But when it comes to breeding new kinds of cabbages and other crops for modern food production, the specifications for farming, food processing, and long distance shipping are first on the breeder's agenda. Nutritional considerations are usually not a priority.

NUTRITION: LAST IN LINE

Today in the United States it is the needs of the food industry that determine what will be bred, not nutritional value. The nutrient level in most agricultural crops is secondary to how well the crop performs and holds up in the food-making process. Before a vegetable or cereal grain ever reaches the consumer, it must run a gauntlet of agricultural and manufacturing processes. Farmers, millers, food manufacturers, feedlots, and shippers all have their demands met well before consumers do. And in this field-to-table journey what matters most is not so much the nutrient value or taste that food will have at the end of the process, but how much abuse a vegetable or cereal grain can take along the way.

Crops in the field must first meet the tests of yield, uniform growth, and simultaneous maturity. After this, their fruit or kernels must be able to withstand the rigors of mechanical harvesting, repeated handling, and various kinds of transport from one point to another. Next come the trials of steaming, crushing, or canning. In some cases, the raw agricultural crop must "store well" or "travel well," or be good for freezing or frying. And genes are the keys to meeting each of these steps in the food-making process; the genes that control the field-to-table characteristics of every crop from broccoli to wheat. In this process, the genes that matter are those of yield, tensile strength, durability, and long shelf life. However, the genes of nutrition—if considered at all—are for the most part last in line.*

Take, for example, the tomato. In terms of its overall raw nutrient content for a group of ten vitamins and minerals, the tomato ranks about sixteenth among major fruits and vegetables. For its vitamin A content the tomato ranks sixteenth, and for vitamin C, about thirteenth. Yet in terms of its actual vitamin and mineral contribution to the average American diet, the tomato ranks at the very top of the list because it is found in such a wide variety of processed foods—from pizza pie and ketchup to Bloody Marys and spaghetti sauce. Not surprisingly then, a lot of processed tomatoes are grown in the United States; about 70 percent of all tomato acreage is for processing tomatoes. The remaining acreage is for fresh market tomatoes used in salads and for other more immediate needs.

Today, the processed tomato business in the United States is booming; the per capita consumption of tomatoes and tomato products has risen 15 percent in the last ten years. Americans are eating tomato-based Mexican and Italian foods like never before, and the

*Government and scientific reports generally confirm the view that other traits—those primarily related to yield and marketability—are typically sought well before those of nutrition. For example, in a 1973 report from a National Academy of Sciences study group called the Task Force on Genetic Alterations in Food and Feed Crops, it is noted: "Nearly all plant breeding programs in the U.S. emphasize yield, uniformity, market acceptability, and pest resistance. Plant breeders have lacked the resources to extend their evaluation to factors of nutritional importance for reasons of time, effort, cost, technology, and lack of defined goals. Nutritional quality has not been recognized as a distinct dimension in plant breeding programs." And in a January 1979 memorandum to former U.S. Secretary of Agriculture Bob Bergland, the National Plant Genetic Resources Board noted: "Most plant breeding has been directed toward the improvement of yielding ability and other agronomic traits that affect yields either directly or indirectly. . . . Increased attention needs to be given to nutritional quality and food safety."

demand for concentrated tomato products such as pastes and ketchup is soaring.

As a result of forty years of breeding tomatoes for the processing industry, and the agricultural operations that supply it, a number of specific genes have been identified that control one or more of the traits of this plant and its fruit. Such "genes of commercial importance," as they are called, have been bred into most of the tomato varieties used for processing. One of these is the "u gene" for uniform ripening, which also eliminates the "problem" of unripe, green shoulders appearing on the tomato. Another is the "sp gene" (for self-pruning), a gene discovered in Florida in 1914 which insures that tomato plants will grow in a compact, determinate fashion—so determinate in fact that tomato varieties incorporating this gene terminate their branch growth at approximately the same distance from the center of the plant. This gene has facilitated the work of plant breeders like Gordie C. Hanna of the University of California at Davis, who during the 1950s and 1960s developed the "VF-145," a tomato variety suitable for machine harvesting and which also incorporates some genetic traits for high yield, concentrated ripening, and disease resistance. The VF-145 and its progeny are still used extensively throughout California.

In developing such tomato varieties suitable for the processing industry, some of the tomato's nutrients haven't fared so well. "In our work with the quality of processing tomatoes," wrote University of California tomato breeder M. Allen Stevens in 1974, "we have found that some of the characteristics considered desirable for mechanical harvest are in opposition to those needed for high vitamin C." Similarly, if the genetic characteristics for good, red tomato color and high vitamin A content are brought together successfully, other problems which are unfavorable for tomato production may develop. Working to maintain the crimson gene in some tomato varieties can result in a decrease in Beta carotene. Other factors can reduce germination, seedling vigor, and crop establishment; sometimes cause brittle stems, increase susceptibility to certain foliage diseases; and through these effects, generally reduce yield. And anything that reduces yield is out of the question: yield is, quite plainly, profit.

Since 1975, tomato growers have been getting more tomatoes from fewer acres. In fact, in recent years, an average 9,000 fewer acres each year have been required to meet the ever rising demand for processing tomatoes, indicating a steady increase in tomato yield and other productivity factors. Tomato growers, processors and buyers all benefit from increasing yields. In fact, they have become accustomed to getting more yield on less land, and they expect that trend to continue.

Cathy Cryder-Sower of the Asgrow Seed Company breeds tomato varieties for the canning and processing industry in San Juan Batista, California. Asgrow sells seed for processed tomato varieties all over the world, and Cryder-Sower travels quite a bit, often doubling up her responsibilities as both a salesperson and plant breeder.

In a telephone conversation with one prospective customer, she described some of the tomato varieties Asgrow has for sale: "Murietta . . . peels beautifully; number 649 . . . lacks firmness; Pacesetter 502 . . . will mechanically harvest." Other varieties are described by their form—round, square, or pear-shaped.

Cryder-Sower explains that in her tomato-breeding program at Asgrow, "the absolute major priorities are processing quality and disease resistance." By "processing quality" she means "peeling qualities," a high solids content, and high viscosity. "Disease resistance," she explains, means resistance to nematodes and a fungus known as fussarium wilt, race II—the latter a major problem in California. Also on her list of breeding priorities are "high yield," the ability to withstand mechanical harvesting, and plants that are easily grown. Nutrition is not mentioned.

As for the influence of tomato processors in California, Cryder-Sower notes that "the canner is in the driver's seat." She explains that canners publish a list of approved tomato varieties, and often stipulate which varieties can be grown for experimental purposes as well as the number of acres to be grown in such varieties. She says that canners will "blackball" some varieties as totally unacceptable, and that this practice has been "peculiar to California since 1970" and the demand for new high-yield varieties.

THE GENES OF NUTRITION, HEALTH, & SAFETY

In the American diet, fruits and vegetables alone account for about 90 percent of vitamin C, 50 percent of vitamin A, 30 percent of vitamin B6, 25 percent of magnesium, 20 percent of thiamine, and about 18 percent of riboflavin and niacin. One cooked, medium-sized broccoli spear, for example, provides about 25 percent of the U.S. Recommended Daily Allowance (RDA) for folacin, an essential B-vitamin involved in the maturation of red blood cells. A single fresh banana provides 45 percent of the adult RDA for vitamin B6, essential for protein metabolism. One cup of frozen collard greens provides 23 percent of the adult RDA for magnesium, a mineral element that serves as a catalyst in several hundred biological reactions.

Many consumers, however, often assume that one kind of cabbage or tomato is pretty much like any other, even though, when it comes to vitamin and mineral content, all crop varieties are not created equal. And plant breeding can and does make a difference.

When plant breeders and seed-company officials are asked why they don't produce nutritionally improved crop varieties, they respond that consumers are more interested in shape, color, and consistency in cooking than they are in nutrition. Glenn Kardel, seed stock manager and onion specialist for the Asgrow Seed Company, says "consumers won't buy ugly vegetables," and adds that people don't always buy what's good for them. "Although General Motors can produce a car that's completely safe, people will not buy it if there's no style," says Kardel. He points out that in 1956, Asgrow introduced a yellow tomato that had increased carotene value. That tomato produced a juice that was "more toward orange juice" in color, but consumers wouldn't buy it. According to Kardel, Asgrow invested a lot of money in the yellow tomato experiment and was left holding "a big bag of seed" when consumers turned up their noses. He calls it their "yellow tomato fiasco." Asgrow also tried introducing a larger, yellow, "high-volume" Italian pepper in the U.S. market, but that didn't sell either.

A number of factors contribute to the lack of attention to food-quality considerations in breeding programs, says tomato breeder M. Allen Stevens, now with the Campbell Soup Company. One of these is that breeders have not been pressured to be actively concerned with nutrients. "The very slight demand for vegetables and fruits with high nutritional value," says Stevens, "has been no match for the great demand for a plentiful supply of low-cost crops." In other words, cheap food and producing food in volume are the forces that drive modern food production, not nutrition.

However, there is no question that plant breeding, and certainly the powers of genetic engineering, can alter the nutritional levels in crops. It is clear from plant breeding practices in the past that vitamin and mineral levels in crops can be raised or lowered. The key question is whether such nutrient levels should be tampered with at all, and if so, by whom and for what purpose? Should nutritional considerations in new crop varieties always be placed behind those of yield, harvestability, processing performance, and appearance?

Over the years in the United States, the consumption of fresh fruits and vegetables has generally declined, while the consumption of meat and processed foods has increased. Recently, however, there has been increasing consumer interest in the nutritional and health value of fresh fruits and vegetables and a rise in their consumption. Moreover, government studies and dietary guidelines have periodically suggested that Americans reduce their consumption of meat and increase consumption of fruits and vegetables.

In June 1982, for example, the National Academy of Sciences (NAS) released a comprehensive report entitled *Diet, Nutrition, and Cancer* which emphasized the importance of eating fruits and vegetables high in vitamins C and A. The NAS study noted that vitamin C in foods can inhibit the formation of cancer-causing substances and also seems to lower the risk of cancers of the stomach and esophagus. Fruits and vegetables high in a vitamin-A precursor called beta-carotene—such as carrots, spinach, and broccoli—were also found to be associated with a reduced risk of certain cancers, including those of the lung, breast, bladder, and skin. According to the NAS report, vegetables in the cabbage family, such as broccoli, cauliflower, kale, and brussels sprouts contain natural cancer-inhibiting substances.

The American Cancer Society launched a national campaign for an anti-cancer diet in February 1984, and among its dietary guidelines were those suggesting a reduction in total fat intake and an increase in the consumption of high-fiber, whole-grain cereals, fruits, and vegetables. Not to be outdone, the American Heart Association followed in May 1984 with its dietary plan to help Americans lower their blood fat levels, including recommendations to eat less red meat* and increase the consumption of fruits and vegetables.

Studies and dietary recommendations such as those of the National Academy of Sciences, the American Cancer Society, and the American Heart Association underscore the importance of cereals, fruits, and vegetables in the diet, and, indirectly, the raw nutrient levels in such crops.†

We know, however, that nutrient levels in food crops can be inadvertently altered through plant breeding, and presumably, genetic engineering. But should plant breeders and genetic engineers now try to improve the nutritional levels in important crops? Or should they guard against "nutritional erosion" in crops due to breeding them for harvesting or processing goals? Moreover, some crops contain naturally occurring toxicants that can also be increased or reduced in the process of genetic alteration. This last effect has so far been notably rare, but there are thousands of such substances in plants, and a few have the ability to kill. "We simply know very, very little about plant composition," says Columbia University nutritionist Joan Gussow. "Plants are made up of thousands of chemicals, therefore all plant breeding is going to reduce or increase certain chemicals, many of them unmeasured, many of them perhaps even unknown." Should plant breeding and genetic engineering then be monitored for nutritional and toxic-constituent changes in crops?

Curiously, the federal government once did consider monitoring

*Were a decrease in meat consumption to occur in the United States, the vitamin and mineral content of certain vegetables would become even more important than they are now. For example, meat is a major source of zinc, an element involved in metabolism, digestion, and the action of insulin. Some vegetables, such as cowpeas and spinach, are also sources of zinc. One cup of boiled cowpeas, for example, can provide 20 percent of the adult RDA for zinc.

†In fact, some scientists have suggested that if fruit and vegetable consumption declines or even remains constant in some societies, because of the preponderance and availability of processed foods, or because price increases for fresh fruits and vegetables may prevent a marked increase in their consumption, then one potential means of increasing the dietary contribution of fruits and vegetables would be to grow crops higher in the prime nutrients. Vegetable scientists John F. Kelly and Billy B. Rhodes, for example, have suggested that we "encourage the use of high beta-carotene cultivars of sweet potatoes, carrots, and tomatoes; the preservation of ascorbic acid in fruits, tomatoes, and green vegetables; and the increase in protein quality and quantity in legumes and root crops."

changes in both the toxicant and nutritional constituents of new crop varieties, but after a three-year controversy the whole affair came to a rather ignominious ending. Yet there is still a federal regulation on the books—little known and even less talked about—but a regulation nonetheless. Now with genetic engineering on the near horizon, it may be time to reexamine how this government action came about and the controversy this regulation stirred up. The story begins in the late 1960s in Ontario, Canada.

THE LENAPE POTATO CRISIS

On a Sunday afternoon in the summer of 1969, a middle-aged Canadian named Gary Johnston settled down to an undistinguished meal of beer and potatoes. Johnston was then a plant breeder and professor of horticulture at the University of Guelph in Ontario who had been experimenting at the university greenhouse with a new American potato variety named Lenape, developed by the U.S. Department of Agriculture. Cross breeding the American potato with some Canadian varieties to see what he could come up with, Johnston had grown a few of the Lenapes in the greenhouse that yielded a small crop of tubers. These potatoes, about the size of half dollars, looked pretty good to Johnston, so he decided to take them home for dinner. He cooked them with butter and a few slices of ham, and along with the beer, he ate them.

Later that evening, Johnston developed a blinding headache and became terribly sick to his stomach. All through the night he was very sick, with chills and vomiting, but when he recovered in a day or so, Johnston dismissed the illness as "some kind of bug." A few days later he happened to mention his illness to a university colleague, Ambrose Zitnak, a biochemist who also worked in horticulture; when Johnston described the symptoms of his sickness, Zitnak said it sounded very much like alkaloid poisoning from the potatoes, which Zitnak had seen before in other work in Canada. Glycoalkaloids, Zitnak explained, are present in all potatoes, but at high concentrations can cause severe illness.* Johnston said he

*Glycoalkaloids are also found in eggplant, peppers, and tomatoes. In some instances, primarily in Europe, high glycoalkaloid concentrations in potatoes have been associated with intestinal disorders and even death to humans and livestock.

never dreamed that his illness might have come from the potatoes. The two scientists then set out to test that hypothesis.

Johnston planted some of the Lenape potatoes in test plots, and after a period of growth, Zitnak began testing the harvested tubers when they were about the size of quarters. The biochemical analysis revealed that the potatoes were very high in glycoalkaloids—containing as much as 40 to 50 milligrams per 100 grams of fresh weight, with a few containing as much as 60 to 70 milligrams per 100 grams. Anything beyond the 15 to 20 milligram range is not advisable for eating.

Johnston and Zitnak had the Lenape potatoes grown and tested in plots all over Canada, and they found a similar pattern. But when they moved to publish their findings, they discovered that there was some resistance to letting the world know about these potatoes. The editorial board of the *Potato Journal* refused to publish the Johnston/Zitnak paper. Recalls Johnston: "There was no way that the *Potato Journal* was going to publish an article about poison potatoes!" The economic ramifications of a potato scare would be devastating for the whole industry, said the editors of the *Journal*.

Later in 1969, Johnston and Zitnak released their results at a small gathering of international potato experts held at the University of Guelph. Not long after that report, the editorial board of the *Potato Journal* agreed to publish the Johnston/Zitnak paper if the authors would "take the scare tactics out." But it wasn't until July 1970 that their article finally appeared.

However, by December 1969, word of the poisonous Lenapes had made its way to the American Embassy in Ottawa, and from there to Washington, D.C. A few clips of the incident appeared in the American press, and then all hell broke loose. The U.S. Department of Agriculture began to feel the heat, but resisted any admission of a potato problem.

In American evaluations and field tests, the Lenape had done well in yield when compared to the Kennebec and Katahdin, two other popular potato varieties. The Lenape was also resistant to some potato diseases such as late blight, mild mosaic, and tuber necrosis. But in 1967 when the USDA had first released the Lenape, there was great excitement about its potential use for making potato chips. In its formal release, the USDA headlined the Lenape as "A New Potato Unusually High in Solids and Chipping Qualities."

Because of these qualities and a very low sugar content, the USDA concluded that the Lenape "should be very valuable for the processing of potato chips, especially in production areas where the crop matures under high environmental temperatures." And one such production area was Pennsylvania, home of the Wise Potato Chip Company. The USDA had singled out the company in its announcement for its help in evaluating and testing the Lenape potato.

By the time the Johnston/Zitnak findings on the Lenape began to penetrate USDA, potato chip companies such as Wise and others, as well as potato seed growers, were already using or planning to use the Lenape potato. In fact, the new potato had been planted for seed in as many as ten states, from California and Montana in the West, through Pennsylvania and Maine in the East. Some Lenape potato chips had also been manufactured. Thus, the idea of recalling the potato was resisted in many quarters. Finally, however, in February 1970, the USDA and the Agricultural Experiment Station of Pennsylvania issued a joint statement withdrawing the Lenape from further agricultural use.

"The USDA and the Pennsylvania Agricultural Experiment Station have withdrawn the name of the Lenape variety of potatoes," the announcement said. "This variety is no longer recommended for planting, and no basic seed stocks will be released in the future." Other parts of the announcement explained the problem, while attempting to minimize the possibility of any adverse public reaction. The Lenape, continued the announcement, "was found to contain approximately twice the level of glycoalkaloids carried by commercial varieties of potatoes." Although glycoalkaloids are "a normal constituent of all potatoes," explained the USDA, "some instances have been reported in Europe where excessive amounts have caused intestinal upsets in humans."

"Lenape-variety potatoes are not suited for boiling or baking," warned the announcement, "and in addition, the consumption of quantities of whole Lenape potatoes so prepared might produce discomfort or even illness. It is this possibility that has occasioned withdrawal of the variety."

However, USDA further reported that because "the glycoalkaloids are concentrated in the peel, "chips made from the flesh of the variety do not exceed safe levels."

The Lenape potato incident raised some new and disturbing questions about the age-old practice of plant breeding. Plant breeders were good and honorable people, of course, doing the painstaking work of developing new crop varieties to help feed a hungry world. But something had gone wrong in the government's screening process. How could a new variety of potato that might make people sick get out into the market so easily? Some people began asking questions about naturally occurring toxicants. Could such toxicants be present in other food crops as well? Could they be increased through plant breeding as they were in the Lenape potato? And what about important traits in food crops—the vitamin and mineral content of vegetables, or the protein level in cereal crops? Could these also be affected by plant breeding? The answer to all these questions was yes, and as was soon discovered, hardly any screening or evaluation of such changes was being conducted anywhere in the country, either by government or industry.

THE FDA RILES THE PLANT BREEDERS

At about the same time the Lenape potato controversy surfaced in Washington, the U.S. Food and Drug Administration (FDA) was involved in a White House–ordered review of certain food additives, touched off by findings that cyclamates were carcinogenic. The FDA, it seemed, had a long list of substances—mainly chemical food additives—that were "generally recognized as safe" until cyclamates showed up among them. This supposedly safe list of substances was a product of the policies and politics of the late 1950s.

In 1958, President Dwight Eisenhower signed the Food Additives Amendment to the Food, Drug and Cosmetic Act. That amendment changed the way the federal government regulated food additives. Prior to that time, all food additives were presumed safe unless proven otherwise; after 1958, additives had to be proven safe before they could be marketed. However, there was a catch. The Food Additives Amendment also created a new category of substances that were exempted from the Act, and not subject to pre-market FDA clearance. These exempted substances, hundreds of them, came to be known as the GRAS substances—"generally recognized as safe"—and eventually constituted the GRAS list of accepted food additives.

The first GRAS list was compiled in the early 1960s, but it was soon learned that almost anything—including substances such as cyclamates—could be proposed for addition to the list unless challenged by the FDA. In 1969, however, the cyclamate time bomb exploded, prompting President Richard Nixon, under considerable public pressure, to call for a complete review of all GRAS substances. It was in the process of that review (which still continues), that the FDA decided on its own that new varieties of food crops should be included as a category of substances to be scrutinized.

In December 1970, the FDA proposed that "foods that have had a significant alteration of composition by breeding or (genetic) selection" be included for review and regulation under the new GRAS process. It was the first time the FDA had ever attempted to regulate raw agricultural crops rather than finished food products made from crops. Shortly after the proposed FDA regulation appeared in the *Federal Register,* representatives of the food industry, seed industry, land grant colleges, and the USDA all complained that they had not been consulted and that the proposed regulation was too broad.

"The first reaction of most plant breeders," reported USDA scientists L. P. Reitz and B. E. Caldwell to the Crop Science Society in 1973, "was one of indignation, even rage. . . . Didn't FDA know we had the interest of the public at heart and a keen sense of responsibility? The inclination was to wrestle free from this oppressor, fight back, restore our independence and good name, and get on with the breeding. Some 'hot' letters were written; some emotional statements were made; even resolutions were drafted. Fortunately, the cooler judgments of the National Canners Association, the American Agronomy Association, the Crop Science Society of America, and USDA were available."

What the "cooler judgments" of these and other agricultural interests amounted to was a three-year campaign, between 1971 and 1974, to wear down the FDA people who had proposed the "plant breeding rule."

In explaining why his agency had proposed the rule, Alan T. Spiher, who was charged with heading up the GRAS review process for the FDA, said that "as [plant] breeding stocks are manipulated in order to develop a redder tomato, a shorter maturation time, or a variety more resistant to blight, mold, rust, or insects . . . we need to monitor [such changes] to assure ourselves that other, less desirable changes have not taken place."*

Taking tomatoes as an example, Spiher pointed out their importance as a nutritionally significant source of vitamin C, but that "data show that a recently developed new variety intended for mechanical harvesting, and the concomitant marketing techniques developed for this variety, deliver to the consumer a tomato having about 15 percent less vitamin C content." Thus, he said, "the need to keep a close watch on this kind of development is evident."

In a May 1971 meeting with food and agricultural interests, Spiher noted that nutrient labeling requirements for food products made it important that changes in crop varieties would not produce fluctuations in nutrient levels. He also indicated that changes in low-level toxicants in all new food crops should be monitored. Spiher assured his listeners that the FDA wasn't interested in extensive guidelines, and only wanted to be informed of any "significant" nutritional or toxic change in a new crop variety before it was released. However, he said, the burden of responsibility was on the industry, and particularly the plant breeder, whether employed in a public institution or a private company.

The FDA regulation became final in June 1971, and the agency defined "significant decreases in nutrients" as a decrease of 20 percent or more; for toxic constituents, increases of 10 percent or more were deemed significant and proposed for monitoring. At that time, however, not much was known about either toxic constitutents or the nutritional composition of food crops. "There was no body of knowledge of the total nutritional value of the various vegetable varieties," recalls Edwin Crosby, then with the National Canners Association. Nor was it well known what toxicants were present in commercial crop varieties or at what levels.

By December 1971, it was clear that more information was needed.

*There was some argument from agricultural and plant-breeding interests that raw agricultural commodities were "foods" and not "food ingredients," and therefore should be exempt from the Nixon-ordered review. FDA's Spiher stated why it was important to treat the plant-breeding process in the same way a chemical additive would be treated: "We also have to recognize that the mission of the Food and Drug Administration is to monitor the use of all food ingredients, and the introduction of deleterious levels of alkaloids or other chemicals into a raw agricultural commodity through breeding or selection is just as significant as adding an alkaloid as a chemical entity to the cooking pot."

The food and seed industries wanted to know which nutrients and toxicants should be monitored, and in which crops—all or only certain ones? Spiher said that the FDA wanted to focus only on those crops and nutrients that were particularly good sources of nutrition. As for toxicants, Spiher explained that the FDA wanted to monitor those toxicants that posed especially serious problems, such as glycoalkaloids in potatoes. In October 1972, a joint USDA, FDA, and food-industry task force was appointed to establish some recommendations and guidelines for nutrient and toxicant monitoring. After sixteen months of work, a set of recommendations was drafted and released for wider review.

The task force recommended only one specific action in the area of naturally occurring toxicants: that the glycoalkaloid content of existing potato varieties not be exceeded. For all other crops they suggested that plant breeders pay "special attention" to any potential increase in toxic constituents when using exotic and wild varieties in their breeding programs. However, the task force recommendations in the area of nutrient monitoring drew attention and met with heated resistance.

The task force recommended that nine crops—wheat, white potatoes, carrots, sweet potatoes, oranges, cabbages, tomatoes, peanuts, and dry beans—be monitored for changes in one or more nutrients.* They arrived at this list by concluding that any crop that provided at least 5 percent of the average national intake of any nutrient should be monitored. In addition, if any crop constituted

*The food crops identified as candidates for monitoring by the USDA/FDA Task Force, with their significant nutrients, are listed in the table below:

Food Crop	Nutrient				
Wheat	Protein	B6	Thiamin	Niacin	Magnesium
White potatoes	—	B6	Thiamin	Niacin	C
Carrots	—	A	—	—	—
Tomatoes	—	A	—	—	C
Sweet potatoes	—	A	—	—	—
Dry beans	—	—	Thiamin	—	Magnesium
Oranges	—	—	—	—	C
Cabbage	—	—	—	—	C
Peanuts	—	—	—	Niacin	—

Source: F. R. Senti & R. L. Rizek, "Nutrient Levels in Horticultural Crops," *Horticultural Science,* vol. 10(3), June 1975.

at least 10 percent of the average intake of a specific nutrient by a particular group of people for reasons of income, age, sex, regional supply, or ethnic preference, that crop should be monitored as well. Of the nine crops chosen, only dry beans fell into this latter category because of their popularity among Mexican-Americans. By July 1974, the task force's recommendations were the subject of hot debate.

The American Seed Trade Association, the Crop Science Society of America, and the National Council of Commercial Plant Breeders quickly lined up in staunch opposition to the recommendations, claiming that if implemented, they would jeopardize large research investments in plant breeding; that such an FDA intrusion would be a "first step" in applying nutritional requirements to new products; and that such regulation might eventually lead to a ban on nutritionally deficient crops or the foods made from them. Alan Trotter, a plant geneticist with the Asgrow Seed Company, and H. M. Munger,* a professor of plant breeding at Cornell University—who were both members of the task force—also opposed the proposal. On the other side was Frederic Senti, a senior USDA official who chaired the task force for two years, and who said, "these guidelines will simply enable us to determine what is happening to the nutritional value of our food. [They are] a useful and reasonable first step."

Another member of the task force, Stuart Younkin, then with the Campbell Soup Company, also supported the FDA regulation and the task force recommendations. So did the National Canners Association (NCA), whose Edwin Crosby said, "as the industry moves to nutritional labeling, we will have a difficult time in meeting labeling requirements if advertently we begin growing a new plant variety which has been substantially altered in [its] nutrient content by breeding." Food processors were beginning to realize that the task force's proposals for monitoring nutritional levels in food crops

*Cornell's Munger later wrote in 1979: "The idea of breeding more nutritious fruits and vegetables has great popular appeal, but a realistic analysis indicates that it has limited potential for improving the U.S. diet. It seems to me that breeding which has made fruits and vegetables more economical to produce and market, attractive to look at, and appetizing to eat has done, and will probably continue to do, more to increase nutrient intake than could have been done by breeding for higher percentages of nutrients."

shifted some of the burden from the food industry to seed companies and plant breeders.

Speaking for consumers, Rodney Leonard, executive director of the Washington-based Community Nutrition Institute, supported the proposed recommendations. "The guidelines represent a healthy trend," he said. "We should not be eating vitamin- and mineral-fortified cardboard." Despite this, the opponents of the guidelines successfully outflanked those supporting the task force's findings.

Weighing in heavily in opposition to the task force's recommendations was a National Academy of Sciences (NAS) study group called the Task Force on Genetic Alterations in Food and Feed Crops. In an earlier August 1973 report, this NAS study group stated: "The FDA regulation indicating that selected plant cultivars should not differ significantly in nutrition or toxicity . . . is unsound, impractical and regressive.

"The relative lack of baseline data and analytical capabilities makes it impossible to meet the . . . guidelines for new plant varieties. . . . [The] regulations at present cannot be sensibly enforced by FDA, nor can they be taken seriously by those who introduce new genetic materials."

This NAS group also argued for a two to three year NAS study of the whole issue, including a cost-benefit analysis. Although that study never started, its mere suggestion seems to have helped put the whole idea of monitoring food crops on the back burner.*

*Although the NAS study group opposed the task force's recommendations and the FDA regulation, it did have important things to say about nutrient levels in crops. "It is desirable to avoid a significant decrease of those specific nutrients in plants recognized as good sources of these nutrients," said the study group in its 1973 report. "Where plants are not a significant source of specific nutrients, there is little need for concern with variations in content. Specific crops for which an improvement in one or more selected nutrients would be of practical nutritional significance should be identified and authoritative recommendations made to plant breeders to achieve increases in these nutrients whenever feasible, while maintaining adequate acceptability of the food.

"Till now in plant breeding programs the nutritional goals have either been ignored or are secondary to yield and pest resistance. The relevant nutritional characteristics of new varieties should in the future be identified prior to their commercial introduction. *Nutritional characteristics could become on [a] par with yield and pest resistance* [emphasis added]. This would encourage the application of appropriate regulatory standards to prevent progressive depletion of the nutritional quality of the national supply of food and feed and encourage its improvement."

The FDA-USDA Task Force recommendations for monitoring important crops for nutritional alteration were never adopted, and to this day no guidelines exist. FDA's 1971 regulation on nutrient and toxicant alternations in food crops is still on the books, but has not been actively enforced and information on whether nutritional or toxic changes have occurred in crops since that time is scant at best.*

NUTRITION & THE NEW GENETICS

Today, however, even as FDA tries to come to terms with its obscure plant breeding regulation, it is becoming clear that the newer genetic technologies will have potential impacts for the quality and safety of food, and the agency will have to become more directly involved in genetic monitoring. "As [recombinant DNA technology] becomes more of a factor in the production of food," said FDA's acting commissioner Mark Novitch in a March 1984 address to the National Food Policy Conference in Washington, "we will be paying close attention to microbiological safety, to nutritional quality, to the formation of new, unexpected com-

*In a July 1984 response to a letter from the Environmental Policy Institute inquiring about the history and current status of the FDA plant breeding regulation, Dr. Sanford A. Miller, Director of FDA's Center for Food Safety and Applied Nutrition, replied to several questions as follows:

Q: What is the history of FDA actions with regard to substances reviewed under this regulation since 1971? How many of these kinds of nutritional and/or toxic changes have been brought to the attention of FDA?

A: We have dealt with many inquiries regarding new varieties on a case by case basis. We have never maintained a list of inquiries on this topic and do not have the capability to generate such a list.

Q: How many times has FDA exercised its authority to review crops or other substances for compliance with this regulation and what were the results?

A: To my knowledge, FDA has never established a mandatory monitoring program for the purpose of checking the level of nutrients or toxicants in new varieties of food substances. I do know that potatoes were monitored on a voluntary basis, but this was done by USDA.

Q: What enforcement actions have been taken or formally recommended by FDA regarding this regulation?

A: I am unaware of any enforcement actions taken by FDA regarding the substances in question.

pounds, and to unpredicted substances in genetically-engineered food products." But whether the FDA will have the funding or the trained personnel to keep up with such monitoring is another question.

Meanwhile, among biotechnology companies and major corporations now involved in genetics research, there is much talk about "quality improvements" in food crops, including nutritional and even flavor improvements. Monsanto boasts of "food plants with enhanced nutritional value" a decade or more away. Phytogen, a Pasadena-based biotechnology company, is working on "a more nutritious potato." And Calgene's president Norm Goldfarb talks of "better nutritional quality" being one of his company's crop research goals. In a 1982 Calgene brochure entitled "The Second Green Revolution Is Here Today," the company explains: "Bioengineering crop plants for better flavor or protein content from the human point of view will substantially increase net product value with minimal cost to the plant."

Yet some genetic engineers—while talking of nutritional enhancement and improved varieties—are pursuing essentially the same kinds of goals with individual plant genes as traditional breeders have done in the past working with whole plants—namely, developing crop varieties to meet the needs of agricultural mechanization, food processing, and long-distance shipping. "Consumer needs" have also been mentioned as a concern of some biotechnologists, yet it is unclear exactly what role genetic engineering will play in meeting those needs.

"Genetic engineering related to the development of improved plant varieties for the food processor and the consumer" say biotechnologists William R. Sharp, David A. Evans, and Philip V. Ammirato writing in the February 1984 issue of *Food Technology*, "is the big genetic engineering research opportunity in the crop sciences. This [technology] makes it possible to breed the raw plant material according to the specifications prescribed by the processor or, in the case of fresh market products, by the consumer." Sharp, Evans, and Ammirato are, respectively, scientific director, associate scientific director, and manager of developmental genetics for the DNA Plant Technology Corporation of New Jersey. "DNAP," as it is called, is a biotechnology company founded, in part, to work on the genetics of the tomato.

THE "HIGH-SOLIDS" TOMATO

Tomatoes, if left to their own devices, are about 95 percent water. Tomato processors, however, aren't interested in water. What they want for their soups, ketchups, and pastes is a tomato with a "high-solids" content. The more tomato you have in a tomato, they reason, the less water you need to grow it. Thus, a tomato that doesn't require much irrigation water, and which doesn't carry excess water weight when processed and shipped, is a tomato that will save growers and manufacturers money while producing more soup or ketchup. Plant geneticists say it is possible to increase a tomato's solids yield some 20 percent simply by increasing its solids content by one percentage point—all with the help of the right genes.

"The economic benefit derived from an increase in tomato solids alone clearly justifies the new plant genetics efforts," says Dr. George H. Kidd, a genetics consultant with the L. William Teweles Company. "An increase in tomato solids from the current level of 5 to 6 percent would add almost eighty million dollars to the annual value of processing tomatoes in the United States alone." Accordingly, a number of biotechnology companies, such as DNAP, as well as major corporations such as H. J. Heinz and Campbell Soup, are knocking themselves out to come up with the perfect, "high-solids" super tomato.

The Campbell Soup Company was one of the first major food-processing corporations to take the plunge into biotechnology. In 1981, the company was gently pushed into a biotechnology venture when two of its plant geneticists resigned to start the DNA Plant Technology Corporation. Today, DNAP operates in a donated Campbell's research facility with twenty former Campbell employees on the payroll and $10 million in Campbell backing. Campbell went along with the venture because DNAP's scientists might just come up with the perfect tomato, or at least new superior tomato varieties that Campbell could use in its soups, V-8 juice, spaghetti sauces, and other tomato-based products.* The soup company now

*Campbell is no stranger to plant breeding, seed selection, and plant genetics. The company holds one of the world's largest private collections of vegetable seed, and is especially well-stocked in tomato seed of all kinds. In the early 1970s, for example, Campbell's Eldrow Reeve, then vice-president for vegetable research, told a Congressional committee, "Our bank of tomato genetical material presently

owns about 60 percent of DNAP and has specified contracts with the new company to conduct research on wild tomato types. But Campbell and DNAP are not alone in the high-solids tomato race.

Heinz, the nation's largest ketchup manufacturer, is placing its bets on Atlantic Richfield's plant genetics subsidiary, the Plant Cell Research Institute of Dublin, California. In December 1982, Heinz signed a five-year agreement with the ARCO subsidiary to conduct tomato genetics research. In this work, ARCO's plant geneticist, Jon F. Fobes, plans to use wild tomato varieties obtained from the Peruvian Andes which have a solids content of 13 percent. With genetic engineering techniques, Fobes hopes to move those genes responsible for that "high solids" characteristic into commercial tomato varieties. And high-solids genes aren't the only genes that Fobes, ARCO, and Heinz are interested in using in commercial tomato varieties; Fobes is also looking at the Peruvian tomatoes for their cold-tolerant genes, their drought and disease resistance genes, and genes that may allow commercial tomato varieties to be irrigated with saline water.

Like the Heinz-sponsored research at ARCO, the Campbell/DNAP work will try to incorporate valuable production and processing traits—high-solids content among them—into new tomato varieties. DNAP's William Sharp, also one of the company's founders, says his company is already field-testing a tomato with a 15 percent solids content. "We don't think they [Heinz/ARCO] have anything like this," he observes. On the West Coast, David L. Zollinger, executive vice president of the California Tomato Growers Association, says that his organization "would welcome" the high-solids tomato. Meanwhile, at Campbell headquarters in New Jersey, Chairman R. Gordon McGovern says of his company's latest genetics efforts, "We'd like to get a magic balance all in the same tomato if we can." McGovern says that Campbell wants the kind

contains 2,249 items." And the company proudly points to its past accomplishments in tomato breeding. "Campbell has spent decades developing nutritious and sturdy tomato varieties," boasts the company in one of its annual reports. "We've boosted yields from six tons to twenty-four tons an acre. We specially bred the tomato for mechanical harvesting to boost productivity even further. Then we developed a tomato paste so our production schedules aren't hamstrung by the seasons." Over the years, Campbell's has developed more than thirty new varieties of tomatoes, dating from research which began at the turn of the century.

of tomato that will produce "a can of tomato soup so good the consumer just has to have it."

By September 1984, Campbell had signed another biotechnology research contract with Calgene to develop "proprietary high solids processing tomato varieties with optimum taste, color, and texture," and two additional contracts with DNAP, one of which was aimed at "developing fresh market tomatoes with superior characteristics." Of these latest research investments Campbell Vice President for Research and Development James R. Kirk said: "In the processing area, a tomato with increased solids, greater flavor, better texture and color, and extended-season availability will enable tomatoes to be processed into a wide range of Campbell food products at lower costs. The development of a superior fresh-market tomato available in supermarket produce departments year-round for the first time ever, will help meet rising consumer demand for fresh produce."

Although Campbell makes much of its "nutritious products" in corporate advertising and other literature,* nutrition does not appear to be a priority in its biotechnology programs. For example, when DNAP's David Evans ticks off the tomato traits his firm is hoping to develop for Campbell, improved nutrition is not one of them. Evans says that the tomato plants his company is developing for Campbell should produce fruits that look and taste appetizing, and which have a firm texture and skin for easier handling and

*Questions about Campbell's claims for some of its products, as well as the accuracy of the nutritional information the company conveys to the general public, arose in an October 1984 action brought by New York Attorney General Robert Abrams against Campbell's for misleading advertising. Abrams charged that some of the company's ads in its "Soup Is Good Food" campaign made misleading comparisons between Campbell's soup and other foods. For example, in the middle of one print ad with the headline "All Calories Are Not Created Equal," Campbell's made the following claims:

> ". . . . Most Campbell's Soups are dense in nutrients, so you get a higher level of nutrition with fewer calories than you get from many other foods.
>
> "Because of their nutritionally hardworking calories, many Campbell's Soups are more nutritious than other good foods. Calorie for calorie, Campbell's Tomato Soup has more Vitamin C than carrots or apricots."

Abrams charged that this comparison was misleading, because carrots and apricots are not good nutrient comparison sources for vitamin C, and in fact, are only mediocre sources of vitamin C. The company's attempt to elevate tomato soup's vitamin C content by comparing it to apricots and carrots was not a fair one in Abrams's view. And although Campbell's agreed to stop such practices in future ads, it denied that its ad campaign was misleading.

shipping. In addition, the new DNAP tomato plants should produce foliage that is low and compact to aid in mechanical harvesting, and have a uniform ripening trait so that tomatoes produced in the field are synchronized with the processing factories.

The new genetics of the tomato now being pursued by Campbell, Calgene, DNAP, Heinz, ARCO and others is not unique, and will soon be coming to other vegetables, fruits, cereal crops, and livestock. The Kellogg Company, for example, is sponsoring research at the Agrigenetics Corporation for the development of high-yielding, hard starch corn varieties, and Agrigenetics itself claims to have developed a high-protein wheat strain ideally suited to the baking industry whose flour will absorb water 30 percent faster than that from other varieties.

In the food-making process, business is interested in certain kinds of mass; the cheapest possible aggregation of raw product mass at the beginning of the process—as in tons of "high-solids" tomatoes—and the cheapest possible mass output of finished "food units"—as in loaves of bread or cans of soup. Put simply, bulk matters more than quality. And as plant breeding priorities in the United States and elsewhere have amply demonstrated in the past, it is tonnage and processing fit that matter most; the genes of yield, durability, and tensile strength which are the most important commercially.

Although there is the promise of nutritional and dietary improvement through the specificity and diversity promised with genetic engineering, there is also the very real possibility that the agricultural genes of first selection and capital backing will be those that serve the industrial apparatus of food production rather than those that improve the quality of food. The emphasis, it seems, is to use genes to serve the *business* functions of food rather than the *dietary wholesomeness* of food. And in that process, consumers will be last in line.

CHAPTER

9

VANISHING VEGETABLES

One must cultivate one's garden.
—*Voltaire*

Every summer, more than 38 million Americans—a number roughly equivalent to the combined populations of Texas and California—plant a vegetable garden of some kind. Each year, in fact, much of the 1.7 million acres of American land devoted to such gardens—from backyard patches to urban window boxes—is planted with the hope of producing fresher, better tasting vegetables; food that doesn't have to make the long trip to market, and vegetables that don't have to be mechanically harvested.*

"When you grow your own," says Nancy Bubel, author of *Vegetables Money Can't Buy, But You Can Grow,* "a whole new world of real food becomes available to you." Yet even in the privacy of one's own backyard the long arm of corporate America can be found in the genetic material of the millions of seeds, fruit trees, berry crops, and other plant materials sold to home gardeners every year. For increasingly the character of this "home-grown" industry, once dominated by family-owned companies, is being corporatized and consolidated like the farm seed industry.

*According to the National Association for Gardening, Americans choose to vegetable garden primarily because it saves them money, produces fresher, better-tasting vegetables, and is an enjoyable and healthful activity. Others also garden to avoid pesticides and food additives, and still others do it for peace of mind and the opportunity to work with the soil.

In the United States, although there are hundreds of companies involved in the sale of garden seed and plant materials to home gardeners, a relative handful of companies actually breed and produce the material sold to consumers. And in certain sectors, like the mail-order seed business, fewer than five companies dominate the business. In recent years, it has been these companies that have been the targets of corporate takeovers and investment. In fact, the nation's largest garden seed company, the W. Atlee Burpee Company of Warminster, Pennsylvania, has been owned by General Foods and ITT, and is now in the process of being sold by ITT to yet another company. Another of America's major garden seed houses, the Joseph Harris Seed Company of Rochester, New York, is now owned by LaFarge Coppee, a giant French construction corporation now moving into new markets. And a number of other American garden seed and horticultural suppliers have been acquired or consolidated in recent years.* "When I look around at my friends in the business," says William J. Park, owner of the George W. Park Seed Company of Greenwood, South Carolina, "I realize I'm the last of the privately owned major mail-order gardening seed companies."

In Europe, a similar pattern has developed. In England, for ex-

*In addition to the corporate ownership of Burpee and Joseph Harris seed companies, two other companies prominent in the mail-order seed business formerly owned by Amfac—the Gurney Seed & Nursery Company of Yankton, South Dakota, and the Henry Field Seed & Nursery Company of Shenandoah, Iowa—are now operating as one company. Also in the home garden seed business, several "packet seed" houses have been acquired by foreign corporations, including the nation's largest such firm, the Northrup King Company (owned by Sandoz), the Ferry-Morse Seed Company of Mountain View, California (owned by Limagrain, a French company), and the Fredonia Seed Company of Fredonia, New York (acquired by the Stokes Seed Company of Canada). In the horticultural supply business, including both retailers and wholesalers, the Weyerhaeuser Corporation has been quite active in recent years acquiring four companies: Wight Nurseries of Cairo, Georgia (retail ornamental); Shemin Nurseries of Greenwich, Connecticut (garden and floral supplies); Hines Wholesale Nurseries in California and Texas (national landscape supplier); and Oakdale Nurseries of Florida (tropical houseplants). In 1983 and 1984, the General Host Corporation spent $72 million to acquire two regional nursery chains: Frank's Nursery & Crafts in the Midwest and the thirty-store Flower Time, Inc., operating in the Northeast. The nation's largest rose breeder, the Jackson & Perkins Company of Medford, Oregon, is now owned by R. J. Reynolds. And Stark Brothers Nurseries, one of the nation's largest mail-order fruit tree and horticultural firms, has acquired Interstate Nurseries in Iowa.

ample, seed industry analyst Pat R. Mooney finds that three corporations—Kema Nobel and Cardo of Sweden, and Royal Dutch-Shell—through their seed company subsidiaries, now account for about 80 percent of the garden seed market.

Major corporations are acquiring garden seed companies for the same reasons they have bought farm seed companies. "It's mostly to get into the basic production of food and fiber," says William J. Park. "That's why chemical and pharmaceutical companies buy seed companies, along with the genetic engineering possibilities that came along recently." Adds one Burpee spokesman, "Gardening companies are very attractive businesses because one thing is for sure: we all have to eat." Garden seed companies—especially the ones that operate in national or even large regional markets—can be quite profitable operations. In the mail-order business, for example, there are few middlemen, which means larger profits for the seller. Moreover, regular orders from home gardeners provide lucrative sources of cash-in-the-mail for parent corporations.

But many gardeners who order their vegetable seed by mail from one of Burpee's or Joseph Harris's catalogues may never know that their favorite seed company is owned by a multinational corporation like ITT or LaFarge Coppee. Usually, there is no mention of the parent corporation in the catalogue. According to some seed industry spokesmen, however, nothing has changed. "As far as we're concerned," said Charley Wilson, a manager at Joseph Harris when that company was owned by Celanese in 1979, "there's no effect whatsoever. We go on just exactly the way we always have." Despite this, others argue that dramatic changes have occurred, and that more are on the way.

VANISHING VEGETABLES

"Multinational agrichemical conglomerates are buying out family-owned seed companies, dropping their collections of standard vegetables and replacing them with the more profitable hybrids and patented varieties," says Kent Whealy of Decorah, Iowa, who heads a unique organization of backyard gardeners and seed savers called the Seed Savers Exchange. "In many cases," continues Whealy, "the deleted varieties represent the life's work of several generations within these families, and are extremely well adapted to local weather, pests, and diseases. Far from being obsolete or inferior, I believe they are the best home garden varieties that we will see." As Whealy looks into the future, he sees the trend of vanishing varieties accelerating, pointing out that "More vegetable varieties will be dropped in the next three years than have been dropped in the last twenty."

Whealy recalls that at the turn of this century, there were lists of vegetables "in which bean varieties alone covered six standard-sized sheets of paper with single-spaced typing," that pea varieties required four pages, and onions two and a half. Today, he says, it is estimated that less than 20 percent of those 1900-vintage vegetables can be found in commercial seed catalogues. And, he explains, "whenever a variety is dropped from commercial availability—unless an individual or seed bank decides to make a concerted effort to keep [it] alive—it is on the road to extinction."

Garden writer Carolyn Jabs notes that the seed catalogue of the Joseph Harris Seed Company, "listed more vegetable varieties in 1900 than it does today." And that pattern, she adds, "is typical of other companies." But the American Seed Trade Association sees matters differently. "If there was a demand for an old variety," says ASTA representative Bob Falasca, "a seed company would continue to carry it. In most cases new varieties replace the old because consumers find them more desirable."

Seed companies, even those that do a healthy business in the garden trade, are after efficiency and economy in their seed production operations, and this is reflected in the kind and number of varieties they offer for sale. Kent Whealy points out that "Most seed houses these days simply can't afford to offer any vegetable that sells fewer than five hundred packets a year, so the 'small-time' cultivars are simply discontinued."

Clearly, it is less costly to breed and handle fewer popular varieties; the overhead is lower and the profit margins are obviously better. The trend in the garden trade is toward product homogenization rather than product diversification; toward developing fewer popular varieties of seeds and plants that appeal to national markets rather than more different varieties that can be sold in regional and local markets.

"Seeds are not like cars," says Carolyn Jabs. "Nature thrives on messy diversity. Growing conditions differ from garden to garden, even from state to state. The bean that flourishes on loam soil with hot, dry summers may be altogether different from the one that likes muck and moisture. The seed companies can't accommodate such discrepancies so they select a few all-purpose varieties, common denominators that will grow everywhere with reasonable success."

Some seed companies also improve their efficiency by selling the same varieties to home gardeners that they sell to the commercial vegetable growers who supply food processors, supermarket chains, and national markets. Because most modern vegetables varieties are developed for the commercial grower and not the home gardener, explains Jim Johnson, author of the *Heirloom Vegetable Guide*, considerations such as uniform size and shipping durability receive priority in breeding programs. "These qualifications matter little to the home gardener," says Johnson. "We want wholesome, tasty, productive varieties."

"Of course we breed for the commercial growers," says Joseph Harris of the Joseph Harris Seed Company. "That's where the money is." But Harris also argues that home gardeners have benefited at the same time. "Today's peas," he says, "taste better than the old varieties. We have fabulous tomatoes, radishes that don't get pithy, and almost all [of our] vegetables are resistant to at least some diseases."

Leroy Schmidbauer, a New Jersey tomato connoisseur and seed collector who has preserved thirty-five tomato varieties for their "old-time" tomato flavor, objects to some of the newer commercially-produced hybrids. "The new hybrids may be disease resistant and they may ship well, but they don't have the slightly acid taste of a real tomato." New hybrids like the Big Boy "are no better tasting than old varieties like Rutgers or Ramapo," says Schmidbauer. "The tomatoes handed from generation to generation are often the best," he says, "but you really have to look for them." And that's what Kent Whealy's Seed Savers Exchange is all about.

SEED SAVERS MAKING A DIFFERENCE

Seed savers are not, for the most part, mainstream backyard gardeners. They comprise a tiny faction of today's 38 million American gardeners and represent perhaps the most extreme phalanx of the nation's gardening movement. They are activist gardeners who are, figuratively and literally, taking matters into their own hands.

Seed savers grow and save their own vegetable seed, and in some cases exchange it among like-minded souls across the country.* For many of these gardeners, saving seed and protecting old varieties of vegetables is more than a hobby; they are dedicated to preserving and continuing the old varieties for reasons which often serve the larger purposes of biological diversity, exquisite taste, nutritional quality, and cultural significance.* They are serious about what they do and they are forming "growers networks" around the country to produce more seed for exchange to keep these rare and endangered older varieties alive and in circulation for the future.

The Seed Savers Exchange, one of the largest organizations of its kind, began as a tiny, informal letter-writing and seed savers group, and within a few years, became a nationwide organization. In 1983, Whealy's exchange had 500 participating members offering seed for exchange or sale. "Over the past seven years," says Whealy, "our organization's members have offered access to the seed of 3,000 unique vegetables, and have sent off enough samples to make an estimated 250,000 plantings of vegetables that were usually not in any catalogue, and in many cases were quite literally on the edge of extinction."

One of the projects now consuming Whealy and his exchange is a computer-based inventory of all nonhybrid vegetable varieties available in the United States and Canada. With such an inventory, it will be easier, says Whealy, to determine "what is being dropped

*Although the activities of the Seed Savers Exchange are focused on preserving and perpetuating traditional vegetable varieties, there are other organizations working to preserve, protect, and/or exchange fruits, nuts, cereals, and livestock breeds, as well as local and regional groups collecting and saving seed of local or cultural interest. For example, the North American Fruit Explorers of Hinsdale, Illinois is a membership organization and exchange which promotes traditional and hard-to-find varieties of fruits and nuts; Native Seed/Search of Tucson, Arizona works to preserve heirloom vegetable seeds and other rare native plant materials of the American Southwest and northwest Mexico; and The American Minor Breeds Conservancy of Hardwick, Massachusetts works to promote and conserve minor, rare, and endangered breeds of agricultural livestock. For other examples of such organizations see *The 2nd Graham Center Seed and Nursery Directory*, published by the Rural Advancement Fund, Pittsboro, North Carolina.

*According to Kent Whealy, there is a single gene in green beans that determines stringlessness, and well over 99 percent of the snap beans grown in this country are stringless. "Now imagine what would happen if one disease zeroed in on that particular gene," says Whealy. "People would be scrambling frantically to find a bean without the stringless gene . . . one of the old string varieties. And it's likely that only seed preservation programs like ours would still have those rare cultivars."

and how quickly." Using such a list, organizations like the Seed Savers Exchange can buy up endangered vegetable varieties and begin distributing them to seed savers throughout the country, keeping them in use for the present and producing seed for the future.

OUTLAW VEGETABLES?

Some home gardeners and seed savers became involved in the 1979–80 Congressional fight over the amendments to the Plant Variety Protection Act (PVPA) when they thought that the seed-patenting law might lead to a prohibition on growing certain vegetable varieties. During that debate, backyard gardeners wrote hundreds of letters urging their representatives and senators to oppose the bills then before Congress. The amendments were drafted to extend patent-like protection for new varieties of celery, cucumbers, tomatoes, okra, carrots, and peppers—six vegetable species Congress excluded from the original act in 1970.*

The letters of protest figured prominently in slowing congressional acceptance of the amendments, and also contributed to the scheduling of a second hearing on the topic in 1980 by the House Agriculture Committee. "I have received more mail on this issue than any other agriculture issue in recent memory," wrote California Congressman George E. Brown, in a November 1979 letter to agriculture subcommittee chairman Kika de la Garza. Brown asked for further hearings on the matter.

New York gardener Carolyn Santelli explained in her letter to one Congressman that, "As an enthusiastic home gardener, I believe that home-grown produce is more nutritious, inexpensive and economical than mass-produced, non-local and out-of-season produce . . . I purchase my seed from small seed companies who specialize in varieties which sell in this area—most often, these are older, well-established types that are often not available from larger, national companies. They adapt well to the climate and are healthier and more dependable than newer varieties I have tried."

Santelli urged the congressman to oppose the bill because she believed that patenting new varieties of vegetables would "encourage the growth and eventual takeover of large seed companies," and "discourage the preservation of old seed varieties by pushing

*See Chapter 4 for more details on this subject.

small companies out of business." She also suggested that the seed-patenting law might encourage the banning of certain vegetables for legal reasons, as was then the apparent practice in some European nations.

In England, a "common catalog" of accepted vegetable varieties is used as a legal document to enforce plant patents. Excluded from the accepted list are near-relatives of patented varieties. "In an effort to cull out 'weak' food crop species," reported *The New York Times* in August 1980, "Great Britain has ordained that some 500 vegetable seed species can no longer be sold or be catalogued in any national seed list (for the) official reason . . . that many of the banned species were similar to other vegetable varieties but did not have the heavier weight, brighter colors or commercial appearance of their near-relatives."

Opponents of the bill amending the American seed patent law argued that the common catalogue system was "a logical next step" in enforcing seed patents in the United States. USDA officials and the American seed industry denied that any such system was in the offing for this country. Moreover, they argued, no such provision was contained in the 1980 PVPA amendments being considered by Congress.* Yet even the rumor of such a system moved many gardeners to write their Congressmen in opposition to the bill.

Because no European-like provision was actually contained in the 1970 PVPA law or the 1980 amendments to it, and because the bill amending the Act was regarded as a minor, noncontroversial measure, the gardeners' objections to it were dismissed by most politicians and consigned to the "emotional" fringe. Congressman Kika de la Garza, for example, who chaired the subcommittee considering the bill, said of the opponents' position: "These people are fighting the wrong war on the wrong battlefield. They're emotional. What this bill does is allow little bitty seed companies to patent a seed so that the big companies won't steal it from them the next year. The first thing I verified myself was whether this bill was going to hurt my little chili plants and my little tomato plants in my back garden, and the answer was 'no!' "

But others, like seed saver John Withee, a cigar-smoking, retired

*Some objectors to the seed-patenting bill suggested that a clause be added that would prohibit the American adoption of such a European system of patent enforcement—a clause that Congress was unwilling to include in the final bill.

medical photographer from Boston, did not think that the PVPA or its amendments were harmless measures. "They're trying to tell us this little amendment they're working on down there in Washington won't mean anything to people like me who are trying to save the old, heirloom varieties from dying out. But I don't think it's harmless. This is just a case of the camel getting its nose under the tent. Pretty soon we'll have the whole beast in here with us." Withee, and others like him, viewed patenting laws as strengthening the position of the larger seed companies as well as corporations moving into the genetics business.

GARDEN GROUPS JOIN THE FRAY

During the PVPA debate, a few gardening organizations became involved in the seed-patenting issue, including the National Association for Gardening (NAG), a Burlington, Vermont organization with 60,000 members. In June 1980, Kit Anderson, director of the NAG's Botanical Gardens Project, appeared before a Senate agriculture subcommittee. Noting that four of the six vegetable species proposed to be added to the seed patenting law—tomatoes, peppers, carrots, and cucumbers—were "among the ten most popular home garden crops," Anderson asked for an objective evaluation of the law's impact. "The actual imact of the 1970 patent legislation on American backyard gardeners . . . farmers, and all those trying to fight inflation by raising their own food," she said, "has not been explored." Anderson wanted to know what effect the law had on the seeds available to consumers, and how it affected seed prices and varieties. She wondered whether all consumers were being provided with "high quality seeds and a wide selection of types suited to their particular needs and geographical region."

In a follow-up letter to Vermont Senator Patrick Leahy after she testified, Anderson noted, "There are some serious questions about recent trends in crop breeding and seed marketing, and until we can see that the PVPA is not a direct or indirect cause of these trends we cannot support the legislation as written." Anderson's concerns, and her organization's request for a study, however, were not regarded as important enough to be answered before the amendments were adopted. The bill was signed into law by President Jimmy Carter in December 1980.

Another gardening organization that became involved in the PVPA debate, although not as an opponent of the law, was the Rodale Press of Emmaus, Pennsylvania, which through its monthly magazine, *Organic Gardening,* reaches more than one million gardeners. In a February 1980 editorial, Robert Rodale argued that the issue was not so much one of the pros and cons of seed patenting, as of the need to conserve crop genetic resources for the future. He presented the arguments on both sides of the patenting question, and did not take a position on the pending amendments. He did say, however, that further hearings then scheduled on the bill "should also address the need for more support for seed exploration, identification and conservation."

Following up on the subject in a May 1980 *Organic Gardening* article entitled, "The Real Scoop on the Plant Patent Controversy," Rodale staffer Anthony De Crosta wrote, "neither the current law nor the bills before Congress directly threaten the backyard grower." In fact, De Crosta noted that gardeners "could benefit from the plant-patenting laws since breeders would have an incentive to develop and market more open-pollinated varieties of tomatoes, peppers, carrots, cucumbers, okra, and celery." He added, however, that Congress had heard from few of the nation's home gardeners. "Although the hearing was labeled a 'public' meeting," wrote De Crosta, "most gardeners didn't even know it took place until months later. The people most affected by the pending amendments had the least input."

De Crosta, like Rodale, stressed that the loss of genetic diversity in crops was perhaps the more serious problem, but he also mentioned the emerging influence of multinational corporations, which, he said, "By controlling the seed companies . . . have the potential to control the food-producing resources of this country." De Crosta, also pointing out that plant-breeding research is expensive and time-consuming, noted that "if big companies are the only breeders that can afford to invest the time and money in new varietal development, then they would possess the exclusive rights to many crops that you plant in your garden."

Despite the outcome of the 1980 amendments to the PVPA and the gardeners' involvement in that fray, there is one clear political potential for garden interests in future such fights. At nearly 40 million strong, gardeners of one kind or another are found in practically every congressional district of the United States. Without question, they would represent a formidable political bloc if ever moved to speak forcefully on any issue, and with the help of organizations such as the Rodale Press, the National Association for Gardening, the Garden Club of America, the Seed Savers Exchange, and others, could become an important political force in a range of food, resource, and environmental issues—and especially on those that begin with the genetic envelope called the seed.

GARDENS OF CORPORATE DELIGHT

Whether the seed patenting law has helped major corporations tighten their grip on the backyard garden trade is still an open question among some observers, particularly economists who say that other market forces are perhaps more responsible for economic concentration. Yet by 1984, one thing was clear: the seed-company subsidiaries of six major corporations—Upjohn, Sandoz, Lubrizol, Limagrain, LaFarge Coppee, and ARCO—had garnered more than 230 vegetable-seed patents among them, well over half of those that had been issued. Moreover, some of the corporations that already owned garden-seed companies, or that were becoming involved in biotechnology research, were also broadening their involvements in the home gardening market.

Under ITT's ownership, for example, there has been some "business synergy" between the products of O. M. Scott & Sons of Marysville, Ohio, and those of the Burpee seed company. In 1971, ITT acquired O. M. Scott—the nation's largest home lawn care company—for $108 million, the sixth largest merger that year. Today, ITT has integrated some of Scott's product lines with those of Burpee. The latter company—as a 1983 *Newsweek* advertisement entitled "Seed. Feed. Succeed." explains—is now selling fertilizer to the home gardener. "A good home garden," reads the advertisement, "is like a good homemade soup. If you use the right ingredients, there's nothing more satisfying."

"At Burpee we supply the right ingredients . . ." says the ad. "Let's start with the seed. Burpee Seeds have always been bred with the home gardener in mind. You can choose seeds bred for a big yield in a smaller space, disease and heat tolerance, a shorter growing season, a sweeter taste . . . even smaller flowers." The advertisement continues by boasting about "the Burpee tradition of quality," and then comes the pitch for fertilizer.

"Now, seedlings need food to grow on," says the ad. "Good Food. That's where Burpee can also help—with Burpee Grow Fertilizers. Just sprinkle some on and you can rest assured your plants will get the right nutrients at the right time. . . . So if you want more flowers, more vegetables and fuller shrubs (and who doesn't), make sure your garden's eating right with Burpee Grow Fertilizer."

Other corporations, such as American Cyanamid, Chevron, Ciba-Geigy, Mobay Chemical, and Stauffer Chemical are also in the home gardening business. In 1981, for example, the sales of lawn fertilizers, pesticides (indoor and garden), lime, and other chemical-based garden products were $1.2 billion, up 9 percent from 1980, and pesticide sales to home gardeners are expected to continue to rise through 1985. "We see more activity with insects, and we hear more reports of insect problems," says Philip J. Young, director of the Spectrum Home and Garden Products Department of Ciba-Geigy, a company that also owns several farm seed companies.

Chevron's Ortho Consumer Products Division, the nation's largest seller of home pesticides, is also in the business of publishing and selling gardening books. "1982 was the best year in our history," says Ortho's Daniel P. Hogan, Jr. The company's garden books are priced at $5 to $6 each, but their real benefit to Ortho, according to *Chemical Week*, comes with the "hefty collateral sales of garden products when readers return to the store for materials needed to carry out projects the books suggest."*

Companies directly involved in the vegetable-seed business—such as Upjohn's Asgrow Seed Company, Sandoz's Northrup King, ARCO's Dessert Seed Company and others—are now using biotechnology to develop new vegetable varieties, to find shortcuts to hybridization, and to make new vegetables and vegetable seeds suited

*In fact, in Chevron's 1983 annual report it is noted: "Ortho's new hardcover book, *The Complete Guide to Successful Gardening,* sold out its first printing of 90,000 copies; an additional 60,000 were ordered for the holiday gift-giving season. The five hundred-page book is also being offered as a Book-of-the-Month Club selection." Chevron also explained that Ortho published 11 new softcover books in 1983, bringing the total number of titles for the company to fifty-three. "More than 15 million Ortho books have been sold in the last decade," says Chevron.

for world markets, commercial agriculture, and food processing. For companies like Northrup King that deal in both the commercial trade and the home garden trade, much of this new seed will probably be used in both markets. Yet the genes of first priority used in the making of this seed will not, in all likelihood, be those preferred by the home gardener, but rather those of commercial importance to large-scale growers, food processors, and shippers.

The seed and seed improvements that do come to the home gardener by way of biotechnology, however, are likely to be more expensive. In the years ahead there will be more hybrid vegetable varieties available for the home gardener, and there may also be a new kind of encapsulated, pre-germinated seed—a seed that will guarantee to produce plants within a day or two rather than a week or more, as is normally the case.

Plant Genetics, a California-based company working with a biotechnology process known as "somatic embryogenesis," can take tissue from a stem of a celery plant, for example, and reduce it in laboratory culture to cells, which when stimulated by hormones, become what are called "somatic embryos"—or seeds without their coats. "Soma" is the greek word for body, and "somatic cells" are those from the body of plants and animals, as in the cells of stems and leaves in plants. Somatic cells are distinct from germ or reproductive cells, in that they have, in the case of stem and leaf cells in plants, differentiated to perform "stem and leaf tasks." Yet, even individual somatic cells in the stems and leaves of plants contain the genetic material in their DNA to make an entire plant, or clone of the parent, and this is what Plant Genetics is making with its pre-seed embryos. For plant breeders that have had a particularly difficult time hybridizing some crops by traditional cross-breeding techniques, or producing true hybrid seed for such crops, somatic embryogenesis offers a way to produce those crops. Plant Genetics scientists have also devised an artificial seed coat—a transparent organic jelly coated with a biodegradable polymer—to encase the embryos, called GEL-COAT.*

*Plant Genetics is not the only company working with artificial seed technology. Upjohn, Hoffman-La Roche, Agrigenetics, Cetus, and DNA Plant Technology Corporation, among others, are also researching somatic embryogenesis and/or seed encapsulation techniques.

The motivating commercial interest for "synthetic or artificial seeds," as Plant Genetics' gel-coated embryos are often called, comes from the vegetable industry, which wants uniform germination in its seed and the assurance that every seed will become a plant. Time will tell, however, whether artificial seeds, and the "instant gardens" they promise, will also carry the genes of vegetable varieties of interest to the home gardener. One potential problem with this technique for home gardeners and everyone else, however, is that it may foster genetic uniformity. "The reason we want to make artificial seeds," says Keith Redenbaugh of Plant Genetics, "is because natural seeds have a lot of variability. In clones, all the material is uniform."

Yet, there is no reason why biotechnology could not be used to diversify and expand the range of varieties and options available to both commercial growers and home gardeners. Biotechnology could, for example, be used to produce new garden-crop varieties genetically honed for superior nutritional qualities, taste, and pest-resistance. But whether that happens will depend, to some extent, on how loudly home gardeners raise their voices to demand such varieties.

In the meantime, for people like Kent Whealy, John Withee, and other seed savers, the more important mission is saving the old, nonhybrid seeds rather than depending on the seed or biotechnology industries to devise "genetically superior" new ones.

CHAPTER

10

INVITING DISASTER

See that farm down there?" Mr. Simplot asked, pointing out the airplane window. "Gotta be the world's largest potato farm. Thirty thousand acres. So big you couldn't walk across it in a day.
—*Robert E. Rhoades,*
"The Incredible Potato"

J. R. Simplot, the seventy-four-year-old founder and principal owner of the J. R. Simplot Company of Caldwell, Idaho, is often given to boasting about his potato farm. And rightly so—Simplot's "potato farm" is one of the largest privately held agribusiness operations in the United States. Worth an estimated $1 billion in annual sales, Simplot's operation includes thousands of acres of potatoes along the Snake and Columbia rivers of Idaho, Washington, and Oregon, at least eight potato-processing plants, assorted feedlots, fertilizer plants, and other facilities. In addition to his own operation, Simplot is also involved with the Boeing Corporation in a partnership called SIMTAG farms, a 40,000-acre irrigated wheat and potato operation along the Columbia River. But J. R. Simplot didn't need the help of Boeing to get where he is today; all he needed was the potato.

Simplot is the undisputed "potato king" of the United States, known in the northwest as "Mr. Spud." Simplot produces so many potatoes, in fact, that he and a partner were accused in 1976 of pushing prices down on the commodity markets by selling contracts for 100 million pounds of potatoes. The Commodity Futures Trading Commission charged him with selling contracts he did not have (a charge Simplot denies), and in 1978 barred him from futures trading until 1984.

Like a character from a Horatio Alger story, J. R. Simplot first became involved in the potato business in the early 1920s after he dropped out of the eighth grade and began sorting potatoes by hand and doing manual labor along Idaho's irrigation canals. At the age of fifteen, he bought and later sold seven hundred hogs, the income from which he used to buy some land and equipment to begin farming. He soon moved into potato processing, and was one of the first to use electric potato sorters. By the age of thirty, Simplot was a millionaire and fast becoming one of Idaho's most notable tycoons.

Today, J. R. Simplot lives on a hilltop just outside of Boise, surrounded by a 120-acre estate complete with stables, a training arena, and a collection of fine horses, including a few Tennessee walkers. His home is marked from afar by a gigantic American flag mounted on a sixteen-story flagpole. The Greek paintings, carved jade figurines from China, and gold fixtures of leaping fish that decorate Simplot's home provide some hint of his $500 million in personal wealth. But stretching over thousands of acres of land in Idaho, Washington, and Oregon is the source of Simplot's real wealth: the potato, and especially, one variety of potato known as the Russet Burbank.

"Early one morning," wrote Robert E. Rhoades, who visited Simplot's operation for *National Geographic* magazine in 1981, "Mr. Simplot and I lifted off from Boise, Idaho, in his private plane, headed northwest. Somewhere over Oregon, which is fourth in United States potato production, the airplane dipped. As far as I could see were perfect 130-acre circular fields, designed to accommodate . . . massive pivoting irrigation systems."

THE REACH OF ONE POTATO

The potato is a member of the genus *Solanum,* in which there are 2,000 species. All potatoes used in the United States, and most of those grown commercially throughout the world belong to one species, *Solanum tuberosum.* Today in the United States, twelve varieties from this one species constitute 85 percent of the nation's potato harvest, but one variety—the Russet Burbank—is by far the dominant variety.

In 1982, for example, nearly 40 percent of the potatoes planted in the United States were Russet Burbanks. In Idaho that year, 95 percent of the potatoes grown were Russet Burbanks; in Montana, 93 percent; and in Oregon, 72 percent. Yet, these acres and acres of the same kind of potato—in this case genetically *identical* potatoes since they are reproduced vegetatively from a single plant variety—could create a big problem should one disease or insect key in on the Russet Burbank, just as the Southern Corn Leaf Blight wiped out 15 percent of the nation's hybrid corn crop.*

In the wake of that crisis, the National Academy of Sciences (NAS) took a hard look at the genetic situation of many other crops, including the potato. In the 1972 report that followed, *Genetic Vulnerability of Major Crops,* the NAS found that 72 percent of all American potato acreage was planted to just four potato varieties, adding that the varieties then in use were "very closely related" genetically.

Since the NAS wrote its report, the genetic situation of the potato hasn't improved much. The acreage share of the top four varieties has dropped to 60 percent, but the Russet Burbank's share has grown a full 12 percentage points—from 28 percent of America's total potato acreage in 1970 to 40 percent in 1982.

UNIFORMITY ABOUNDS

The National Academy of Sciences' 1972 report concluded that most crops—not only the potato—were "impressively uniform genetically and impressively vulnerable" to disease and pestilence. And while some scientists argue that genetic diversity in the nation's agricultural crop base has improved somewhat since 1972, others still see a dangerous level of genetic uniformity throughout American agriculture. Appearing before a U.S. Senate Agriculture panel in 1977, Dr. J. Artie Browning, a plant pathologist at Iowa State University, observed that "Corn and soybeans in Iowa today are more genetically homogeneous than they were in 1970."

*See Chapter 1 for more details.

In certain regions of the United States today, there are tens of thousands of acres of farmland planted to continuous blocks of the same kinds of wheat, corn, cotton, and other crops. Although there are more than 250 varieties of wheat available, ten varieties dominate the landscape. In 1981, six varieties accounted for nearly 40 percent of America's wheat acreage. The situation is similar in other crops: four varieties account for 65 percent of the nation's rice acreage; six for 42 percent of the soybean land; three for 76 percent of the snap beans planted; two for 96 percent of the peas; and nine for 95 percent of the peanuts.

Despite the painful lesson of the 1970 Southern Corn Leaf Blight, 43 percent of the nation's hybrid corn acreage is planted to varieties derived from six inbred lines. Dr. Sam Levings, a corn geneticist at North Carolina State University, says it is "very likely that 50 to 70 percent of the hybrid corn in the Northern Corn Belt" contains the genetic material of one parent line called B-73. "There are only four or five prominent inbred groups used in about 90 percent of the current domestic [corn] hybrids," adds John Dillon, research director for the Renk Seed Company of Sun Prairie, Wisconsin. He says that "The seed corn industry is probably working with the narrowest genetic base in its history." Because of this, Dillon sees "a threat of considerable damage by insects, diseases or growing conditions to which our popular hybrids are potentially susceptible."

Even the once-ubiquitous and varied apple has dwindled to a few commercially favored varieties. "Over the years and centuries," says Fred Corey of the International Apple Institute, "there have been more than 8,000—maybe even 10,000—apple varieties named and recorded in history." But of the one hundred or so varieties—old and new—now being grown in the United States, he says, only "a dozen or fifteen varieties constitute 95 percent of our . . . commercial production."

In livestock, the situation is not much better. The Holstein cow constitutes about 70 percent of the nation's dairy herd. In the cattle industry, Angus and Hereford breeds account for more than 80 percent of all registered breeds. In the hog industry, cross-breeding among eight purebred lines—Berkshire, Chester White, Duroc, Hampshire, Landrace, Poland China, Spot, and Yorkshire—provides some 90 percent of American pork production. In fact, some estimates find that the Duroc and Hampshire breeds together account for 60 percent of all hogs in the United States.

The broiler industry relies extensively on the Rock Cornish hen. The White Leghorn and its derivatives provide most of America's eggs. And broad-breasted white breeds account for most of the nation's turkey meat.*

LEARNING THE HARD WAY

The history of modern agriculture is a history full of disease- and insect-caused crop disasters, many brought on by the use of a few popular crop varieties. In 1845, Ireland's primary potato variety, the Lumper, had no resistance to a potato blight that raged for three years, resulting in the Great Irish Potato Famine which took the lives of more than one million people and forced another million and a half to emigrate.

In 1916, the red rust of wheat destroyed 200 million bushels of the American crop and another 100 million bushels in Canada. In North and South Dakota that year, nearly 70 percent of the crop was destroyed. By 1917, Herbert Hoover's Food Administration had put into effect a voluntary program of two "wheatless days" per week. People ate cornbread because of the wheat shortage.

Rusts of wheat have plagued mankind since the dawn of domesticated agriculture. In the first century B.C., the Romans held a feast around April 25 in which they sacrificed a red dog to the god Robigus in the hope that wheat rust would not appear. Wheat rusts hit the North American breadbasket several times before the 1916 calamity, and a wheat stem-rust wiped out 65 percent of America's Durum wheat in 1953 and 75 percent in 1954. Most of the Durum wheat planted at that time was of the Steward variety. "The wheat rust epidemics of modern times," wrote the National Academy of Sciences in its 1972 report, "are clearly genetically based, in that as resistant varieties become available, fungus mutates to a form that attacks the new variety, and an epidemic ensues."

But wheat isn't the only crop that has been periodically ravaged by disease or insects. Between 1900 and 1908, two-thirds of all Bartlett pear trees in California succumbed to fire blight. An epidemic of root rot and kernel smut devastated America's sorghum

*The white-breasted turkey is the epitome of what some scientists refer to as "overdomestication." The females *must* be artificially inseminated since the big-breasted males cannot productively mate with them.

crop in 1924. Nearly half of the tomato crop in the Atlantic Coast states fell prey to late blight in 1946, the same year that Victoria blight swept through 86 percent of the nation's oat crop. In the latter epidemic, at least thirty oat varieties, all with a common parent, accounted for the genetic sameness that led to almost a total crop failure.

Outside the United States, despite the "export" of modern agriculture and new Green Revolution crop varieties, disease and pestilence have also taken a grim toll. While the Southern Corn Leaf Blight was ravaging the American corn crop in 1970, leafhoppers were dispensing Tungro disease on Green Revolution rice throughout the Philippines and a leaf rust was devastating genetically uniform coffee plantations in Brazil. Wherever agriculture is practiced, disease and pestilence are always probing for a point of entry.

A basic tenet of biology is that living organisms have a certain capacity to evolve and adapt to changes in their surroundings—whether those be changes in temperature, predators, or the characteristics of some new crop variety. Stated in genetic terms, this means that as environmental changes exert different "selection pressures" on a given population of organisms, certain genes previously expressed in low frequencies in those populations become favored because of their usefulness in the "new" environment. Viruses, fungi, bacteria, and insects have short life cycles, and so can rapidly adapt to new crops and livestock—particularly those favored varieties or breeds used by man again and again. Because of their capacity for rapid genetic adaptation, plant pests and the organisms that cause agricultural diseases such as wheat rusts have overcome just about everything science has thrown at them, from pesticides to stronger crop varieties.

One way in which man has made the handiwork of disease and insects easier, however, and at times more predictable and more severe than it might otherwise be, is in the breeding of crops and animals to fit the strictures of modern agriculture.

THE GENES OF MONOCULTURE

Higher plants contain upwards of 10 million genes—more than ten times the number found in man. This vast sea of genetic information controls the way a plant performs. Some plant genes act alone in controlling a specific trait; others act in combination with other genes to control particular traits and biological reactions inside and outside of plants. The overall genetic composition of an individual plant—the sum total of its genes, called its "genome"—has evolved with time in response to changing environments.

When man first began to domesticate food plants, selecting those most suited to cultivation for a variety of reasons—good seed germination, growth rate, harvestability, the taste or color of the fruit, or size of the grain—he also in effect began to select genes and alter the genome of individual plants. Because the basis of his selection was the ability of certain plants to do well in an agricultural environment as opposed to a wild environment, man was, in other words, ensuring the perpetuation of those genes that governed good agricultural performance, rather than those that governed survival in the wild.

As man continued to select new crop varieties for yield and other desirable agricultural traits, the new varieties became far removed from their wild origins and began to carry new arrangements of genes selected for agricultural performance. Stated in vastly simplified terms, "old" arrangements of genes—those which evolved in complex plant communities—were gradually replaced by "new" arrangements of genes selected for agricultural production. Furthermore, as plants became more and more domesticated, and chosen for their productivity, the genes needed for survival in the wild were increasingly displaced or "lost." Such plants—now agricultural crops—then became completely dependent on man for their survival. A modern hybrid corn variety, for example, is virtually helpless in the wild; it has no mechanism for dispensing its seed. In fact, some plant geneticists believe that corn would become extinct if it were not cultivated by man. According to the Cetus Corporation's agricultural research director Winston Brill, corn "is a crop plant that has been bred for years for the highest yields. It is a machine of man, rather than a creature that can survive on its own."

With the advent of modern plant breeding, and particularly crossbreeding and selective pollination, the process of altering the "evolved genome" of individual plants was greatly accelerated. With such techniques, man began to scramble the carefully assembled genetic heritage of plants, rearranging their genetic instructions; saving and enhancing the instructions for high yield, for example, while often muting or discarding the instructions for disease resistance. In this process, some of the old, naturally honed and long-evolved genetic characteristics of plants, such as disease resistance,* were lost or greatly weakened in a process scientists call "genetic erosion."

Reporting for *Scientific American* in 1981 on the genetic situation of the world's wheat crop, for example, geneticist Moshe Feldman and biologist Ernest R. Sears noted that "the range of genetic variation of the cultivated wheats has decreased drastically in recent years," and that the continuing erosion of the wheat gene pool "not only reduces the possibility of further improvements in productivity but also makes the world wheat crop increasingly vulnerable to new diseases and to adverse climatic changes." Feldman and Sears also explained that the genetic variability of the cultivated wheats, which had accumulated over 10,000 years of cultivation, "has been diminished by the introduction of modern, scientifically planned breeding practices."

Jack Harlan, a plant geneticist who has written extensively on crop evolution and genetic diversity, has noted that plant systems in domesticated agriculture are "always much simpler than natural ecosystems." Moreover, says Harlan, the natural defenses of plants to disease and pests, which have evolved over evolutionary time spans, "tend to be extremely complex and stable," while the defenses developed in breeding nurseries "tend to be simple and unstable."

The generally better level of disease and pest resistance found in wild plants and those in primitive agriculture is due to both environmental and genetic factors. In the wild and in primitive farming, plants are often intermixed with one another, as well as with weeds, which "hide" them from pests and disease. In addition, being locally adapted, wild species and primitive crops usually have a broader base of genetic resistance to disease than do modern crop varieties; that is, they usually have genetic tolerance or resistance to more than one disease common in their natural region. Thus, in primitive agriculture, some crop damage from disease or pests might occur every year, but there is little danger of a catastrophic wipe-out. Yet

*It is important to note, however, that wild relatives of crops, while they have genetic resistance to one to three diseases that are common in their native region, they are often highly vulnerable to many diseases that are infrequent in wild environments.

as man further domesticated his crops, he continued to change both the genetic and environmental mechanisms of disease resistance, by breeding crops for specific traits and planting them in large numbers of their own and often identical kind.

As time passed, man began to replace nature's regime of biological diversity among plants and animals with his own regime of agricultural predictability, and, increasingly, genetic uniformity. Then came a steady progression of improved farming technology. In 1701, Jethro Tull invented a machine drill for sowing farm seed; by the 1730s farmers had begun to grow crops in straight rows. Cyrus McCormick unveiled his first reaper in 1831, and John Deere followed in 1837 with his steel plow. In advanced societies, horses gave way to tractors after World War I, and single-crop agriculture and continuous cropping were not far behind.

But monocultures of corn or potatoes or any other crop soon gave rise to matching monocultures of damaging insects, viruses, fungi, and new diseases. To stay one step ahead of the ever-adapting microbes and pests, man became a "treadmill plant breeder," developing new crop varieties that would stop the advance of a pest, but only until it adapted to that variety. But when pesticides came on the scene, the role of the plant breeder began to change.

UNDER THE CHEMICAL CANOPY

With the rise of monoculture, the need for "crop protection" methods to minimize the destructive effects of diseases and insects became great. In the late 1800s, the discovery that chemicals could be used to fend off crop pests ushered in a new era of agriculture—and, most importantly, a new era of plant genetics. Agricultural chemicals gave plant breeders the "freedom" to focus almost exclusively on the genetics of yield. At that time, when societies everywhere stood helplessly before the onslaught of disease and insects, the initial foray into the use of chemical pesticides for agricultural and public health reasons was seen as a great step forward. Yet this development served to further remove the breeder from the "genes of the past"; the real sources of plant diversity and disease resistance.

For a short time in the United States, marked roughly from the establishment of the land-grant universities in the late 1880s, to shortly after the turn of the century, "biological methods" of pest control seemed to be in the ascendancy. It was in the late 1880s, for example, that University of Illinois entomologist Stephen A. Forbes coined the word "ecology" (that branch of biology dealing with the interrelationships of organisms and their environment) and stressed the application of ecological principles in the control of agricultural pests. With the scientific community's recognition of Gregor Mendel's laws of heredity in 1900—which had gone virtually unnoticed since his publication of them in 1866—crop selection and plant breeding were undertaken with great interest, and a number of horticultural and cereal crop varieties resistant to diseases were produced and put into use.

Breakthroughs in plant breeding for disease-resistance were made earlier than those for insect-resistance. Between 1899 and 1909, for example, one of the first plant breeding programs directed at developing disease-resistant crops gave rise to varieties of cotton, cowpeas, and watermelon resistant to *Fusarium* wilt. By comparison, the first successful breeding program for insect resistance came to fruition only in 1942, when the first Hessian-fly-resistant wheat variety was developed.

However, the brief flurry of crop breeding for disease resistance at the turn of the century, and the early attempts employing biological methods for controlling crop diseases and pests, gradually gave way to newer technologies. The biological methods initially used in agriculture were cumbersome and not always completely effective, while plant breeding for disease and pest resistance was painstaking and time-consuming. As chemical pesticides became available, the approach to controlling agricultural diseases and insect infestations began to change dramatically.

Writing for the President's Council on Environmental Quality in 1979, Dale Bottrell summed up the change in approach to pest management that occurred between the late 1880s and 1920s: "The ecologically oriented approach . . . shifted to control by chemical pesticides as effective materials became available," noted Bottrell. "Pesticides were often more effective, were much simpler to use than the more complex and labor-intensive nonchemical approaches, were cheaper, gave greater yields, and provided readily available and inexpensive insurance to the user. Their use displaced many of the earlier control techniques, such as cultural and biological control, pest-resistant crop varieties, and habitat management." Moreover, said Bottrell, "The new chemical pesticides could be used by themselves and could achieve higher levels of pest control: they greatly simplified [such] control, and the earlier integrated pest control schemes were viewed as obsolete."

Chemical pesticides first made their appearance on American farms in 1887, when an arsenic compound called "Paris green" was used to control the Colorado potato beetle. An accidental discovery in Europe, also in the 1880s, led to the development of one of the first widely used pesticide sprays.* In later years, spray attachments for tractors and cultivators were adopted, becoming more sophisticated in their range and use with time. In the United States, the first aircraft application of pesticides was made in Ohio in 1921, and led to a practice that became widespread after World War II.

Some of the earliest pesticides were fashioned from naturally occurring minerals and plant products—compounds of arsenic, copper, lead, manganese, zinc, and other minerals, as well as pyrethrum from dried chrysanthemums and rotenone from some leguminous plants. However, these simple pesticide compounds soon gave way to a plethora of synthetic pesticides that were the commercial outgrowth of World War II chemical-warfare research. Between 1947 and 1960, the production of synthetic pesticides underwent a fivefold increase—from 124 million pounds to 638 million pounds. By 1981, it had reached 1.8 billion pounds.

With the arrival of cheap pesticides, crop diseases and insect problems were increasingly left to the chemical industry. Plant breeders, meanwhile, began to focus increasingly on developing crops for higher yield. And while crop yields did increase dramat-

*In the Medoc area of southwestern France it was the custom of vineyard keepers during the 1880s to spatter their grape vines near roads and footpaths with a poisonous-looking substance to discourage passers-by from pilfering grapes. One passer-by, Pierre Marie Alexis Millardet, a professor of botany at Bordeaux, noticed that the vines near the path he walked one October day in 1882 still retained their leaves, while the other plants had lost theirs to a mildew that was then causing great problems in the French vineyards. After questioning the vineyard's manager, Millardet discovered that the substance used on the vines was a mixture of copper sulfate, lime and water, which later became known as the "Bordeaux mixture," one of the first widely used fungicides. The Bordeaux mixture was used for nearly sixty years on all kinds of crops, and was a harbinger of other pesticides to come. The first "knapsack spraying machines," essentially portable canisters strapped on the backs of farm workers, were used to disseminate the Bordeaux mixture in French grape vineyards duing 1886 and 1887.

ically, so did the use of pesticides. In fact, since 1945, the yields of most major crops have doubled, whereas the use of insecticides has grown tenfold. In many ways, chemical pesticides were used to compensate for the lack of genetic resistance that might have been bred into crops.

For a period of nearly thirty years, breeding for disease and insect resistance was generally neglected; and in each specialized area of pest concern—whether for insect, fungus, nematode, or mite—a corresponding insecticide, fungicide, nematicide, or miticide was developed and introduced, allowing plant breeders to pursue the genetics of yield and other economically valuable traits.

This is not to suggest that all plant breeding for disease and insect resistance stopped dead in its tracks. Some plant breeders continued to work on these problems, and some new crop varieties carrying such resistance were produced. Yet the emphasis in crop research following World War II was more on chemically assisted agriculture than on developing and using genetically resistant crop lines. Land-grant universities began to receive money from the chemical industry, and research priorities began tilting in that direction. During the 1950s and 1960s, many of the new crop varieties developed were those of super yield rather than super pest-resistance. This was the heyday of the hybrid crop variety, booming agricultural yields, and increasing pesticide use. Some breeders would later call it the era of "spray and pray" plant breeding.

"The development of insect-resistant crops has received less emphasis," said the Congressional Office of Technology Assessment (OTA) in its 1979 report *Pest Management Strategies in Crop Protection*, "partly because of the easy availability of economical and effective insecticides." Plant scientist Mano D. Pathak of the International Rice Research Institute generally echoes that view: "Until the mid-1960s, the study of host-plant [genetic] resistance as a method of insect control received little attention, except in a few cases where other methods of insect control were not practical." Such efforts, he explains, "require close collaboration among entomologists, plant breeders, geneticists, and often biochemists, as well as several years to develop a resistant plant variety." Not surprisingly, this course of action "appeared cumbersome," he says, "when compared with using chemical pesticides to knock down pest populations."

Fungicide sprays and coatings for seed have had a similar though less dramatic impact on breeding for disease-resistant crop varieties.

In 1975, the National Academy of Sciences pointed out "some plant breeders feel that attention to the incorporation of resistance to certain diseases was relaxed when fully effective [fungicide] seed treatments became generally available at reasonable cost."

To this day, new crop varieties are often selected and tested in chemically protected breeding environments. In California, for example, new varieties of strawberries are selected and field tested in soil that has been fumigated with the chemical chlorpicrin—even though, wrote the National Academy of Sciences in 1975, the "selection and testing [of crops] . . . in a protected environment" may mean the creation of a crop variety "that depends upon continued [chemical] protection."

With some crops, there has been greater reliance on pesticides than with others. In the Texas cotton fields, for example, insecticide use following World War II climbed to a 1964 peak of 20 million pounds annually. Such widespread use of insecticides had a direct impact on the kind of cotton varieties Texas cotton breeders turned out during those years.

"Following World War II," says the OTA in its report, "cotton breeders used the 'insecticide umbrella' to develop cotton varieties with superior yield and fiber qualities." Dr. Perry Adkisson, vice president for Agriculture and Renewable Resources at Texas A & M University, recalls, "It was like magic. Farmers planted longer-fruiting cotton and made unheard-of yields under an umbrella of insecticides." But the magic soon ended. "With the development of insecticide resistance in the mid-1960s," notes OTA, "the insecticide umbrella ruptured, and the entire production system began to change."

Cotton insects such as the boll weevil, budworm, and bollworm became resistant to DDT as early as 1962. Soon thereafter, Texas plant breeders gradually began to stress the development of crop lines that were either tolerant or resistant to various diseases and insect pests. However, there have been some reports noting that breeders in Texas have had difficulty in getting new, "low pesticide" varieties released for use.* Today, although earlier-maturing cotton varieties have helped to alleviate some of the insect problem, and

*There is one report that a Texas cotton breeder developed a cotton variety that used 75 percent less water, 80 percent less pesticide, and 50 percent less fertilizer, yet still had trouble getting it released.

disease-resistant varieties have also been developed, Texas cotton farmers still depend on insecticides to control the boll weevil and cotton fleahopper.

In the case of wheat, there are a few examples from the 1920s of successful breeding for insect resistance, such as that for Hessian fly resistance, and some breeding aimed at disease problems such as wheat rusts. After the devastating 1916 wheat-rust epidemic in the Great Plains, for example, a "wheat-rust laboratory" was established by USDA in Minneapolis to begin working on the development of new rust-resistant crop varieties, and no major losses occurred after that date (although rusts did continue to impact the nation's spring wheat crop periodically though 1955). In 1979, however, OTA found no basis for complacency. The "vulnerability of wheat to leaf rust in the Great Plains is high," it said, "and the diversity of resistance to this disease is inadequate. Virulence exists in the leaf rust population of the United States for all useful resistant cultivars. Therefore, a major epidemic could occur any year."

The problem of developing reliable disease and insect resistance in crops lies not so much with the crop itself as with the disease organism or the problem insect. Each new genetic twist the breeder makes in a crop variety invites an adaptation on the part of the disease organism or insect pest. Thus, while a new wheat variety may be resistant to one known strain of Hessian fly, it may not be resistant to all known strains, or to a new, developing strain. Pesticides that killed indiscriminately solved part of this problem, but created others.

As happened in the Texas cotton fields and elsewhere in the early 1960s, disease organisms and insects began adapting to pesticides. At last count, there were at least four hundred species of insects, mites, and ticks worldwide with strains resistant to one or more pesticides. But insect resistance to pesticides wasn't the only problem facing plant breeders and the chemical industry. By the early 1970s, the economic realities of freewheeling pesticide use, with its true energy costs, began to hit home. The Arab oil embargo of 1973–74 revealed the expensive energy underbelly of pesticide production as prices soared. In addition, the 1970–71 Southern Corn Leaf Blight had already touched off a reexamination of conventional breeding programs, with more emphasis being placed on breeding for disease and insect resistance. But shifting back to a genetic emphasis in breeding for disease and insect resistance after nearly thirty years of breeding under the chemical umbrella hasn't been all that easy.

During the 1950s and 1960s, new hybrids and other high-yielding crop varieties were turned out by plant breeders each year. Farmers became accustomed to higher yielding varieties and expected more of the same each year. More importantly, when it came to a choice between a new, high-yielding variety without disease resistance and one that had a little less yield but also carried disease or insect resistance, the farmer would typically opt for the high-yielding variety without resistance.

Farmers quickly learned that in some crops the genetic traits for yield were inversely linked to those of disease or insect resistance*; that with improved disease or insect resistance in some varieties, yield would decline, or "standability" (stalk strength) would be affected, or some other farmer-sought trait would be weakened or eliminated. Learning this, and being pushed from all sides to maximize their crop yield (i.e., by low prices, federal farm programs, or simply to pay bills), farmers have often been wary of varieties with "good pest resistance." This accounts for the fact that even though there are disease- and insect-resistant varieties on the market today—some with yields that do not suffer—farmers are nonetheless reluctant to use them.

Today in the United States, there are approximately 150 crop varieties resistant to one or more kinds of disease, 150 varieties resistant to nematodes, and over 100 varieties resistant to some twenty-five types of insect pests. Impressive as this might sound, however, there are many diseases and insect pests for which no resistant crop varieties exist. In the Midwest Corn Belt, for example, there are at least thirty insect pests and fifty disease pathogens that can attack corn. So far, disease-resistant corn varieties have been

*The National Academy of Sciences has written that the genetic basis for insect resistance in many crops "is complex and multiply determined." The NAS also notes that the genetic basis for insect deterrence "may, in specific cases, be negatively correlated with agronomic traits." A similar relationship is often true for disease resistance and yield. In corn, for example, when yield is increased, certain diseases such as corn-stalk rot do more damage because the vegetative tissues of the corn plant stalk are weakened in the process. In this case, the genes for yield and tissue strength are somehow negatively associated with one another, a problem that is not always overcome in breeding programs.

developed for about twenty-two of the most damaging corn diseases—but the record for developing new insect-resistant corn varieties is generally not as good as that for disease resistance. Corn is the nation's second largest consumer of insecticides, accounting for a hefty 25 percent share. In the Corn Belt, insecticides are the primary means used to control rootworms,* army worms, and cutworms.

Today, pesticides are used extensively on Great Plains wheat to fight off insects such as army worms, cutworms, aphids, and grasshoppers. They are also used to kill greenbugs that attack sorghum. The first greenbug-resistant sorghum variety for Texas was developed only in 1975, even though that pest has damaged Texas sorghum fields since the early 1960s. Insecticide use in Texas to control the greenbug comprised 60 percent of harvested acreage in 1976. Other crops are also still dependent on pesticides for protection.

In 1972, the National Academy of Sciences said of potatoes: "At the present time we rely on chemical protectants more heavily than we should." Yet five years later, the OTA was still reporting that "no potatoes with resistance to insects are available commercially in the Northeast."

In the southeastern United States, chemical pesticides have been the chief means for controlling soybean insects, nematodes, and several plant diseases. As of 1979, for instance, there were no insect-resistant soybean varieties on the market, although a few are now being developed. In states such as Alabama the root-knot nematode and the soybean cyst nematode are particularly vexing problems (recent Alabama Agricultural Experiment Station surveys indicate that 28.6 percent of the soybean fields in Alabama are infested with cyst nematodes, 23.3 percent with root-knot nematodes, and 9.7 percent have mixtures of both), and until recently both have been almost entirely controlled by two chemical nematicides—EDB and DBCP. These pesticides, however, were recently banned by the U.S. Environmental Protection Agency for public health reasons, and that prohibition revealed how dependent soybean growers were on

*In the Corn Belt, the western corn rootworm has become especially problematic, expanding its range in an ever-enlarging circle from a small infestation that began in southern Nebraska in 1960, to more than eighteen states by 1978. Now resistant to the pesticide heptachlor (which became illegal by 1979), the western corn rootworm continues to expand its range by 140 miles a year. As of 1984, no rootworm-resistant corn varieties had been developed.

these pesticides as well as how little breeding had been undertaken to develop nematode-resistant varieties.

"In the past," reported Auburn University researchers D. B. Weaver, R. Rodriguez-Kabana, and D. G. Robertson in 1984, "soybean producers have relied mainly on nematicides to control nematodes. The most effective nematicides were the fumigants DBCP and EDB. These chemicals were easy to apply and provided good nematode control at a moderate cost." Yet in the wake of EPA's action banning these chemicals, they explained, "no economical and effective nematicide" was available, and the only alternatives were using other crops in rotation with soybeans or soybean varieties with genetic resistance. But, they concluded, "There is a critical need for resistant varieties to assure profitable yields in the absence of effective nematicides." (In testing results of ten new soybean varieties yet to be used widely by farmers but purported to have resistance to nematodes, these researchers found three which exhibited good resistance.)

In California too, tomato breeders have, until recently, left the problem of nematodes to the chemical industry. Only when the EPA began to remove certain harmful fumigation chemicals in the 1970s did tomato breeders begin to pay more attention to developing nematode resistance in their new varieties. In fact, says the OTA of the California vegetable industry as a whole: "breeding for resistance to pests and diseases as a primary means of pest control has never received the recognition and funding that it deserves."

PUBLIC DECLINE, PRIVATE NEGLECT

One development complicating work on disease- and pest-resistant crop varieties has been the deemphasis on public plant breeding programs at the nation's agricultural experiment stations and USDA laboratories. In 1972, the USDA terminated its research program on the wheat-stem sawfly, a major insect pest in the Great Plains. "In the absence of this research effort," says the OTA, "the resistant cultivars presently grown are expected to be replaced with susceptible cultivars that have other improved agronomic characteristics" (i.e., yield, harvestability, etc.). Because of this development, says OTA, "infestations of wheat-stem sawfly are expected to increase."

Plant-breeding programs at USDA laboratories and state agricultural experiment stations have generally been responsible for developing wheat varieties resistant to the Hessian fly as well as hybrid corn lines resistant to the corn borer. Yet between 1973 and 1977 there was a 24 percent decline in the acreage planted in Hessian fly-resistant wheat in Kansas and Nebraska. Meanwhile, in parts of the Corn Belt the OTA found an increase in the use of corn hybrids more susceptible to the corn borer with resulting increases in infestations and insecticide use.

In South Dakota in 1978, wheat farmers lost an estimated $25 to $50 million worth of their crop to the Hessian fly when the insect suddenly infested over 1.2 million acres in a part of the state where it did not normally occur. The OTA suspects at least part of the problem lies in the release by private seed companies of new high-yielding wheat varieties (which are replacing those developed at public universities) that do not carry resistance to the Hessian fly. "Insect resistance has not been a significant component of commercial breeding programs," observes the OTA, "and none of the new commercial wheats have resistance to Hessian fly."

In the face of government deemphasis of important plant breeding programs such as those for the Hessian fly, wheat-stem sawfly, and European corn borer, it was assumed that commercial seed companies would do the work needed to produce new disease- and pest-resistant crop varieties. Yet, says the OTA, "Experience indicates that this was not a correct assumption."*

Clearly, any trend that deemphasizes the use of genetically based pest resistance in crops—whether the result of a decline in public support or commercial neglect—will cause an unnecessary use of pesticides, and perhaps prolong the era of pesticide dependence. Although more attention has now turned to breeding and genetically engineering crops for disease and insect resistance, there are still questions about the adequacy of current breeding programs, the kinds of genetic resistance being incorporated into crops, and the ownership of seed companies by chemical and pharmaceutical interests.

*Interestingly, in 1982–83 field tests of ten new soybean varieties for resistance to nematodes at the Alabama Agricultural Experiment Station at Auburn University it was reported: "The public varieties Foster, Kirby, and Braxton had the best performance. . . . Proprietary varieties RA 701 [Ring Around Seed Company], A7372 [Asgrow Seed Company], and Agratech 67 [Gold Kist] were generally inferior to the public varieties."

Moreover, there can be no doubt that chemical pesticides have played a major role in how plant breeding programs—both public and commercial—have been conducted in the recent past. It just might be that the nation's nearly forty-year immersion in synthetic pesticides has cost American agriculture and the American public decades in real plant-breeding progress, millions of dollars in distorted research priorities, and still unknown costs to public health and safety.

ELITE GENES & THE GENETIC SHUFFLE

The history of crop agriculture in America is analogous to a pyramid whose present-day base of numerous crop varieties points backward to a few points of genetic origin; to a handful of varieties that spawned countless offspring. Popular crop varieties—often carried to this nation by immigrants or developed by farmers and early plant breeders—were widely planted, and some were used as parent lines to make new varieties. One wheat variety, Marquis, was introduced into the United States in 1912, and subsequently became the parent of a long line of other wheat varieties, including Ceres, Hope, Reward, Marquillo, Reliance, Thatcher, Sturgeon, Comet, and Tenmarq. Some of Marquis' genetic traits, then, were passed on to each of these other varieties and to their offspring whenever they were used in breeding programs.

When crop varieties like Marquis and their offspring are used repeatedly in breeding programs, they become known to scientists as "elite varieties" or "elite germ-plasm." Plant geneticist Jack Harlan explains that Marquis and some of the other early wheats were "worked . . . over and over by wheat breeders. Pure lines were selected, crossed with other lines, more lines selected out, and the process repeated." This pattern of using a few popular crop lines in cross-breding programs, says Harlan, "is a model repeated many times over in American agriculture." This means that a limited pool of "elite varieties" has been used to fashion most of the crop base we have today—a pool that still remains a primary source for new crop varieties.

Modern plant breeders persist in the use of such elite breeding materials, even for traits such as disease and insect resistance. In a 1981 survey of some 100 American plant-breeding programs, 95 percent of wheat breeders, 83 percent of corn breeders, and 79 percent of soybean breeders said they turned to their "elite, adapted" breeding pools to find pest resistance for new crop varieties. Most plant breeders do not like to "outcross" to "unadapted" varieties because this often reduces the yield of the variety they are trying to improve.

For many breeders, the continued reliance on elite breeding pools is not a matter of great concern; they point to the "thousands of varieties in reserve" that they can use in numerous combinations and crosses to fashion new and resilient crop varieties. Or they note that "variety lifetimes" are shorter today, with old crop lines being replaced by new ones before any disease or pest can adapt to them. Yet some configurations of genes, common to many crop varieties, are exactly the same and may determine susceptibility or resistance to a given disease or pest. The crucial difference, therefore, is not in the number of available varieties or the length of their lifetimes, but in the genes found in those varieties.

SINGLE GENES & THIN LINES OF DEFENSE

Plant breeders generally recognize two kinds of disease and pest resistance in plants: vertical and horizontal. In somewhat simplified terms, vertical resistance is a one-gene/one-pathogen type of resistance, while horizontal resistance is a multiple-gene/multiple-pest type of resistance. A vertically resistant crop variety would be almost completely resistant to the predominant strain of a particular pest for a given period, while a horizontally resistant crop variety would, for the most part, resist all strains of a pest species at some level for as long as the resistance genes remained in that particular variety.

Vertical resistance more or less beckons the adaptation of the pest and its mutation into new strains, encouraging virulent adaptations and epidemics. Horizontal resistance tolerates the pest and usually stands up to its different adaptations, encouraging stability and tolerable pest populations (although sometimes at economically unacceptable levels). By way of analogy, vertical resistance is somewhat like a chemical pesticide manufactured to kill one specific type of pest. Just as the pest can become immune to the pesticide, so can it genetically overcome the resistance produced by a single plant gene. Vertical resistance breeding then—like pesticides—practically guarantees the development of "superbugs." And just as pesticide-resistant bugs send the pesticide maker back to his laboratory again and again to reformulate his compounds, vertical resistance breeding will inevitably consign the plant breeder to an endless process of "treadmill plant breeding." Not so with horizontal resistance.

However, from the perspective of the plant breeder or genetic engineer, vertical resistance—since it is typically a one-gene phenomenon—is much easier to work with, identify, and incorporate into new crop varieties than is the multiple-factored horizontal resistance. Thus, in modern plant breeding, in order to combat specific diseases or pests as they develop, horizontal resistance is gradually bred out of plants (however unintentionally), while vertical resistance is bred into them. Horizontal resistance, then, is bred out of crop varieties *because* vertical resistance is bred into them. Today, years of vertical resistance or other single-trait breeding have weakened or pushed horizontal resistance out of many modern crop varieties.*

"Vertical resistance shines brightest," says plant scientist J. E. Van der Plank, "in the breeders' nurseries, on experimental stations, and in the trial plots laid down to test new varieties in comparison with old. It is here that a new variety is chosen." On the other hand, he explains, horizontal resistance "is at its dimmest in breeders' nurseries, on experiment stations, and in trial plots. It shines brightest where it should, in the broad acres of successful varieties . . . [but] that is not a quality for which a new variety is at present chosen."

Thus, vertical resistance is found throughout most modern crop varieties, including, for example, Hessian fly-resistant wheat, phylloxora-resistant grapevines, and aphid-resistant corn, even though it is guaranteed to fail sooner or later.

In fact, according to OTA's 1981 report, *Impacts of Applied Genetics,* there is some evidence which suggests that the use of

*For example, plant scientist J. E. Van der Plank writes: "If new varieties are selected while they are being protected against disease by vertical resistance, it is ordinarily not possible to ensure that enough horizontal resistance is present as well." And also, "if horizontal resistance was previously selected for, it is likely to be dissipated progressively in new varieties bred under protection of vertical resistance."

single-gene or vertical resistance breeding in some crops has resulted in a corresponding increase in the frequency of virulence genes in pathogens which attack crops. This has been reported in Australia, for example, in breeding for stem rust of wheat, powdery mildew of barley, leaf mold of tomato, and downy mildew of lettuce. Many pathogenic organisms have the ability to transfer genes among themselves asexually, increasing the rate at which they can develop virulence genes in their populations, and hence, adapt to single disease-resistance genes in the new crop varieties which they attack. And this is one reason why some new crop varieties last as few as five years before they are overtaken by a new strain of pathogen, leading to the charge among some critics, that plant breeders employing this technique are, in effect, breeding new populations of pests as well.*

If used widely, single-gene breeding for other popular traits in crop varieties—whether yield or some other characteristic—can also set the stage for disease outbreaks. In fact, it may take only a single susceptible trait found throughout a crop to invite catastrophe. And, as the National Academy of Sciences warned in 1972, the penchant for incorporating single-gene traits in our most popular crop varieties may be setting the stage for serious disease epidemics: "A given technological advance in crop production," wrote the Academy, "often rests on small numbers of genes. If one of these genes is incorporated into many varieties, the crop becomes correspondingly uniform for that gene. If a parasite with a preference for the characters controlled by that gene were to come along, the stage would be set for an epidemic."

The 1970 Southern Corn Leaf Blight was made possible by a single genetic trait (in this case, in the cell's cytoplasm) that made practically every hybrid corn variety then in use susceptible to a

*In its 1981 report, *Impacts of Applied Genetics,* the U.S. Congressional Office of Technology Assessment explains, for example, "There is some evidence that pathogens are becoming more virulent and aggressive—which could increase the rate of infection, enhancing the potential for an epidemic," citing the findings of R. C. Shattock, B. D. Janssen, R. Whitebread, and D. B. Shaw in "An Interpretation of the Frequencies of Host-Specific Phenotypes of *Phytophtora infestans* in North Wales," *Annals of Applied Biology* 86:249, 1977. "Furthermore," says OTA, "pressures brought about in the evolutionary process have developed such a high degree of complexity in both resistance and virulence mechanisms, that breeding approaches, especially those only using single-gene resistance, can be easily overcome."

new strain of fungus. Despite the effects of that blight, similar type traits for male-sterile cytoplasm—which is used in the making of hybrid seed—are now found in sorghum, millet, sugarbeets, onions, cotton, and cantaloupe. Perhaps more surprising is that new hybrid corn varieties with male-sterile cytoplasm, the same trait that led to the 1970 Corn Blight, are again being sold in the United States.

BIOTECHNOLOGY TO THE RESCUE—MAYBE

All the shortcomings of past and present plant breeding—the worries about disease and insect vulnerability in crops, the problem of incorporating more genetic diversity into the nation's crop pool, the leveling off of yield due to shallow breeding pools, and other concerns—now can be overcome through biotechnology and gene splicing, say today's genetic engineers. Every biotechnology company and corporation now vying for a share of the agricultural genetics market, predicts that new and better ways of ensuring disease and insect resistance in crops, as well as higher yields, are just over the horizon.

Typical of some of the promises being made are those of Calgene, a California biotechnology company. In one of its recent brochures Calgene states that "plant diseases and other pests destroy almost a third of the world's crop production" and that "intensive modern agricultural methods accelerate the evolution of new pest types, requiring plant geneticists to more rapidly breed genes into commercial cultivars from resistant relatives." Calgene claims that "bioengineering will speed this transfer process to keep pace with the pathogens, expand the gene pool to unrelated organisms, and even create synthetic resistance genes in the test tube."

Biotechnology techniques such as tissue culture screening, protoplast fusion, and gene splicing offer vast new possibilities for crop alteration. Plant breeders working with traditional cross-breeding methods have been bound by the natural laws of reproduction, forced to work within one particular plant species or another. Now, with gene splicing, the entire plant kingdom becomes one giant, open-ended gene pool in which genes can be moved about freely from one species or genus to any other. This means, for example, that the genetic traits of the oak tree—a species that is not bothered by the rusts of wheat—could conceivably be moved into commercial wheat varieties to make them permanently resistant to rust.

David Padwa, chairman of the Agrigenetics Corporation, talks of making crops "immune" to disease and insects. Plants might also be engineered to emit natural chemicals that repel insects and ward off disease. Yet, as in so many other areas, the promise and reach of biotechnology in the pest-resistance area is not without its risks.

It is possible, for example, that with high-speed and precise genetic technologies at their disposal—capable of moving "commercially desirable" genes into large numbers of widely used crops and/or livestock—science and industry could make mistakes of "genetic uniformity" more quickly than they have in the past. And again, the 1970 Corn Leaf Blight comes immediately to mind.

With classical plant-breeding techniques, a breeder has to go through a painstakingly slow process of cross breeding, selection, further breeding, and final testing before he can successfully move the genetic characteristics he wants into a new crop variety. The entire process, from start to finish, might take as long as ten years. Now, with the presumed ability to leap species barriers, to screen millions of plant cells at a laboratory workbench in a few hours, and to create new varieties and breeds with techniques such as protoplast fusion, microinjection, and gene splicing, beneficial and not-so-beneficial changes can be incorporated into crops and livestock in a matter of days or weeks. And these changes can then be spread throughout the world in a matter of weeks through normal trading and shipping. *The Global 2000 Report* noted, for example, that the culturing of plant cells could lead to the mass cloning of crops. "Should this happen," warned the report, "the world may face even greater genetic uniformity in crops."*

Yet the forces driving genetic engineers, like the plant breeders before them, will be economic ones: the forces of yield, uniformity,

*When Norman Borlaug developed his Green Revolution wheats in the 1960s, the genetic characteristic that made them successful, giving them their semi-dwarf stature, was the dwarf gene obtained in crosses made with the Norin 10/Beavor wheat type. By the early 1970s, Borlaug's new wheat varieties (which did have resistance to several types of rust disease) were growing on 26 million acres in Asia, Europe, Africa, North America, and South America. In 1972, the NAS warned of the "potential danger in having a single semi-dwarf genetic type gain such widespread popularity." By then, there were already signs that the wheats with the dwarf gene were becoming susceptible to the fungus *Altervaria triticiva* in India's Punjab, and in Mexico were falling prey to a leaf blotch caused by a species of *Rhyncasporium*.

and marketability. The traits that plant breeders sought somewhat blindly in the past are likely to be the same ones that genetic engineers will seek with their eyes wide open in the future. And at least initially, it will be the single-gene traits that will come most easily. As for improving disease and insect resistance in crops, might not genetic engineering therefore simply extend the practice of vertical-resistance plant breeding, thereby hastening the adaptation of new strains of disease and insects, and in that process, put us on a faster treadmill than the one we are already on?

In addition, one of the biggest questions about the promises of biotechnology for fending off disease and insect problems in agriculture is the role that the chemical and pharmaceutical industries will play in bringing such advances to market; the very same industries whose pesticide products once made it so easy for plant breeders and entomologists to avoid the more difficult but smarter biological and genetic options.

MORE DIVERSITY THAN IN NATURE?

Genetic engineers say that with their new techniques, the agricultural gene pool will be broadened. Winston Brill, director of research at the Cetus Madison Corporation in Wisconsin, paints a very promising future. "A whole range of very specific plant-genetic modifications can now be considered," he says, "with the use of methods that may someday generate a genetic diversity not naturally present in cultivated plants." And David Padwa echoes Brill on this point, saying that "these new technologies will do unprecedented things. You will be able to find more variability than you can in nature."

Nonetheless, the stark reality confronting genetic engineers and all plant breeders, whether they talk about increasing genetic diversity, making crops immune to disease, or improving yield, is that the elite gene pools now in their laboratories or breeding nurseries are inadequate for making the spectacular improvements they promise. Biotechnologists and breeders alike are now discovering that they need new sources of genetic diversity in their current breeding pools. Indeed, it will be such diversity that will be the key to commercial success.

Because of this, the entire agricultural genetics industry is today engaged in a journey into the past to recover those genes that were left behind in mankind's headlong rush for yield, uniformity, and modernity. Biotechnology companies, major corporations, and national governments are all trying to collect, save, and in some cases, own, the genes of the Old World—the genes that may insure their economic success or the security of their political futures. But most of this valuable Old World genetic diversity isn't found in the United States or North America, or, for that matter, in most other parts of the developed world.

THE GENES OF THE PAST

In the distribution of plant and animal species throughout the world, biological evolution has been more generous to some regions than to others. Owing in part to natural developments such as the Ice Age, much of Europe, the Soviet Union, the United States, and Canada are relatively deprived botanically when compared to those areas of the world located in the tropics, subtropics and regions such as the Mediterranean basin. There are, for example, about eight times as many living species in the Amazon River system of South America, than there are in the Mississippi River system of the United States. And on the volcanic slopes of one Philippine mountain, there are more woody plant species than in all of Canada.

The treasure house of botanic variety found in regions such as the Philippines came about because it is one of the world's great crucibles of biological life; a birthplace to many species of plants, animals, and microbes. In all, there are twelve such regions recognized for their biological origins and diversity—nine major ones and three minor ones—called the "Vavilov centers" after the famous Russian agronomist and geneticist N. I. Vavilov. After years of exploration in the 1920s, Vavilov concluded that certain specific regions of the earth's surface, due to combinations of climate and topography, were the centers of origin for virtually all of man's important crops.* Corn, for example, originated in the Jaliscan Plateau region of central Mexico, the soybean in China, and the

*The twelve Vavilov centers are: Ethiopia, the Mediterranean, Asia Minor, Central Asia, Indo-Burma, Siam/Malaya/Java, China, Mexico/Guatemala, Peru/Ecuador/Bolivia, Southern Chile, Brazil/Paraguay, and the United States. This is not to suggest that these twelve centers comprise the only regions of important biological diversity; there are also other areas with species of plants that could prove to be important sources of food, medicine, and industrial products.

tomato in South America. In the United States, only the sunflower, blueberry, cranberry, and Jerusalem artichoke are considered native.

Today, the twelve Vavilov centers are among the most important botanic gene pools for world food security. The rich biological diversity found in these centers can serve as an important genetic bulwark against the constant threat of disease and the unforseen mutations of pests. The Vavilov centers are also among the most important sources of potential food crops, medicines, and beneficial microbes. Yet these centers of priceless diversity are being threatened by population growth, development, and modern agriculture. Says author Norman Meyers in *The Sinking Ark:* "It is not unrealistic to suppose that, right now, at least one species is disappearing each day," and that "by the late 1980s we could be facing a situation where one species becomes extinct each hour." Meyers believes that by the year 2000, one million species could be lost to all forms of habitat encroachment and development.

Besides the Vavilov centers themselves, and often in the same general regions, are the areas of Old World agriculture where many plants were first domesticated thousands of years ago. Crop varieties that have evolved under primitive and traditional systems of agriculture—many of which are still in cultivation in the Third World—are called "land races." These land races are not limited to the Vavilov centers or the Third World, but many of those that remain are found in these regions; the United States, for example, once had thousands of land races used by American Indians. Yet like the Vavilov centers, the remaining land races and areas of Old World agriculture are under pressure from population growth and modern agricultural methods.*

The loss of genetic diversity throughout the old regions of agri-

*Even in the United States, valuable genetic diversity, often within fifty miles of American agricultural research facilities, is being lost through sheer neglect and failure to collect and preserve it. Native American agriculture, for example, was, and in some regions still is, a rich source of botanic diversity. Yet, since the time of Columbus, according to ethnobiologist Gary Nabhan of Tucson, Arizona "at least six plant species, and hundreds of land races in eighteen other species, have been lost from Indian communities." Nabhan believes that the United States "is not working hard enough on the remaining native land races that we now have." He says that "rather than greedily lurching over others in the Third World, I think we should clean up—diversify—our own house first." (For Nabhan's recent study of Papago Indian agriculture in Arizona and Mexico, see *The Desert Smells Like Rain,* North Point Press, 1982.)

cultural domestication is being accelerated by the introduction of high-yielding crop varieties that displace the traditional land races. Sorghum land races widely used in South America have been displaced by Texas hybrids. Almost the entire population of wheat varieties native to Greece has become extinct, some displaced by imported varieties. Ancestral strains of Turkish wheat, the genetic forebears of North American red wheat, have all but disappeared from that country. And in India, Mexico, and the Philippines, Green Revolution wheat and rice varieties have displaced local varieties.

Despite this, it has in recent years become increasingly apparent that the old land races of crops and the genetic resources found in many parts of the Third World are highly valuable to the modern agricultural systems of the industrialized world. For example, the genetic material found in one Ethiopian barley variety resistant to yellow barley mosaic now saves American farmers an estimated $150 million a year, and some Turkish wheat genes resistant to stripe rust have saved American wheat farmers $50 million annually since the 1960s. Ironically, some modern crop varieties strengthened by native strains found in Third World countries have wound up being sold to farmers in those countries by American and European seed companies.

In some cases, this circularity of genetic material is an inadvertent and unintentional occurrence, since twenty years or more may pass between the time that a native strain is collected from some country and the time that some of its genes wind up in an improved commercial variety sold in that country. Yet in other cases the practice is quite controlled, and is generally done more by design than by accident. A hybrid corn variety suited to Iowa, for example, can be adapted with some native genes from country "x" or country "y"—where the photo period, soil conditions, and local climate are similar to those in Iowa—and can then be made available for sale in that country. Similarly, a high-yielding hybrid bean variety from Michigan can be adapted for sale in Guatemala by crossing it with some native Guatemalan lines so that it looks like a local variety but sells for a higher price. In these instances, the changes are analogous to a high-performance General Motors automobile engine (high-yield genes) being fitted into a Guatemalan or Turkish chassis (crop size or local disease-resistance genes). And in many cases, it is the native genes that make the "model change" and the sale possible.

However, the free flow of priceless genetic materials from the "gene-rich" Third World to the "gene-poor" industrialized world may soon end. Some Third World governments are now beginning to impose restrictions on the collection of their genetic resources. The Mexican government has restricted the collection of some wild plant species in that country. Ethiopia has adopted legislation that restricts the export of native plant materials; and Indonesia and Bangladesh prohibit the export of tea plants, partly to protect their position in the world tea market. And there is also a growing movement in some "gene-rich" Third World nations to devise financial compensation and/or royalty plans for the use of their resources.

As Third World governments and the United Nations grapple with ways to accommodate the genetic-resource needs of the industrialized world while protecting the interests of smaller and less advanced nations, the emergence of a multi-billion dollar biotechnology industry has now intensified the search for new sources of genetic material—varying from exotic plants to rare species of fungi. Scientific teams from universities, seed companies, and chemical corporations are now scouring the most isolated Andean valleys and remote tropical jungles for the "genes of economic importance"—a development that may have major impact on how genetic resources are collected, preserved, and utilized.

According to James Murray, a biotechnology consultant with the Policy Research Corporation in Chicago, it is "unrealistic" to attempt to preserve all existing biological diversity. Murray believes that current priorities and methods for germ-plasm conservation are unrealistic because they do not reflect the new possibilities of genetic engineering. A better approach, says Murray, might be to determine for each group of organisms, such as microbes or plants, what traits, rather than which species, will be economically important, and then seek to preserve the species that have those traits.

But a genetic collection and preservation strategy based on one group's idea of economically valuable traits—let alone species—will inevitably conflict with those of other interests. Moreover, given the current economic frenzy and heated international competition among corporations and national governments in the area of biotechnology, the political and economic pressure on Third World governments to accommodate some kind of genetic-resource collection will be considerable. This issue, not unlike the "common heritage of mankind" aspects of the fledgling Law of the Sea Treaty—where nations attempted to devise a system for the equitable access and allocation of ocean-based resources believed common to all mankind—promises to be one of the most difficult and controversial facing the world community during the next decade.*

DON'T WORRY, IT'S ALL IN THE BANK

Gene banks for storing seed and other plant genetic materials now exist in some sixty nations. Yet despite the implied assurance that this "banking" of genetic diversity will provide the genetic wherewithal to improve crop productivity and fend off disease-borne catastrophes, there are nagging questions about the adequacy of such collections, who will have access to them in times of crisis, and concerns about their long-term security. Take, for example, the National Seed Storage Laboratory (NSSL) in the United States, which is part of the National Plant Germplasm System.†

Located on an earthquake fault in Fort Collins, Colorado, roughly equidistant between a nuclear power plant and the Rocky Flats plutonium facility, the NSSL is not exactly the Rock of Gibraltar. Opened by the federal government in 1958 with an initial budget of $450,000, the NSSL today has some 240,000 "seed accessions"

*For some of the recent political infighting among nations on this question in the United Nations and elsewhere, see, for example, Cary Fowler's "Report on the 22nd Session of the Conference of the Food and Agriculture Organization of the United Nations, November 5–23, 1983, Rome, Italy," prepared on December 13, 1983, available from the National Sharecroppers Fund, Pittsboro, North Carolina. See also Pat Roy Mooney's "The Law of the Seed: Another Development and Plant Genetic Resources," *Development Dialogue*, The Dag Hammarskjold Foundation, Uppsala, Sweden, 1983, pp. 1–2.

†The key parts of the NPGS are (1) The Germplasm Resources Laboratory at Beltsville, Maryland; (2) three Plant Introduction Stations at Glenn Dale, Maryland; Savannah, Georgia; and Miami, Florida; (3) four state-federal regional plant introduction stations located at Pullman, Washington; Ames, Iowa; Geneva, New York; and Experiment, Georgia; (4) the state-federal Potato Introduction Station at Sturgeon Bay, Wisconsin; (5) the National Seed Storage Laboratory at Fort Collins, Colorado; and (6) a large group of federal, state, and private plant germplasm curators located throughout the United States. In addition to these collections, there are also twelve clonal repositories planned for vegetatively propagated fruits, nuts, and other crops, five of which are now in operation at Corvallis, Oregon; Davis, California; Miami, Florida; India; California; and Mayaguez, Puerto Rico.

or collections in its confines. One *New York Times* reporter who toured the NSSL in 1981 found it so crowded that seeds were "piled on the floors in brown cardboard cartons and sacks."

More than 20 percent (and some charge as much as half) of the NSSL's collection has not been cataloged, and some seed stored in the facility has lost its ability to germinate. Other crop varieties stored at the NSSL have been "grown out" in the wrong environments, leading to the possibility, for example, that seed collected for dry-land cultivation may end up evolving—through successive growouts under irrigated conditions—toward a use for which it was not originally saved. Moreover, some seeds mutate in storage, and evolve "storage traits" rather than agricultural traits.

At the NSSL, there have been occasions when refrigeration equipment needed to preserve seed viability has broken down, and sloppy maintenance and inadequate storage procedures have also plagued some of the government's regional seed-banking facilities. Seed-storage officials charge that they have been victimized by a lack of federal funds, supported at the $10–$12-million level, which has gone unchanged for nearly fifteen years.

In April 1981, the General Accounting Office of the U.S. Congress issued a report criticizing the USDA's handling of the nation's germ-plasm system. Although the GAO noted that genetic vulnerability and variability are critical issues that need national attention, it found the nation's germ-plasm system "inadequate" to manage United States resources or to meet current research needs.

Another potentially serious problem that could be stalking the nation's supposedly "safe" stores of plant-genetic material are seed-borne viruses. Plant pathologist Richard O. Hampton of the USDA says that after seven years of research he has found evidence of "known or probable" seed-borne viruses in the government's seed collections for seventeen major crops, including corn, beans, soybean, and wheat.

"When undetected and unrecognized," says Hampton, "seed-borne viruses in the germ-plasm of seed banks can accompany crop genes into breeding programs." When this occurs, such viruses constitute "biological time bombs," he says, threatening new crop development, adjacent breeding materials, and surrounding crops. Hampton also points out that when such infected material is used in breeding programs, the viruses can be passed undetected to the offspring, eventually creating a problem that may require costly eradication measures.

A second problem, explains Hampton, is that virus strains not yet common to American agriculture "can be introduced and stored in germ-plasm collections, ready to spread and become established as soon as the germ-plasm is distributed for use" in new breeding programs. There are a few methods that can now be used to detect seed-borne viruses in stored seed, but given the NSSL's poor funding record and other more pressing maintenance duties, the extent to which they will be used remains to be seen.

At the international level, the seed banking situation is often worse than it is in the United States. J. T. Williams, an FAO official who is also executive secretary of the International Board for Plant Genetic Resources in Rome, explains that most crops of economic importance are "barely represented" in many nations' germ-plasm collections. Williams notes, for example, that although there is a "very large collection" of rice seed held by the International Rice Research Institute (IRRI) in the Philippines, it is "relatively deficient in wild material." And after a survey of world wheat seed collections (not all countries were willing to release their information), Williams found the taxonomic range of the wheats in storage to be "completely inadequate." The wheat collections Williams surveyed were so bad, in fact, that he asked whether the nations involved should "start again and go about it in a much more scientific manner."

Similarly, Dr. Te-Tzu Chang, head of IRRI's seed collection program, also finds little to be complacent about when it comes to seed banking facilities around the world. "Few of them have comprehensive collections of their indigenous germ-plasm," he notes. "Most have many leading commercial varieties, but are deficient in less important, traditional, and primitive varieties. Foreign introductions are duplicated in many collections. The lack of refrigerated facilities for seed storage in most tropical countries, the accompanying need to renew seeds every year, and the lack of trained personnel have placed such a heavy burden on most national centers that valuable accessions continue to disappear, or become badly mixed."

In addition to the poor record of seed banking both in the United

States and internationally, there have also been some reports questioning long-term seed storage strategies and some scientists who feel there may be better ways to go about preserving biological diversity. "The maintenance of a species apart from its natural environment for significant periods of time is an uncertain means of preserving germ-plasm," wrote the National Academy of Sciences in a 1978 report. Because of the expenses involved and the fear that "within a few generations the gene pool may become modified as a result of artificial and systematic selection pressures," the Academy found such long-term storage less than ideal. And others agree.

"Seed banks may help plant breeders today," says University of Wisconsin botanist Dr. Hugh Iltis, "but preserve varieties over a thousand years? Forget it. The only thing we should do now is work like mad to help the countries with the richest flora to set aside natural parks that cannot be touched." Yet for Third World governments faced with burgeoning populations and a need to expand their food production, that option may not be realistic, especially if such preserves are viewed as servicing the needs of the industrialized nations.

OF UNIFORMITY & THE McDONALD'S POTATO

Despite the known dangers of genetic uniformity in world agriculture and the strategies aimed at preserving and using genetic diversity, the economic forces of yield and uniformity still drive the market. In the years ahead, these forces, embodied in the actions of corporations and national governments, will likely determine the genetic condition of world agriculture. Take, for example, the McDonald's corporation and the Russet Burbank potato.

"People think all potatoes are alike," says McDonald's spokesman Bill Atchley, "but they aren't. A Russet Burbank potato has a distinctive taste and higher ratio of solids to water, which makes for crispier fries." McDonald's, in fact, has its own set of french fry specifications built around the Russet Burbank potato, and is not in a hurry to change them. For example, 40 percent of all McDonald fries must be from two to three inches long; another 40 percent must be over three inches; and the remaining 20 percent can be under two inches. For this order, the long slender Russet

Burbank fills the bill, and most of McDonald's Russets, about 80 percent in fact, come from none other than J. R. Simplot.*

Over the years, McDonald's business has been growing dramatically, and the company is now adding franchises all over the world, taking the Russet Burbank wherever it goes. McDonald's has already been successful in moving the Russet Burbank into such foreign countries as Tasmania, from where the company supplies its Australian outlets. And it is now moving the Russet Burbank into the Philippines with an eye toward using it in other tropical countries. "If we can grow these potatoes in the Philippines, we'll learn a lot about how to do it in other tropical countries," says McDonald's Bill Atchley. Europe, however, is another story.

Russet Burbanks do not grow well in Europe, and McDonald's is not even sure that they will. Nevertheless, for ten years, the company has tried to move the Russet into Europe, despite the availability of hundreds of European potato varieties. However, Europe has a long memory when it comes to potatoes and potato diseases, and it is also very protective of its own farmers. The Common Market prohibits potato imports.

When McDonald's attempted to have the Russet Burbank introduced into Holland in 1981, the potatoes had to sit in quarantine for eight months before they could be given trial plantings. But the potatoes then proved vulnerable to a European potato virus, and were therefore unacceptable for European agriculture. So now McDonald's is trying a backdoor approach, attempting to have the Russet Burbank adopted in Spain, which is in the process of joining the Common Market, and—McDonald's hopes—with Russet Burbanks growing in its soils.

McDonald's isn't the only major corporation that has developed a vested interest in one particular crop variety, but its worldwide enthusiasm for the Russet Burbank does illustrate the lengths to which some interests will go to perpetuate the use of one favored variety. Such one-variety favoritism—sometimes elevated to the force

*In the early 1960s, J. R. Simplot patented the process of making frozen french fries. A few years later he hit paydirt when he signed a big supply contract with McDonald's creator Ray Kroc, who was then just beginning his hamburger chain. The McDonald's/Simplot contract—based on Simplot's supply of french fries made from Russet Burbank potatoes—helped make Simplot the "potato king" of the United States.

of law as with the required use of Acala cotton in California—is being repeated for other crops and livestock all over the world. Powerful economic interests of the stature of McDonald's, J. R. Simplot, and greater, are building their futures on specific crop varieties and livestock breeds too. Biotechnology may only make such genetic favoritism even more possible in the future.

Yet, in the McDonald's case, what appears a genetic godsend and economic bonanza for that company today, could well become an economic nightmare for them tomorrow—not to mention farmers and other involved support businesses—should one tiny organism find a genetic window of virulence in the Russet Burbank potato. Should that happen, McDonald's—and other commercial interests like it who are building entire economic systems on the back of one or a few packages of genes—will have contributed mightily to the spread of a genetic epidemic. And for that, they would be culpable, and perhaps should be liable.

Today, biotechnology companies, chemical corporations, and major agribusiness operations are all seeking the genetic resources that will be used in tomorrow's food and agricultural systems, whether they be the genes used to produce new varieties of high-yielding wheat or those that make a particular strain of microbe scavange more effectively for nutrients in the soil. These agricultural products of tomorrow, and their genes, will be mass produced and distributed worldwide over the broadest possible market for the longest possible time. Under such circumstances, genetic uniformity may be spread far more widely, and become much more intractable than it has ever been in the past.

CHAPTER 11

MAGIC MOLECULES, CLEVER CHEMISTRY

The marriage of genetics and chemistry makes exciting things possible in agriculture.
—*Ralph Hardy, E. I. Du Pont de Nemours & Company*

The main research complex of E. I. Du Pont de Nemours & Company, the Du Pont Experiment Station, is located north of Wilmington, Delaware in a pleasant country setting just off State Highway 141. Mature dogwood trees in May flower line the left-hand side of the driveway leading to the older portion of the complex. The entrance here is controlled by a guardhouse and a small office building, and the entire 150-acre research station is surrounded by a cyclone fence topped with barbed wire. While Du Pont police guards are pleasant and cheerful, a white sign with black lettering mounted on a traffic light behind a small guardhouse in the driveway warns: "Vehicles and/or occupants entering and leaving are subject to search and/or inspection." Visitors entering on foot must pass through the office to the right of the guardhouse, have the proper clearance, and wear a Du Pont lapel tag. Cameras are not permitted on the premises unless prearranged with the company.

From a modest office in one of the red brick buildings at this complex, Dr. Ralph Hardy directs a thousand or so Du Pont scientists in the "life sciences"—which broadly include agrichemicals, pharmaceuticals, and diagnostics. Hardy is a likeable, slightly rumpled fifty-year-old biochemist who has worked with Du Pont for more than twenty years. He comes across as part scientist, part businessman, and part professor. Hardy holds a Ph.D. from the University of Wisconsin and has extensive research experience. Before he rose to the lofty position of life sciences research director, Hardy spent a lot of time as a bench scientist and research supervisor, studying the nitrogen and carbon cycles in plants like the soybean. He has also studied photosynthesis, chemical uptake, and chemical partitioning in plants.

In the late 1970s, Hardy was the man Du Pont turned to when it was decided that the life sciences and biotechnology were the way to the future. It was then that he became research director of Du Pont's formidable life sciences division, and almost immediately began recruiting specialists in DNA research for a biotechnology program, the first such program estalished by a major American chemical company. By 1980, Du Pont had the largest in-house genetic engineering team of any American chemical maker.

Up the hill from Ralph Hardy's present office is a brand-new $85 million life-sciences research center, which will eventually house an additional seven hundred Du Pont scientists. Another building not far away has been upgraded with new offices, modern laboratories and plant growth rooms. These facilities represent part of the Du Pont commitment to long-term research in the biosciences, an area in which the company's budget increased 17 percent in 1980, 20 percent in 1981, and 25 percent in 1982, and now amounts to more than $200 million a year.

When the conversation in Ralph Hardy's office turns to questions about Du Pont, biotechnology, and agriculture, he becomes somewhat professorial, generously tutoring the layman about the new science in which his company is engaged. Du Pont sees two possibilities for biotechnology in the agricultural future, says Hardy: the first involves chemical applications and the second genetic applications. Correspondingly, from the vantage point of the company's research program, there are two "product endpoints," as Hardy calls them: chemical products such as pesticides, and more purely genetic products such as seed. Du Pont prefers the chemical path for two reasons: the company has a strong research capability in the chemical area, and it regards the patent or proprietary position of the seed industry as weak. And besides, Hardy says, "Du Pont is not well positioned in [plant] genetics," meaning the company lacks a commercial outlet. While other major chemical corporations have acquired seed companies, Du Pont has yet to show an interest in doing so. More probable for Du Pont, as Hardy explains, is "matching" chemicals to seeds. "That's no secret," he says, "that's been talked about." And with genetic engineering, a whole range of other chemical possibilities opens up for companies like Du Pont. "I have a genetics operation in my shop," says Hardy, "to use as a tool to give me chemical ideas."

Ralph Hardy and Du Pont see a future in agricultural biotechnology because they see world population growing, world food demand spiraling, and the markets for new kinds of agricultural products booming. Hardy foresees world population, growing at two percent a year, climbing to six or seven billion by the year 2000. World affluence is rising, too, according to Hardy, and that means a demand for more meat, which in turn translates into a demand for more crops to feed animals. Yet, because the process of converting plants to animals is inefficient, crop production must increase at a greater rate than world population. If the world is to keep eating meat, Hardy explains, the production of crops like soybeans and other feed grains will have to increase from 150 million metric tons today to about 500 million metric tons by the year 2000. And for companies like Du Pont which provide the ingredients that make crops grow, or the genetic wherewithal to make them "more efficient," that means big business.

In 1928, Du Pont acquired the basis of its agricultural business when it bought the Grasselli Chemical Company of Cincinnati, Ohio, a company that specialized in some early fungicide mixtures and other farm chemicals. By this time, Du Pont's own ammonia department had already moved the company into the nitrogen fertilizer business, and by the 1940s, the company had upgraded ammonia and begun manufacturing and marketing urea fertilizer. In later years, Du Pont would sell urea as a feed supplement, and in a urea-formaldehyde package, as a slow-release nitrogen fertilizer. In the 1970s, Du Pont sold its nitrogen fertilizer business, but was already using its experience with nitrogen compounds to build a pesticide business.

When the company began producing herbicides and other pesticides in the 1950s, it did so with phenylurea compounds. In 1954, Du Pont developed Karmex (diuron), the first weed killer used on cotton, and a key chemical in the company's ascendancy as a pesticide manufacturer. Another phenylurea herbicide, Lorox, became very successful for Du Pont in a number of crops, including soybeans, wheat, parsnips, and potatoes.*

In recent years, Du Pont's agrichemical business has grown at an annual rate of 15 percent, and at least half of its current agrichemical revenue comes from laboratory-fashioned products developed in the last twelve years. Today, the company sells over thirty different agrichemical products in a worldwide market that includes 110 countries. New agrichemical products continue to pour forth from Du Pont labs each year. In 1980, for example, 80 patents were filed for Du Pont crop-protection chemicals in the United States, and equivalent patents were filed for the same products in other countries.

In the early 1980s Du Pont crossed an important scientific threshold in herbicide chemistry producing a new herbicide for wheat called "Glean." Billed by the company as "a new class of chemistry," Glean does its work on weeds at extremely low dosages—at quantities one-tenth to one-hundredth that of other herbicides. Less than one ounce of Glean per acre is all that is required. Yet Glean is a sulfonylurea compound, a product that builds on the company's long involvement and experience with urea molecules. And there are other sulfonylurea herbicides coming from Du Pont, too—Ally aimed at the European wheat market, Classic for the American soybean market, and Londox for rice. Now, with the help of biotechnology and genetic engineering, Du Pont and the rest of the chemical and pharmaceutical establishment are looking ahead to an even more sophisticated era of agricultural chemistry.

*Among Du Pont pesticides there have been many market "firsts": in 1968, Lannate, the first rapid-breakdown insecticide was introduced; in 1970, Benlate, the first broad-spectrum systemic fungicide; in 1974, Vydate, the first systemic foliar insecticide-nematicide for tree crops and row crops; and in 1978, Lexone DF, the first "dry flowable" formulation for a major soybean herbicide.

Du Pont is the nation's largest chemical company and, with annual sales of $23 billion, is one of the world's largest corporations. Yet the company's move into biotechnology in the late 1970s was in part a survival move; born in the cold, hard light of the energy crisis and the decline of its traditional synthetic fibers business. In the mid-1970s, when energy prices soared and the demand for traditional chemical products waned, Du Pont, like the rest of the chemical industry, began facing some very tough decisions and market realities. The economics of producing old-line chemical mainstays such as plastics and synthetic fibers changed dramatically, and market positions began to erode.* As newer competitors with modern chemical plants began production in Mexico, Venezuela, and the Middle East during the late 1970s, many old-line chemical firms began seeking a new direction and new sources of income. Companies such as Du Pont, Chevron, Monsanto, Celanese, ICI, Allied, Dow, and Union Carbide all began to move away from "heavy feedstock chemicals" such as plastics, where they were losing the advantage, and toward "specialty chemicals" such as herbicides, and new research ventures such as biotechnology.

Du Pont, for one, expects to be doing $5 billion worth of life-sciences' business by 1990, with biotechnology and agrichemicals as major contributors in that effort. The company projects the potential market for the new products and techniques of biotechnology to be in the "tens of billions." Du Pont president Edward G. Jefferson says that biotechnology "could be a major source of new

*Du Pont built its business empire on synthetic fibers and plastics. The company's invention of nylon in 1939 ushered in the era of synthetic fibers. In 1951 Du Pont invented Orlon, followed a year later by Dacron. During the 1960s, 40 percent of the company's sales and half of its profits came from synthetic fibers. Until 1974, Du Pont continued to rely on its synthetic-fibers base for much of its income. The manufacture of synthetic fibers is accomplished with oil and natural-gas feedstocks, which are responsible for as much as three-quarters of the cost of a staple fiber such as polyester. The costs for making fibers rose precipitously with the Arab oil embargo, and the recession that followed slashed consumer demand for synthetic fibers such as Dacron polyester, Orlon acrylic, and nylon. Between 1974 and 1975, for example, Du Pont profits in fibers sank from $126 million to $6 million; from 31 percent of company profits to 2 percent. Other chemical companies, including Monsanto and Celanese, were similarly devastated. Monsanto, for example, posted a $50 million operating loss on its fiber products in 1975.

approaches to making drugs and new approaches in the plant and animal kingdoms over the next quarter-century."

THE MEDICIS OF SCIENCE

Du Pont approaches its science philosophically; it is a company steeped in a tradition of heavy research expenditures. Its officials and board members understand the need to give scientists room to flourish—a luxury not every company can afford. The company's philosophy is, as Hardy puts it, "close to the philosophy of the university." It will often "pick an area where science is dynamic, and see what happens," he says. In some respects, Du Pont might be regarded as a modern day patron, like the sixteenth-century Medicis of Florence, but backing the Michelangelos of science rather than of art. One difference, however, is that Du Pont scientists produce substances for the Du Pont trade name, not their own.

Research at Du Pont is synergistic; work in one area such as pharmaceuticals may wash back on another area such as agricultural chemicals or crop genetics. Solving the molecular riddle of "short-livedness" in proteins now used in insulin and interferon might eventually lead to a better understanding of those molecules in the next generation of research, and finally, to the production of new molecules that have applications for agriculture. Among the company's most important projects now under way in its laboratories are those arcane investigations of the microscopic chemical and genetic processes inside plants that control crop growth and regulate agricultural productivity. And Du Pont is looking at agriculture in its broadest possible dimensions.

In 1983, when it acquired Connoco, Du Pont moved into the energy business—where biotechnology could be helpful in the years ahead. "We obviously use a lot of commodity chemicals," says Hardy. "We might, in time, take renewable resources and use biotechnology, or use biotechnology coupled with catalytic chemistry, to produce some of those. The nylon of the year 2000 may well be the cornstarch that is growing at that time or the wood in trees."

Du Pont, of course, is not the only chemical giant to grasp the emerging importance of biotechnology. Most of the world's major agrichemical and animal drug manufacturers are now involved in genetics research* for several apparent reasons, one of which has to do with the way pesticides, crops, microbes, and animal products will be fashioned and packaged for agriculture in the future.

In 1962, Rachel Carson wrote, in her book *Silent Spring,* of a "seemingly endless stream of synthetic insecticides" brought about by the "ingenious laboratory manipulation of molecules [by] substituting atoms [and] altering their arrangements." Today, more than twenty years later, there is a new world of potential genetic manipulation and genetic variation that can be added to the older world of pesticide chemistry. Take, for example, the genetic engineering of crops to make them resistant to the ill-effects of herbicides.

CROPS THAT USE MORE CHEMICALS

Herbicides are chemicals designed to kill plants. They are used in modern agriculture to kill weeds that compete with crops for nutrients, sunlight, and water. In the United States, herbicides now account for nearly 60 percent of all agricultural pesticides, and the market for their continued use is expanding worldwide. In 1981, American farmers applied 625 million pounds of herbicides to their crops and land, a 175 percent increase over 1958 levels.

For some corporations, herbicides have become the single most lucrative agricultural product area.† In 1982, for example, Monsanto sold over $1 billion worth of herbicides, nearly half derived from its weed killer Roundup. Similarly, Eli Lilly's Treflan has brought

*Today, in the United States, for example, nine of the top ten chemical companies—Du Pont, Dow, Union Carbide, Monsanto, W. R. Grace, Allied, American Cyanamid, PPG, and Diamond Shamrock—are involved in the new biogenetic frontiers of agricultural research. So are at least thirteen of the world's top twenty pharmaceutical firms, including Pfizer, Bayer, Merck, SmithKline Beckman, Eli Lilly, Abbott Labs, Hoechst, Schering-Plough, Upjohn, Sandoz, Ciba-Geigy, and Hoffman-La Roche. (See Appendix for more details.)

†Du Pont, Ciba-Geigy, Eli Lilly, Dow Chemical, BASF, American Hoechst, ICI Americas, American Cyanamid, Chevron, FMC, Rohm & Haas, Monsanto, PPG, Shell, Stauffer, and Uniroyal all produce herbicides that are either currently, or soon to be, on the market. In the late 1970s and early 1980s, however, the herbicide business began to change, as did the science of making herbicides. Herbicides are what are known in the industry as "specialty chemicals," and in recent years, companies with pharmaceutical experience have been gaining the upper hand in

in as much as 12 percent of the company's earnings in a single year. But the use of herbicides with crops—since they are chemicals designed to kill plants—can create problems for farmers.

In Illinois, for example, about 10 million acres of corn are treated each year with the herbicide atrazine. Atrazine is chemically classified as a triazine compound and, with at least eight other triazine herbicides, is now widely used around the world to control weeds. When applied to the soil, triazine herbicides are taken up by the plant roots, and later move up through the stem into the leaves, where they inhibit photosynthesis, causing the leaves to yellow and the weeds to die.

Unfortunately, some triazine herbicides are indiscriminate in their effects, killing crops as well as weeds. Corn, however, can tolerate atrazine because it contains enzymes which detoxify the chemical. Thus, corn is naturally resistant to atrazine's lethal effects. But this is not the case with soybeans, a crop often used in rotation with corn. And atrazine has the unhappy effect of lingering in the soil from one season to the next—a phenomenon known to farmers as "residue carryover." Consequently, an Illinois farmer using atrazine on his corn crop and then following that crop with soybeans is likely to have atrazine-damaged soybeans and reduced yield. Moreover, other crops, such as alfalfa and small grains, are also sensitive to atrazine.

The same kind of problem exists with herbicides such as Du Pont's new wheat herbicide, Glean. Only wheat, it seems, "knows" how to detoxify Glean, or metabolize it into harmless byproducts. Crops such as soybeans, sugar beets, sunflowers, and corn that might follow wheat in rotation will be damaged by residues of Glean remaining in the soil. Says North Dakota farmer John Leppert, "It would be at least four years in North Dakota before a field treated with Glean could be used for some broadleaf crops." But for chemical companies, genetic engineering is changing this "herbicide problem" into a potential herbicide bonanza.

this market because the newer herbicides are based on small, highly active molecules. Pharmaceutical firms, dealing with drugs and antibiotics, generally have more experience with such smaller molecules than do more strictly chemical or energy companies. In the late 1970s, herbicide chemistry was becoming more sophisticated chemistry, and companies such as Monsanto, Eli Lilly, and Ciba-Geigy were leading the way.

During the early 1980s, scientists such as Charles Arntzen of Michigan State University, and others working at biotechnology companies, began to make discoveries of genes that could make plants resistant to the harsher side of certain herbicides.* Arntzen, for example, working with others at MSU and Harvard, discovered that a single gene found in the chloroplast of a plant cell could be used to resist the deleterious effects of the herbicide atrazine. Before long, other companies and scientists were pursuing the same techniques and making similar discoveries.

Another method for making crops resistant to herbicides is to find the genes for herbicide resistance in other plants or weeds and transfer them into the living cells of commercial crop lines using techniques such as micro-injection and gene splicing. Classical plant breeding can also be used, but only when a herbicide-resistant weed is cross-fertile with the desired crop species. Interestingly, in some cases, one "resource" of herbicide-resistant genes are the thirty or more species of weeds that have become resistant to herbicides since the early 1970s. A third method of imparting herbicide resistance to crop varieties involves altering the chemical composition or molecular arrangement of an existing gene or genes already in the crop.

By July 1982, *Chemical Week* wrote of a "slow but steady push" among herbicide makers toward the "genetic manipulation of corn, soybeans, and other crops to make them more resistant to herbicides." "The theory is," explained the magazine, "that farmers would then be willing to use even more of the weed killers, safe in the knowledge that their crop won't be damaged."

Chemical companies with new herbicides on the line, as well as

*Prior to the use of biotechnology techniques and genetic engineering, herbicide resistance in some crop varieties was achieved more or less by chance or by painstaking plant breeding—by observing which varieties could tolerate a given herbicide, and then cross-breeding that variety with others. Tissue culture techniques used in the laboratory greatly accelerated the process of screening plant cells for herbicide-resistant mutants. For example, Monsanto's plant sciences research director, Robert J. Kaufman, explained his company's tissue culture work with alfalfa plants and the herbicide Roundup before a Congressional subcommittee in 1982: "Alfalfa tissue was placed into culture first on solid media and later into liquid culture. In liquid culture, the cells were exposed to a lethal dose of the herbicide Roundup and the survivors were plated out for regeneration into whole plants. These new plants [or variants] were transplanted into the field and treated with Roundup the way a farmer would use the herbicide. Several variants were found to have field resistance to the herbicide."

biotechnology companies looking to make their mark on Wall Street, became especially attracted to the genetics of herbicide resistance. "It's no coincidence that companies involved in herbicides get into biotechnology," says Garo H. Armen, a chemical analyst with E. F. Hutton. "They're trying to cover themselves."

The L. William Teweles Company, a seed and biotechnology consulting firm, was soon reporting that some agricultural chemical companies were developing crops resistant to their herbicides "in the hope of selling the seed and chemical as a pair." Other such companies were seeking herbicide resistance, according to Teweles, "as a way of gaining market share lost after a well-known herbicide has declined in price and popularity [with] the old herbicide sold in combination with a new seed resistant to it." This marriage of herbicides and seed "is a natural," says Teweles' James Kent, which circumvents many existing problems with herbicides while creating "a complementary demand for both chemical and seed."*

In November 1982, Calgene, a California biotechnology company, announced that it had cloned a gene for resistance to the world's most widely used herbicide, Monsanto's Roundup, chemically named glyphosate. By 1984, Du Pont had developed tobacco

*One variation of the herbicide-resistance theme at the seed level comes from Ciba-Geigy and its subsidiary, Funk Seeds International. Ciba-Geigy has developed a "herbicide antidote" or seed safener called Concep—actually a chemical coating—which is applied to sorghum seeds to protect them from the harsher side of herbicides such as Dual, Milocep and Bicep. Ciba-Geigy's seed coating—which blocks the toxic action of the sorghum herbicides on the seed—is a trade-marked process called Herbishield. "By planting Herbishield Seed," says Ciba-Geigy's Funk Seed to farmers, "you will be able to use Bicep, Milocep, or Dual herbicides for more effective control of grassy weeds." Ciba-Geigy sells both "Concep-Safened" sorghum seed as well as the herbicides Dual, Bicep, and Milocep.

Moreover, with the aid of biotechnology and the mass cloning of tiny, pre-seed embryos—i.e., seeds without their coats—new crop varieties may soon be mass produced in embryo form to be sold by pharmaceutical and other companies in drugstore-like capsules, complete with fertilizer and pesticides. Called "artificial or synthetic seeds," companies such as Upjohn, Hoffman-La Roche, Plant Genetics, Agrigenetics, Cetus, and DNA Plant Technology are all working on this new kind of encapsulated seed. For example, Plant Genetics of Davis, California, sees its new encapsulation process—called GEL-COAT—"as a new and more effective delivery system for agricultural chemicals, microbials, seeds, and somatic embryos." The company presently has field testing agreements with FMC, Ciba-Geigy, Chevron, and others, and has experimented with sixty agricultural chemicals in the GEL-COAT system, including herbicides, systemic fungicides, and insecticides.

plants with resistance to its herbicide Glean—plants that were one hundred times more resistant to Glean than were normal plants. The company also cloned the gene for Glean resistance. But Du Pont and Calgene were by no means alone in using biotechnology for such purposes. At least a dozen other chemical and biotechnology companies were also involved.*

Some major chemical and agribusiness corporations began to contract biotechnology companies to develop herbicide-resistant crops. American Cyanamid struck a deal with Molecular Genetics to develop new lines of hybrid corn resistant to a new class of experimental herbicides called imidazolinones. "The new lines of corn," reported *Chemical Week*, "are expected to increase the market for broad-spectrum herbicides, which otherwise have a limited potential when used with current hybrids of corn."

By 1984, Calgene had contracts to develop herbicide-resistant crop varieties with at least three major corporations—Kemira Oy of Finland, Nestlé of Switzerland, and Rhone-Poulenc of France. Calgene spokesmen said the company was also working on ways to move herbicide-resistant genes into crops such as cotton. "Cotton has lots of weed problems," explained Calgene vice president Al Adamson, "and cotton farmers could really benefit from herbicide-resistant varieties."

Chemical companies soon began hiring talented scientists to conduct work in the field of herbicide-resistance. Du Pont hired Michigan State's Charles Arntzen, and Ciba-Geigy hired Mary-Dell Chilton from Washington University to head up the company's biotechnology unit in North Carolina. Ciba-Geigy was hoping to develop, among other things, soybean varieties resistant to its old herbicide atrazine. And before long, even the scientific establishment was singing the praises of herbicide-resistance research. "An atrazine resistant soybean," noted the National Academy of Sciences Board of Agriculture in its 1984 booklet *Genetic Engineering of Plants*, "would be ideal for the Corn Belt."

Not everybody, however, is thrilled about the prospect of

*Among the major corporations and biotechnology companies exploring ways to develop herbicide-resistant and/or herbicide-tolerant crop varieties are Advanced Genetic Sciences, Biotechnica International, Calgene, Ciba-Geigy, DNA Plant Technology Corporation, Du Pont, Eli Lilly, Molecular Genetics, Monsanto, Phytogen, Rohm & Haas, Shell Development Company, and Stauffer Chemical Company.

herbicide-resistant crops. "If we create food crops with herbicide resistance," asks Sheldon Krimsky, a social scientist at Tufts University and former member of the NIH committee on recombinant DNA, "are we not going to reinforce the use of herbicides? Are we not going to reinforce greater chemical use in food production at a time when people are increasingly questioning the agricultural use of chemicals?" Further, if herbicide-resistant crop varieties increased the use or range of certain chemicals in the environment,* might they cause ecological or public health problems?

While some herbicides such as Du Pont's Glean and Monsanto's Roundup are claimed to be among the safest chemical compounds developed by industry so far, not much is known about the long-term effect of these or other herbicides in the environment. Herbicides are generally not regarded to be as toxic as the chlorinated hydrocarbon insecticides used in the 1960s, but they do have side effects. Not much is known about how herbicides completely break down, and according to some scientists, such information is only known for about four of the 150 herbicide compounds presently in use. And herbicides such as atrazine have been found to cause chromosome breakage and other aberrations in plants. In fact, triazines generally are known to be mutagenic to some insects such as fruit flies. Moreover, recent discoveries of herbicides such as alachlor,† atrazine, and paraquat showing up in groundwater in states such as Iowa, Nebraska, and Maryland have raised new questions about

*Among those herbicides for which genetic resistance in specific crop varieties is being sought through tissue culture screening and/or genetic engineering techniques are: atrazine, Bromoxynil, cinmethylin (Cinch), glyphosate (Roundup), imidazolinones herbicides, paraquat, picloram (Tordon), 2,4-D, phenmedipham (Betanal), and sulfonylurea herbicides (Glean, Ally, and Londox). Other herbicides are, no doubt, also being investigated for such purposes, yet specific information on this kind of research is difficult to verify since much of it is proprietary.

†In June 1984, *Pesticide and Toxic Chemical News* reported that EPA might cancel the registration of Lasso because of the discovery of minute traces of the herbicide in Ohio drinking water, which suggested that not all of the pesticide dissolves, as previously claimed. In reaction to that report, Leslie C. Ravity, a prominent Wall Street analyst for Salomon Brothers, withdrew his recommendation for the company's stock, causing a temporary panic in selling. On June 7, the New York Stock Exchange halted trading in Monsanto stock for one hour due to the imbalance in orders. The stock dropped $1.50 per share for the day.

Later reports about the possible EPA review of Alachlor, the active ingredient in Lasso, cited links to cancer in laboratory animals. Lifetime feeding studies with technical grade Alachlor in mice and rats suggest that the material is capable of

herbicide safety. Nevertheless, huge research investments continue to be made in herbicide and other forms of agricultural chemistry, now buoyed by the helping hand of biotechnology.

BIOTECHNOLOGY & THE CHEMICAL CORNUCOPIA

Du Pont's Ralph Hardy sees biotechnology leading to a new era of chemical discovery—and what he calls "a whole new sophistication in agrichemicals." Down the road, he says, biotechnology will lead to the discovery of molecules that regulate biological activity in plants, new types of chemicals that control plant growth, and novel pesticides. "As we understand more about the regulation and coordination of gene expression in plants," says Hardy, "we will design and use chemicals to selectively turn on or turn off genes at critical stages in order to effect beneficial changes in plant growth, development, metabolism and pest resistance." Such new agrichemicals, he says, could be a major indirect impact of biotechnology.

The chemicals Hardy is talking about are known as "plant growth regulators"—chemical cousins of herbicides and insecticides that can be used to regulate, control, stunt, or stimulate specific kinds of plant biochemistry. Some are already used in agriculture to a limited extent.

Apple producers in some parts of the United States spray their crop with a chemical substance that gives their fruit the characteristic "Red Delicious look"—a five-pointed knob on the bottom of the apple.* California growers give their grapes a special chemical

causing tumors in these laboratory animals when it is fed to them in extremely high levels on a daily basis for the greater part of their lifetimes.

In reaction to the possible EPA review, Monsanto's Will D. Carpenter explained: "There is no evidence that Alachlor produces tumors in humans. The risks will be compared by the EPA with the benefits from Lasso," he said. "Lasso effectively controls weeds in major crops including corn and soybeans, increasing crop yield and quality. It is of major economic importance to U.S. agriculture.

"These benefits are known to millions of farmers and documented in economic analyses. The product is being used safely with no unreasonable risks. Any special review, if held, would confirm these results and uphold the product's continued registration."

*One 1984 advertisement in *The Grower* from Abbott Laboratories pitches that company's apple growth regulator, Promalin. The ad, entitled "Designer Apples," features a Red Delicious apple with "Promalin" stitched into it like "Levi" is stitched into a pair of jeans. "Promalin gives you a consistent harvest of big,

treatment to make them larger. And in Europe, farmers use a chemical substance on the stems of wheat plants to stunt their growth and keep them short in order to withstand high winds.

Plant growth regulators constitute a relatively small $255 million market worldwide, but according to Battelle Labs, that market is expected to become a $2-billion-a-year business by the year 2000. And a few companies already have a head start.*

John Senger, general manager of the agricultural technology division for Monsanto, says that plant growth regulants are "the next major wave of agricultural chemicals," and that Monsanto is "heavily committed." Union Carbide's Anson R. Cooke adds: "We believe that compounds can be found that will better enable plants to withstand stress—heat, cold, salinity and drought." Robert H. Becker, agricultural research director for the American Cyanamid Company, says that the potential for plant growth regulators is "almost limitless since there are an almost infinite number of ways to increase the potential of plants." Uniroyal already synthesizes some 1,400 potential growth regulators each year, screening them for favorable activity on a broad range of plants.

At Du Pont, research in plant growth regulators, plant breeding, and plant biotechnology have been integrated, in at least one crop, according to G. D. Hill, director of the company's agricultural chemicals department. The company is taking this approach in an attempt to achieve a 10 to 15 percent yield increase in soybeans, a priority crop for Du Pont. Monsanto is following a similar course integrating

long Red Delicious apples that really look Delicious," says the ad. "The kind you'd be proud to put your label on.

"You'll grow them better, when you give Mother Nature a helping hand with Promalin plant growth regulator," claims the ad. "Promalin gives Red Delicious apples the long calyx lobes that apple fanciers fancy. Makes them longer and heavier—gives you more apple per apple, so you need fewer apples per bin, box, bushel or bag. You get more tray packs, a larger percentage of extra fancies, the choice of the premium export market. So you harvest more profit per acre. . . ."

*Du Pont already produces a "sugarcane flowering suppressant" and a "multiple flower stimulator." Shell produces a "dormancy breaker"; Monsanto, 3M and Pennwalt, a "sugarcane ripener"; FMC and Merck, a "root inducer." American Cyanamid produces a substance that can be used as "height shortener" for wheat; Dow and Agway produce "corn yield enhancers"; and Eli Lilly produces a gibberellic acid used to stimulate amylase content of malting barley. And there are other companies producing growth retardants, "bloom improvers," and chemical defoliants to aid in the harvesting of crops such as cotton.

its research in these three areas hoping to achieve "a synergistic effect that will lead to increased crop yield," according to one company spokesman.*

Agricultural research in the years ahead, like that now being conducted by researchers at Du Pont and Monsanto, will be increasingly guided by what the geneticist and biochemist can work out together. Considering, for example, that there are thousands of pesticide, fertilizer, and even microbial products that can be designed to work with certain kinds of crops, and that such products might be specifically matched to individual genes as in herbicide resistance or plant growth regulation, the range of product possibilities that might match an "inside" genetic instruction with an "outside" chemical or microbial product is potentially limitless. All that is needed to artfully explore the inordinate number of possible product combinations in the genetic, chemical, and microbial realms of agriculture is a computer.†

"The use of the computer rather than the test tube to generate new chemical structures with expected biological effects," says Du Pont's Ralph Hardy, "is in its infancy." This capability, says Hardy,

*Some plant growth regulators have already been used in combination with plant breeding programs, or have been produced to facilitate, or to respond to, mechanization in agriculture. For example, Abbott Laboratories "abscission agent," known as Release, which facilitates the separation of fruit from tree branches, is used widely in the Florida orange groves in combination with tree-shaking machines. The mechanized tomato harvest in Florida is aided by the use of Ethrel, an ethlyene spray made by Union Carbide's Amchem Products that also helps turn tomatoes red. In fact, some Florida tomato varieties—such as the Walter and MH-1 varieties developed at the University of Florida in the 1970s—have been specifically bred to respond to the ethylene spray.

†In their work with pesticide molecules, some chemical and pharmaceutical companies have been using computer-assisted analysis to study molecular properties of potential new agrichemicals. "If you know the mode of action of the molecules," says FMC's agricultural research director Dr. Donald E. Bissing, "you can study their topography [on a computer screen] and see where they can be bound tighter to an active site by changing molecular properties." American Cyanamid is also using computer modeling systems in its research into potential insecticides. Such modeling can determine how one molecule might interact with another for the purpose of determining "enzyme receptor sites," or locations on a molecule where another might be connected. As Robert H. Becker, research director at Cyanamid explains, "we have been able to predict activity on new [molecular] ring structures that would not have been obvious to a synthesis chemist doing traditional research."

"will be coupled with the molecular understanding of biological systems" to produce a "directed synthesis" of new products such as plant growth regulators. The "cut and try" method of the past, he says, requiring chemical companies to synthesize and test huge numbers of chemicals will be replaced by the "design and select" method of the computer, enabling companies to screen and synthesize fewer chemicals to find a viable product more quickly.

But if biotechnology fosters a flood of new chemical and microbial products for use in agriculture, such as plant growth regulators, will the Environmental Protection Agency (EPA) be able to insure that all of these substances are safe for use *before* they are put on the market? Or will industry's ability to churn out these new substances—aided by high-speed genetic and computer-screening techniques—outstrip EPA's capacity to carefully evaluate the safety of such products?*

Although the combination of the computer, genetics, and chemistry makes possible the screening out of undesirable plant growth regulators and other chemicals, as well as the selection of environmentally safe ones, such "design power" also puts companies like Du Pont, Monsanto, and American Cyanamid in a position to build integrated packages of chemical, genetic, and microbial ingredients for use in agriculture. And established companies such as these are not likely to make a quantum leap from old pesticide products to

*While some plant growth regulators may be beneficial and harmless in the environment, others may not be so benign. In July 1984, the EPA announced a special review of Uniroyal's growth regulator Alar (daminozide), a chemical spray used on apples to retard ripening in the field, while also making them redder and increasing their shelf life by two to three months. Used in orchards, the spray allows growers to extend the harvest season, employing fewer pickers over a longer harvest. Used on peanut crops, the same chemical stimulates upright growth in the plant, facilitating harvest.

EPA initiated review of daminozide because it feared the substance could pose a dietary cancer risk in humans consuming raw and processed foods treated with the chemical. "The order of magnitude of risk from this chemical," said EPA pesticide official Michael Branagan in July 1984, "is similar to that of EDB." EPA initiated the review for the chemical when it discovered in tests on laboratory animals that daminozide caused tumors of the uterus, liver, kidney, lungs, and blood vessels. Uniroyal officials, however, insist that the chemical is safe. "We don't believe Alar poses a threat to either the environment or individuals," said James Wylie, Uniroyal's manager of crop protection chemicals. The results of EPA's review will not be known until 1986. Meanwhile, Alar continues to be used.

new genetic ones. Rather, more sophisticated kinds of biochemistry—like plant growth regulators or pesticide-compatible microbes—will be developed to work in conjunction with existing products and specific crop varieties. Biotechnology, then, might not eliminate pesticide toxicity or reduce the use of pesticides in the environment so much as it will change the form of those chemicals and the way they are used in agriculture.

TAKING OVER THE AGRICULTURAL SOFT PATH

Biotechnology not only raises new possibilities for a cornucopia of chemical products on the more traditional side of agriculture, but also in the area of more purely biological products, or what might be called the agricultural "soft path."

Methods of controlling agricultural pests that employ nonchemical means, such as the use of one kind of insect to control other insects, or plants that emit natural chemicals that repel insects or even weeds, are known generally as biological methods of pest control. Rachel Carson, in her 1962 book, *Silent Spring,* was one of the first popular writers to advocate the use of biological techniques in commercial agriculture. Today, a broader category of "soft path," non-chemical pest control techniques might include certain genetically-engineered viruses or bacteria that would kill insects, or crops genetically improved to resist disease or insects, thus displacing the need for chemical pesticides. (Although, to be sure, such products would have to be preceded by the most rigorous kind of scientific review and testing for their own potential ecological and public health impacts. See Chapter 12 for more details.) With the help of biotechnology, some of these biological and genetic approaches to pest and disease management in agriculture may become both environmentally acceptable and economically attractive. And not surprisingly, many of the major chemical and pharmaceutical companies involved in the production of synthetic pesticides and fertilizers are also pursuing the development of new biological and genetic products.

While the pursuit of such biological approaches by major pesticide manufacturers is a commendable development, there is some question about how soon and to what extent such companies will market their new biological product lines. Moreover, just as the major energy companies have moved into alternative energy development buying up the most promising solar energy companies, the "Exxons of agricultural chemistry" have also been at work buying up the few small companies that began to show some signs of promise with biological pest control. Consider, for example, the Zoecon Corporation, a California company initially established to develop biological methods of pest control.

In 1968, Dr. Carl Djerassi, an organic chemist who was then in the process of developing the first human birth-control hormone, met Dr. Carroll Williams, a Harvard entomologist who had discovered and isolated a juvenile insect hormone. At that meeting, Djerassi and Williams decided to establish the Zoecon Corporation, a company they created for the sole purpose of developing nonchemical approaches to controlling insects.

In 1970, Zoecon began research on pheromones, the chemical mating signals given off by insects. Pheromones, synthesized chemically in the laboratory, can be sprayed on fields to confuse insects in their reproductive behavior, and thereby serve as a means of reducing pest populations. Along with this research, Dr. Djerassi hoped to develop a birth-control hormone for insects. In 1971, he successfully produced methoprene, a synthetic mimic of a hormone that disrupts an insect's metamorphosis, locking it in the growth stage between larva and adult, and thus keeping it immature and impotent. Methoprene was the first insect growth regulator to be registered by the Environmental Protection Agency.

In the late 1960s, there was great enthusiasm for Zoecon's newly established venture, and investors were eager to back the company. "We raised $15 million for that corporation in the twinkling of an eye," recalls Dr. Williams. But discovering, registering, and marketing viable products was another story. Zoecon's activities were heavily dependent on research funding, and its biological products had to run the same testing gauntlet at the EPA as chemical pesticides did. Thus, it took three years and $500,000 to register methoprene.

Between 1972 and 1976, Zoecon struggled along in business with methoprene and a few pheromones, managing small sales increases as it went. But its research expenditures were slipping—from a high of $3.2 million in 1973 to $2.6 million in 1976. And for Zoecon, these research expenditures averaged about 17 percent of sales—a much higher amount than most chemical and pharmaceutical companies normally spend on research. Zoecon was also running up against the unique economic reality of biological pest controls: they were selective, often targeted to one insect species, and sometimes used only in one stage of insect development. Such products, therefore, had a very limited market potential. They were not like broad-spectrum chemical pesticides that could be used on a large number of pests and marketed widely, and consequently, were not as profitable.

By 1975, with mounting financial difficulties, Zoecon fired most of its consulting scientists. Two years later the company was acquired by Occidental Petroleum, one of the world's largest energy corporations. According to Carl Djerassi, Zoecon went willingly into the arms of Occidental because of its need for more research funding. For Occidental, on the hot seat with its Hooker Chemical subsidiary and the Love Canal controversy, Zoecon was a potential "good guy" image-builder.

One year after Occidental acquired Zoecon,* it also acquired Ring Around Products, a major southeastern seed company headquartered in Montgomery, Alabama. Occidental then merged the two operations into one unit, hoping that some revenues from the seed business would offset Zoecon's need for time and heavy research spending. Under Occidental's ownership, Zoecon's research budget quadrupled to $15 million annually. By 1978, Zoecon researchers had begun to explore genetic engineering methods for developing new crop lines resistant to such environmental stresses as increased soil salinity. Hybrid cotton and hybrid soybean research were also under way. And Zoecon's work on insect hormones, pheromones, and other "biologicals" continued as well.

By 1982, however, Occidental needed cash to repay money it

*Occidental wasn't the only major company to buy a fledgling biological pesticide company. In 1975, for example, the Amway Corporation bought out Nutrilite Products, one of the first American companies to sell biological pesticides. More recently, other corporations have ventured into the nitrogen bacteria market. In 1982, the Allied Chemical Company acquired Nitragin, a $6-million-a-year Milwaukee, Wisconsin, company specializing in the production and marketing of nitrogen-fixing bacteria. Allied's acquisition of Nitragin was viewed by *Chemical Week* as possibly being "a more profitable strategy for getting essential nitrogen compounds into the roots of corn and other crop plants than the conventional fertilizer business." Part of Allied's strategy includes linking up its in-house genetic engineering work with the seed marketing and distribution capability of its new acquisition. Allied expects that its first genetically engineered product will be a soybean bacteria that absorbs nitrogen more efficiently.

had borrowed to acquire the Cities Services Corporation—a $4 billion transaction—and Zoecon was put up for sale. In January 1983, Sandoz, the world's eighth ranked pharmaceutical corporation, acquired Zoecon for an estimated $80 million. Sandoz already owned four American seed companies and one in Europe, and did not purchase Oxy's Ring Around seed company in the deal.

When Sandoz acquired Zoecon, the California company had already accumulated some 140 pesticide patents on its various insect hormones and pheromones, making it one of the top fifty American pesticide-patent holders. In addition, Zoecon had research and production facilities in Spain, Canada, and Japan. Sandoz appeared satisfied with its new acquisition. "We consider [Zoecon's] high-quality research very important," said Daniel Wagniere, head of Sandoz's agrichemical and seed divisions, "and their products sufficiently different to complement ours." John Diekman, Zoecon's vice president for research and development, points out that Sandoz's American and Dutch seed companies will provide commercial outlets for Zoecon's genetic engineering work on crops.

With a company such as Zoecon under its belt, Sandoz was now positioned to move in any of three directions should agricultural markets shift: the traditional seed and agrichemical products market, the "good guy" biologicals market, or the emerging microbial and biotechnology products market. Moreover, the company was also in a good position to "package" product combinations in any of these areas.

Zoecon's work with pheromones and anti-juvenile-insect hormones complements Sandoz's own work with biological pesticides such as Thuricide and Teknar—the latter of which was used in 1981 on a "fairly large scale" in West Africa in connection with a World Health Organization project—as well as agricultural virus products such as Elcar, which has been sold by Sandoz in Brazil to treat coffee and cocoa crops.

Interestingly, in 1980, Sandoz began a research program to develop a virus that is effective against the coddling moth, an insect that is a significant and economically damaging pest of apples, pears, and walnuts. Scientifically known as "coddling moth granulosis virus," or CMGV, Sandoz produced this virus as an insecticide in the United States in 1981 under an EPA experimental use permit. Shortly after that, however, Sandoz put its viral insecticide program into temporary cold storage because of lack of a market. According to T. Shieh, manager of biological research at Zoecon, at least part of the reason for Sandoz's reluctance in this program was government policy and public attitudes toward pesticides. Neither the chemical industry nor federal or state governments, he says, are "concerned enough" about the "undesirable side effects" of chemical insecticides in the environment. But when these attitudes change, says Shieh, Sandoz will reactivate its viral insecticide program. Yet, Sandoz is still very much in the conventional agrichemicals market, selling insecticides, fungicides, herbicides, and chemical seed treatments in world markets.

Sandoz, of course, is not the only pesticide producer to venture into biological pesticides. Upjohn's Tuco subsidiary produces a bacterial product that is ingested by vegetable worms such as the cabbage looper, tomato hornworm, and imported cabbageworm, attacking and destroying the worms within a few days. Pfizer, Sandoz, Biochem, and Abbott all produce various strains of the same bacteria, known as *Bacillus thuringiensis*. At the same time, ICI Chemicals, Hoechst, and Bayer produce pheromones; Hoffman-La Roche, Stauffer Chemical, Ciba-Geigy, and Phillips have developed juvenile-hormone analogs; and Monsanto is now looking into the possibility of genetically modifying microorganisms that live on the cotton boll so that they will poison insects that attack the boll. Abbott Laboratories is working with a fungus that might be used as a biological herbicide. Biotechnology firms too, such as Advanced Genetic Sciences, Agracetus, Ecogen, the Genetics Institute, Mycogen, Repligen, and the Syntro Corporation, are also working on the genetic modification of agricultural microbes or the development of new biological pesticides.

Yet, despite all of this activity along the agricultural soft path, and the promises of a "pesticide-free" agriculture, it appears that there will be a very considerable period of chemical dependence before there is ever a golden age of benign agricultural biotechnology. Major chemical and pharmaceutical companies, now presiding over the transition of the "old" pesticide era to the "new" genetic one, are not likely to give up their chemically oriented research structures and marketing systems easily.

As agriculture moves into the new era of biogenetics under the auspices of the chemical and pharmaceutical industries, it will also move into a new age of microchemistry and micromolecules—molecules that will trigger certain biological functions in crops and livestock. Genetically keyed chemicals, for example, will help to shift electrons around in the leaves of crops to bring about more effective photosynthesis or to channel more energy to the harvested part of the plant rather than its "less useful" vegetative parts. With such understanding of how "outside chemistry" interacts with "inside genetics," an entire field of soybeans or corn might be administered milligrams of a chemical substance genetically keyed to trigger the beginning of a biochemical process that enhances yield, facilitates harvesting, or insures uniform ripening.

While synthetic chemical use in agriculture will increase in sophistication through the use of biotechnology, some chemical fertilizer use may decline with the ability to genetically alter crops to fix their own nitrogen. On the other hand, the successful hybridization of hard-to-hybridize crops such as wheat and soybeans through genetic engineering would increase the use of fertilizers and pesticides in those crops. Other developments, such as the mass production and druglike encapsulation of preseed embryos, packaged with minute amounts of fertilizers and pesticides, will mean better-targeted, but nevertheless continued, chemical use.

On balance, biotechnology will help major chemical and pharmaceutical corporations maintain their pesticide position in agriculture, while giving them the means to gradually develop and sell a range of geneticlly engineered crop, microbial, and biological products. Chemical-genetic "packages" such as herbicide-resistant crops, chemically coated seeds, and genetically keyed plant growth regulators, may eventually give way to genetically improved crops that are resistant and even immune to diseases and insects, have the ability to repel insects or weeds naturally, and/or the capability of self-fertilization. How fast that happens will depend largely on public attitudes and government policy.

In the meantime, however, the new products of agricultural biotechnology will certainly not be "simple-minded" pesticides applied indiscriminately in tons per acre without regard to ecological niches or environmental consequences. Rather, they will be the carefully honed chemical, genetic, and microbial regulators of agricultural productivity. Such ingredients will be the triggering mechanisms of basic biological processes, and in the end, the instructions that will determine how much and what kind of food appears on the table.

CHAPTER

12

ENVIRONMENTAL ROULETTE

There will always be an Achilles' heel.
—*Arthur L. Koch,*
microbiologist

Steven Lindow doesn't look like the type who would deliberately cause an ecological crisis. Lindow, a thirty-four-year-old professor of plant pathology at the Berkeley campus of the University of California, has blond hair, a neatly-trimmed beard, and could almost pass for one of the Beach Boys. Yet in September 1983, a federal court action and a threatened injunction by a small group of citizen and environmental interests in Washington, D.C., caused Lindow and the University of California to postpone an experiment the citizen groups said might create major ecological problems.

Before the court action, Steven Lindow was about to enter the annals of science history; he was planning to conduct the first out-of-the-laboratory field test of some genetically engineered bacteria. At his university research laboratory, Lindow had discovered that by genetically changing certain bacteria that lived on the leaves of plants, he could forestall the formation of frost on those plants to temperatures as low as 23 degrees Fahrenheit. That fall, Lindow was preparing to test his new bugs on a field of young potato plants in northern California. But in Washington, a few citizen groups worried that these altered microbes might do more than forestall frost damage on potato plants.

In the past, man has introduced numerous non-native organisms, disease pathogens, and "exotic" species into new environments either accidentally or deliberately. Some of these introductions, such as the gypsy moth, Dutch Elm disease, chestnut blight, kudzu vine, Johnson grass, and even starlings and house sparrows, have either become troublesome pests in their new lands, or have caused serious ecological and economic damage. The gypsy moth is a classic example.

In the late 1860s, a French astronomer named Leopold Trouvelot brought three eggs of a relatively obscure variety of European moth, *Lymantria dispar,* into the United States and began breeding the moth in his house in Medford, Massachusetts, trying to develop a disease-resistant form of silkworm. Subsequently, a few of Trouvelot's captive moths escaped from his home into the environment, and from that inauspicious beginning, the ravenous gypsy moth rose to become one of the nation's most disastrous environmental pests.

In the 1940s the moth was thought to be contained within an area east of New York's Hudson River. But by the 1970s the range of the gypsy moth was still expanding, and at an accelerated rate. By 1980, the moths occupied a region of 500 square miles and were spreading rapidly throughout New England. In 1981 alone, they defoliated over 10 million acres of American forests.

Gypsy moths dine on 450 to 500 different plant species, and have adapted to climatic conditions in the United States by developing patterns of egg-laying behavior that insure their survival even through northern winters. Newly hatched gypsy moth larvae are easily dispersed by man and the wind. Despite the fact that some thirty-eight species of birds and a good number of small mammals feed on gypsy moths, and that numerous parasites have been introduced to control their numbers, the gypsy moth today remains out of balance with its environment.

Now, with genetic engineering offering the very real likelihood that new, man-made forms of plants, animals, and microorganisms will be released into the environment—carrying gene combinations that probably don't exist in nature—some people are wondering if such "novel" organisms might not spawn their own "gypsy moth" disasters.

THE VARIABLES & THE UNKNOWNS

As genetic engineers step outside of their laboratories with new microbes, crop varieties, and livestock breeds, they will be entering environments of great biological variety and genetic nuance. In the United States, for example, there are at least 160 species of bacteria, 250 kinds of viruses, 8,000 species of fungi, 8,000 species of insects, and 2,000 species of weeds that affect agricultural crops. This ecological potpourri of living organisms is a dynamic and adaptive part of the agricultural landscape.

"A handful of soil scooped from the fields of a mechanized farm," says microbiologist Winston Brill of the Cetus Corporation, "is the locus of an unruly turmoil of competing microorganisms." In the soil, says Brill, thousands of strains of microbes contend for nutrients and energy, altering, in the process, the chemistry of the soil with the products of their metabolism. Moreover, Brill explains, "the microorganisms themselves evolve in response to stresses imposed by their environment, including the stresses induced by the evolution of their fellow species."

In addition to the environmental variables that operate in the "outside world" of the soil microbe—and other organisms as well—there are countless other variables that operate in the "inside" world of genetics; variables that affect the way these organisms behave. One strain of microbe, for example, contains as many as 3,000 genes; higher crops as many as 100,000. When one gene is removed, or others added in the process of gene splicing, how does that affect the rest of the genes in that organism? Is there an ecological order inside organisms just as there is in the outside environment? Is there, for example, a dynamic balance of gene interrelationships that can be thrown out of kilter when new genes are added or "troublesome" ones deleted?

Scientists working with the genes of plants admit that they don't know much about them. Du Pont's Director of Life Sciences research, Ralph Hardy, says that "Our understanding of plant molecular genetics is still very early," with possibly as few as twenty-five of the 100,000 genes in a plant cell having so far been characterized.

Just how much modern science does not know about genes—and how the scientific establishment will sometimes resist the truth, even when it is staring them in the face—is shown by the work of Barbara McClintock, who was belatedly honored in 1983 with a Nobel Prize for a key discovery she made about genes in the late 1940s. McClintock discovered, in her painstaking cross-breeding of corn plants, that genes—once believed to be stationary in their location and arrangement on chromosomes, and therefore predictable in terms of their association with specific genetic characteristics from one generation to the next—could actually move or "jump" from one chromosome to another, creating changes in the expression of genetically governed traits. McClintock saw such changes manifested in the changing colors of kernels on ears of corn.

McClintock's discovery of "jumping genes" was nothing short of revolutionary, but it flew in the face of the conventional wisdom of her day and was regarded by her scientific peers as heretical. "They thought I was crazy; absolutely mad," recalls McClintock. Her first public report on the discovery was aired at a scientific symposium in 1951 and was received by her peers with cool disbelief. A scientific article she published on the subject two years later drew only three requests for reprints. Incredibly, it took nearly twenty years before the scientific community finally accepted her discovery.

What McClintock learned in her studies of "mobile genetic elements" was that not all genes govern some outwardly manifested trait or some structural characteristic; rather, some genes control the activity of other genes, acting as switches to turn them on and off. This insight is absolutely essential to the understanding of modern molecular genetics, and is a vital foundation for any work in genetic engineering. But McClintock's work also underscores the variable nature of the genetic world *within* organisms, and the still uncharted frontiers of that world. We are now seeing an explosion of information about various forms of newly discovered mobile genetic elements in all kinds of organisms, including microbes, plants, and animals.

Only in 1983 did science first coin the term "promiscuous DNA," describing the apparent natural transfer of DNA sequences from the mitochondria* of a yeast cell into that cell's nucleus. Not long

*Mitochondria are rod-shaped organelles in higher cells that serve as the "powerhouse" for the cell, producing chemical energy.

after that, scientists discovered that genes inside plant cells were also moving around, in this case between chloroplasts* and mitochondria. Scientists studying this particular transfer activity in mung beans, spinach, corn, and peas at Duke University and the Carnegie Institution of Washington concluded that it was not a rare event, but a "general phenomenon." Before this discovery, scientists had assumed that intracellular organelles like chloroplasts and mitochondria were independent of each other. Now they know differently, but how the DNA gets from one organelle to another is still a mystery.

Despite this kind of information and uncertainty, some genetic experiments are being pursued as if the genetic realm were thoroughly understood.

Some scientists say that when a gene is removed from an organism there is "reasonable certainty" that no unintended changes will occur, but others aren't so sure. "Because of the close coupling among genes and gene products," says Cornell University biochemist Liebe Cavalieri, "there is no way of knowing whether this interlaced network has been altered." In fact, says Cavalieri, this genetic situation is so complex that "one does not even know which questions to ask." Cavalieri adds that given the thousands of genes within a single organism, "the permutations among them is virtually infinite." When this internal variability is coupled with the variability of the external environment, the task of assessing what a new genetically altered organism might do when released into that environment is very difficult indeed.

Clearly, gene-splicing experiments with plants and microbes in the controlled environment of the laboratory are one thing, but moving new, genetically engineered organisms as widely used commercial products into the real world of ecological give-and-take—where everything is changing, interacting, and adapting—is quite another.

All new technologies pose some degree of risk to society in their use and, in this regard, biotechnology is certainly no exception. However, some proponents argue that the risk-free use of biotechnology may not be possible. Others say the risks are too high, period.

*Chloroplasts are the chlorophyll-containing organelles in plant cells and some bacteria where photosynthesis occurs.

Still others believe that the environmental and public health hazards of biotechnology can be minimized to the point where there is an "acceptable level of risk." For the most part, the public has not yet entered the debate on risk. But let us return to the case of Steven Lindow and the arguments for and against his frost-inhibiting microbes.

THE "ICE-MINUS" FIGHT

Frost damage in America's orchards, and throughout farm country generally, amounts to about $1 billion a year; worldwide, the toll runs to more than $14 billion annually. Modern commercial farmers have attempted to minimize the damage of late spring and early autumn frosts with everything from oil-powered heaters to huge propeller-size fans that keep warm air from dissipating into the atmosphere. However, these awkward, "Model-T" methods of frost control may soon give way to the "sure and smart" techniques of genetic engineering.

Frost damage to crops and fruit trees occurs as moisture on a plant's surface begins to freeze; as this happens, expanding ice crystals destroy the plant's cells by puncturing and dehydrating them. However, ice forms on plants only around certain kinds of impurities or nuclei, and as it turns out, two kinds of bacteria which populate the vegetative surfaces of most plants—*Pseudomonas syringae* and *Erwinia herbicola*—facilitate the "ice-nucleating" or frost-making process. It is believed that these bacteria contain certain molecules on their cell membranes that react with water molecules, making the bacteria serve as a nuclei for ice formation when the temperature falls slightly below 32 degrees Fahrenheit. Plants that have had *P. syringae* and *E. herbicola* removed from their leaves can survive happily to temperatures of 23 degrees.

In his work with more than one hundred different agricultural crops—including oranges, almonds, beans, corn, squash, tomatoes, and potatoes—Steven Lindow discovered that *P. syringae* and *E. herbicola* were the frost-making culprits, and he began to think of ways to isolate or inhibit their activity. He found that substances such as streptomycin, a common antibiotic drug, killed *P. syringae* and *E. herbicola,* and also discovered that certain compounds used as plant nutrients would disrupt the arrangement of molecules on the bacteria that facilitates the formation of ice crystals, thereby inhibiting their ability to make frost. But Lindow really hit pay dirt when he began thinking genetically.

With two colleagues at Berkeley, Lindow discovered that he could remove the gene in *P. syringae* and *E. herbicola* that governed the arrangement of the molecules on the bacteria's membrane that facilitated ice-nucleation. He tested these genetically altered bacteria on various plants in low-temperature chambers and found that the plants could survive to temperatures of 23 degrees Fahrenheit without frost damage. If such bacteria were as successful outside of the laboratory, Lindow reasoned, many millions of dollars of lost crop productivity could be spared, millions more in frost-protection costs saved, and entirely new options in crop varieties and agricultural practices would become possible simply because of the decreased frost sensitivity of crop plants. First, however, Lindow had to test his altered bacteria in the field; if successful there, they would be worth a lot of money.

In fact, Lindow's research had already attracted the attention of at least one biotechnology company, Advanced Genetic Sciences (AGS), a Connecticut-based firm interested in both the agricultural applications of the so-called "ice-minus" bacteria, and the development of other genetically altered strains of these bacteria that might be used for an opposite purpose—artificial snow-making. Advanced Genetic Sciences was funding some of Lindow's work and looking forward to his field test of *P. syringae* and *E. herbicola*. But because his research was in part federally funded, Lindow had to receive the approval of the Recombinant DNA Advisory Committee of the National Institutes of Health (NIH/RAC) before he could run the test.

After a year of review and evaluation, the NIH approved Lindow's request, and a September 1983 test date was set. But on September 14, a group of citizen and environmental interests based in Washington, D.C.—including Jeremy Rifkin and the Foundation on Economic Trends, the Environmental Task Force, Environmental Action, and Michael W. Fox of the Humane Society—filed suit against the NIH for approving the project. Among other things, the suit charged that the NIH had not conducted an adequate assessment of the potential environmental risks of Lindow's field test, and that the NIH committee that had given its approval contained "no ecologists, botanists, plant pathologists, or population geneticists." "The National Institutes of Health," said challenger Rifkin, "has been grossly negligent in its decision to authorize the deliberate release of the first genetically engineered life forms."

Rifkin is a long-time opponent of genetic engineering and had been active in the genetic controversies of the mid-1970s. In 1983, his book on the subject, *Algeny,* was released, and by June of that year he had formed a coalition with a cross-section of clergy urging a ban on human genetic-engineering research. In September, however, Rifkin's concern was the NIH's Recombinant DNA Advisory Committee (RAC) and its lack of environmental review procedures. "Like the Nuclear Regulatory Commission," Rifkin said, "the National Institutes of Health have attempted to convince the media and the public that its procedures for evaluating and authorizing recombinant DNA experiments in the environment have been scientifically rigorous." Yet, said Rifkin, there had been instances in which the NIH committee had to "send out" for a botany textbook when reviewing a proposal that dealt with a new genetically altered crop variety.

The reaction of scientists and industry to the "Rifkin lawsuit" ranged from strong to guarded. For a time, the Industrial Biotechnology Association toyed with the idea of filing an *amicus curiae* brief on the side of the NIH, but that strategy was abandoned because it might make the new industry appear to be a too-visible, self-serving advocate in the public eye. And there was also some name-calling. One of Lindow's colleagues at the University of California referred to Rifkin as a one-man "super-government" making life miserable for the new biotechnology industry. *Nature* magazine ran a September 1983 story on the lawsuit with the headline "Rifkin's regulatory revivalism runs riot." However, Jeremy Rifkin and the other environmental objectors weren't the only parties raising questions about the NIH review process.

Tennessee Democrat Albert Gore, Jr., then chairman of the House Science and Technology Subcommittee on Investigations and Oversight (and now a U.S. senator), had held hearings in June 1983 on the questions of releasing genetically altered organisms into the environment, how federally funded biotechnology experiments were being reviewed, and how the new products of genetic engineering might be regulated at the point of commercialization.

On the basis of those hearings, Gore wrote to the NIH in October 1983 expressing concern over the RAC's review process. Gore noted concerns among the witnesses he had heard that the RAC's process for evaluating field-test proposals was "flawed," and stated that he was personally concerned that the RAC had already evaluated proposals "without the benefit of either a terrestrial ecologist . . . or a microbial ecologist." Moreover, added Gore, "I am greatly concerned about the absence of a legal basis for the actions of the RAC in reviewing requests for deliberate releases from commercial biotechnology firms." NIH's committee was only an advisory body, and had no statutory basis for regulating the private sector.

In the case of Lindow's experiment, the Rifkin suit suggested the dramatic possibility that the frost-preventing bacteria might be swept into the upper atmosphere, disrupting the natural formation of ice crystals, ultimately affecting local weather patterns and possibly altering global climate. Could California's snow pack be affected, and if so, might that impact the state's elaborate water system, the ability to generate hydroelectric power, and the allocation of water for agricultural irrigation? Although some experts scoffed at these scenarios, ice-nucleating bacteria have been reported in significant numbers in the atmosphere.

Michael W. Fox, a senior scientist with the Humane Society, and another of the plaintiffs in the court action, said that the proposed test "could have adverse effects on animal life since frost kills or arrests the growth of certain microorganisms" that cause disease in animals. Eugene Odum, a nationally recognized ecologist, said in another affidavit that "long and sad experience . . . dictates great caution for any and all proposals for releasing organisms that are foreign, and hence unadapted, to the ecosystem." Odum added that "Introducing microorganisms is especially hazardous because they possess a high reproductive potential and their ecological interrelations with other organisms are not . . . well understood." He also noted that "higher plants, such as trees or crops, are very slow to develop immunity to new microorganisms."

Peter H. Raven, director of the Missouri Botanical Garden in St. Louis, wrote in his affidavit that several hundred endangered and threatened plant species grew in the mountainous area of northern California and Oregon near where the bacteria were going to be tested. "No one can predict," said Raven, "whether [the altered] bacteria [will] reach the populations of these wild plants and endanger them." Raven was also worried that the frost-inhibiting microbes—should they colonize native plant populations—might extend the plants' "photosynthetically active season," altering the natural boundaries of the growing season.

Other scientists, such as Iowa State University economist Robert Wisner, extolling the agricultural benefits of a frost-inhibiting microbe, saw it as extending the growing season for crops in places like Minnesota by ten to fourteen days. "It might stretch the size of the Corn Belt to the north," he said.

Steven Lindow, meanwhile, stuck to his guns, arguing that his experiment posed no threat to the environment. His half-acre test plot near Tulelake, California, was to be surrounded by a six-foot earthen wall. Lindow said he didn't think there was "any possibility this project will have the dramatic effect some people say it will," adding that his group had reviewed it thoroughly and prepared an environmental assessment for the university's review committee. "We just don't foresee this bacteria spreading at all," he said, confident of his field trial.

Lindow explained that for three years he and his colleagues had conducted field trials with chemically mutated bacteria that were similar to the genetically altered ones, with no ill effects on the environment. After three months, he said, the native populations of the bacteria reasserted themselves, while the chemically mutated ones did not spread to nearby plants and soon died anyway. Lindow believes that the excised-gene approach to controlling frost-forming bacteria on plants is superior to the chemical approach, the use of antibiotics, or the use of another method which employs viruses that invade and kill the bacteria. He feels that the genetic route is more precise, less expensive, safer for the environment, and more effective than the other alternatives. All that he and his colleagues have done to the bacteria, he says, is to remove one gene—something that happens often in nature when ultraviolet light hits bacteria.

INCREASING THE ODDS

In nature, the occurrence of a single mutation in a plant or a microbe is a relatively rare event, perhaps one in a million statistically. A second mutation, arising from a successful population of the first mutant—called a double mutation—is even rarer, and perhaps unlikely in nature, put at a probability of one in a billion. But all that changes when scientists begin purposely selecting organisms for their mutations in the laboratory using biotechnology and genetic engineering. What may have been a rare event in a natural environment with competing organisms and other biological pressures can in the laboratory become a predictable, controllable, and repeatable occurrence.

Double, triple, and quadruple mutants "can be made to appear with ease," says Cornell University microbial ecologist Dr. Martin Alexander. Bioengineering, then, can increase the mutational frequency of organisms far beyond that of the natural environment. And such laboratory-bred mutants may express valuable characteristics—commercial or otherwise—that may never have been expressed in nature because of the odds against the survival of new organisms. "We *can* create new genotypes which *do* have survival value even if the potential for the expression of that survival value would not ordinarily have occurred," explains Alexander. But what happens when such laboratory-based mutations are fashioned for use in the outside environment? No one seems to know for sure.

Cornell's Alexander says we don't know much about the survivability of introduced microorganisms in the environment generally. "Organisms will in fact persist," he says, "some will, some will not—but the probability of persistence for an unknown organism . . . is not zero." The problem with genetic engineering, he explains, is that "we have no good information on the fate of new organisms introduced into the environment. And we've neglected gathering information on fates of existing organisms. . . ."

This view is echoed by Dr. Anthony Robbins, former director of the National Institute for Occupational Safety and Health and now a professional staff member of the U.S. House of Representatives Committee on Energy and Commerce. "In most environmental programs," explains Robbins, "we have relied on predictive toxicology to estimate risks and thus safety. Where a chemical is tested in a laboratory and it behaves like other chemicals that cause damage, one can predict what is likely to be dangerous. . . . For chemicals it is likely that toxicological predictions will continue to improve, letting us protect people from hazards before anyone has been exposed. No such battery of predictive tests exists to tell us how novel (genetically engineered) organisms that are released to the environment will behave and whether they will pose a danger in the future. Before we can define the *safety* that we seek, and do it prospectively, we need a well developed science of predictive ecology. . . ."

Don Clay, the EPA's assistant administrator for pesticides and toxic substances, points to the technical problems of gearing up for environmental assessments of genetically altered substances. "There are almost no accepted methodologies for evaluating the safety of genetically engineered products," he observes, explaining that "the risk assessment tools and data we have used for inanimate chemical substances [at the EPA] will not apply in the case of organisms." Yet, at a minimum, says Clay, the EPA "must develop methodologies to evaluate environmental fate, human and environmental exposure, and potential environmental and health hazards of genetically engineered organisms," all of which are "still several years away."

Meanwhile, plants, microbes and livestock breeds are being "genetically enhanced" to perform one or more specific tasks. Genes are being spliced into crop varieties to make them more resistant to a particular disease, or a salty soil, or a particular kind of herbicide. In other cases, as in Steven Lindow's microbes, genes are being removed to alter the microbe's behavior and its interaction with the environment to facilitate one specific activity. Such genetic alterations will make crop plants, livestock, and microbes more precise, more specialized, and in some cases more powerful than they were in their natural state. And even though Steven Lindow's microbes contain one less gene than their natural counterparts, they are, nonetheless, being adapted genetically to grow, survive, and displace their frost-inducing, wild relatives.

Because such organisms are screened and/or designed for persistence in their particular agricultural missions, once they are introduced into the environment, they may not be easily contained. On top of this is the fact that superplants and supermicrobes will be introduced into what ecologists call "simplified environments" consisting of crop and livestock monocultures—environments in which one small genetic change can produce very great and swift consequences.

Moreover, new genetically enhanced organisms are not only being selected and designed for persistence or tenacity in performing a desired task, but they are also being created to perform entirely new roles in the environment. And with the ability to cross species barriers, entirely new "genotypes" will be created. Such new organisms, when introduced into the environment, may confront ecological voids in which there are no competing organisms. This kind of genetic enhancement, explains Colorado State University plant pathologist Ralph Baker, "leads to new ecological balances. The outcome of these new balances has been little studied and is largely unknown."

OF SUPERCROPS & FRAGILE SYMBIOSIS

Although genetically altered strains of crops and livestock may not have the capacity to create environmental problems on the scale that multiplying microbes might, new supercrops and superlivestock breeds may bring about secondary or indirect demands on the environment. Efforts underway to genetically boost "photosynthetic efficiency" in crops, for example, could place heavier demands on the environment through the resources needed to sustain such crops. Most crops convert about 1 percent of the energy they absorb from the sun, but through genetic engineering it might be possible to increase the amount of sunlight absorbed, and so, increase yield. Yet genetically increasing a plant's photosynthetic efficiency may also mean increasing the demands of that plant for water and nutrients, generally resulting in heaver resource-use requirements.*

In the microbial realm, there are some special considerations

*There are also concerns about genetic dabbling in the nitrogen cycle, although these do not pertain directly to agricultural products or applications. Nitrogen is one of the basic ingredients of all life; an essential element in growth of all kinds. In the grand scheme of things, nitrogen is "cycled" throughout the environment in a carefully choreographed biological interchange between atmosphere and earth. In this cycle, nitrogen is facilitated in its use by some plants by soil microbes that "fix" it in a useable form, and by other microbes which "denitrify" it in return to the atmosphere. However, in this cycle, some nitrogen is returned to the atmosphere in the form of nitrous oxide. Scientists have recently learned that nitrous oxide—particularly that being contributed by industry in the burning of fossil fuels and agriculture in the manufacture of synthetic fertilizers—is contributing to the breakdown of the ozone layer, which shields the earth from the harmful extremes of the sun's ultraviolet radiation. But what about the introduction of genetically engineered microbes designed to speed up the process of nitrification, in which ammonium is broken down to nitrates, and nitrous oxide is given off? Such microbes have been proposed for use in connection with waste water management. What if such microbes become established in the soil? Would they significantly add to the amount of nitrous oxide in the atmosphere?

when it comes to introducing genetically modified organisms. It is known, for example, that bacteria can transmit certain genetic characteristics from one species—and even from one genus—to another.* Because of this ability, which may not be unique to bacteria, ecologist Frances Sharples of the Oak Ridge National Laboratory says, "it may be very difficult to keep inserted genes [those that are put into bacteria by way of gene-splicing] isolated in single bacterial strains." Some scientists say that a bacterium genetically altered to kill gypsy moths, for example, could possibly transmit that trait to another bacterium that lives harmlessly inside honeybees.

"The line that distinguishes symbiosis, a beneficial relationship, from pathogenesis is indeed fine," said Du Pont's director of life sciences research, Ralph Hardy, before a Congressional subcommittee in 1983. Hardy noted, for example, the "genetic similarity" between the *Agrobacterium,* a tumor-forming pathogen in plants, and *Rhizobium,* a nitrogen-fixing bacterium that helps the growth of legumes such as soybeans. "Therefore," he told the subcommittee, "means must be established to assure that a new organism created to aid agricultural production . . . does not [become] a pathogenic form."† But whether those means can be established by industry or government, and indeed, what role the federal government should play in overseeing the genetic technologies that will make new organisms for use in the environment, remain unresolved.

*Bacteria do this by means of plasmids, tiny bits of extrachromosomal genetic material. In the late 1970s, for example, it was discovered that plasmids in some strains of bacteria carry genes for resistance to certain antibiotics used widely in livestock feeds, and that these bacteria were transferring that resistance to other species of bacteria by means of plasmids. Moreover, even though there are some differences between the bacterial strains that inhabit or invade humans and animals, the genes for antibiotic resistance can be transferred among and between some of these strains by means of plasmids.

†In a February 1984 report entitled "Environmental Implications of Genetic Engineering," prepared by the staff of the House Science and Technology Subcommittee on Investigations and Oversight, an obscure forestry experiment in New Zealand involving the fusion of two microbes was highlighted as an example of how the fine line between symbiosis and pathogenesis might be crossed. The experiment involved the work of two plant scientists, K. L. Giles and H. C. M. Whitehead, who were studying the association of a particular kind of fungus that lived in a natural symbiotic relationship with the pine tree, *Pinus radiata.* The fungus, named *Rhizopogon sp.,* lived on the tree's roots. Giles and Whitehead thought they could improve tree growth by giving this fungus a nitrogen-fixing ability. Using a technique called cell fusion, the scientists combined the genetic material of the fungus with that of a nitrogen-fixing bacteria called *Azotobacter*

THE POLITICS OF REGULATION

Following the June 1983 hearings held by the House Science and Technology Subcommittee on Investigations and Oversight, and the September litigation stopping Steven Lindow's field test, it became apparent that there were great shortcomings in the NIH environmental review process, and that there were no laws that dealt specifically with genetically engineered substances. Moreover, the history of the NIH's role with the fledgling bioengineering industry is one marked by a gradual relaxation of existing regulations.

The NIH took its first cautious steps toward regulating biotechnology in October 1974 when its "DNA Molecule Advisory Committee" was established to draw up guidelines governing recombinant DNA research. This activity came largely in response to a few scientists who had requested guidance from the National Academy of Sciences (NAS) in 1973. A year later, an NAS-appointed panel called

vinelandii. By fusing the bacterium with the fungus, the scientists hoped the fungus would acquire the bacterial genes and the ability to fix nitrogen. The resulting creations were modified strains of the fungus with an unknown potential for nitrogen fixation.

In the laboratory, Giles and Whitehead applied this new hybrid fungus to some pine-tree seedlings. Later, some of the test seedlings exposed to the new fungus died. A post mortem of the seedlings revealed that the new fungus had penetrated the cells of the root cortex, while no such intercellular growth was found in the seedlings grown with control strains or with wild strains of the fungus. "The [root] cells into which the modified strain had entered," wrote the two scientists in a 1977 report of their experiment, "appeared dead and were empty of cytoplasm. Whether this was due to the presence of the fungus, or whether the fungus had merely entered the dead cells is not known." At any rate, the scientists feared that they might have created a pathogenic microorganism, and destroyed the plants and remaining organisms.

The House subcommittee report, citing the Giles and Whitehead experiment, explained: "In seeking to modify this fungus, the scientists combined two normally nonpathogenic microorganisms. This combination theoretically should have produced a harmless fused organism capable of bringing about the desired result. In contrast, however, one strain of the newly created recombinant fungus was in fact pathogenic." In an earlier October 1983 letter to the NIH, Representative Gore raised the New Zealand experiment, noting that "The incident appears to refute the generally stated contention that nothing can be produced by genetic engineering that is more dangerous than the starting materials." However, both the NIH and the American Society for Microbiology tended to dismiss the Giles and Whitehead incident, charging it to poor study design and lack of good statistical controls. Yet the subcommittee staff believed nonetheless that the Giles and Whitehead incident provided an example of what might happen when using such techniques on plants and microbes in the environment.

for a moratorium on certain kinds of DNA experiments, and in February 1975, American scientists, gathered at the Asilomar Conference in California, called for the "self regulation" of genetic research by the scientific community. A report from the Asilomar Conference became the basis for the NIH guidelines, which were formally released in June 1976, and emphasized the "biological containment" of laboratory experiments.

During 1976 and early 1977, a public controversy erupted in response to the proposed operation of some DNA research labs. At this point in time, biotechnology is a laboratory-based industry, and public concerns are primarily those focused on accidental escape of microbes from those facilities. Local ordinances were adopted in Cambridge, Massachusetts, and regulatory bills were drafted in the New York and California legislatures. In 1977 and 1978, some sixteen bills for regulating genetic research were introduced in Congress, including one from the Carter Administration in which the NIH played a formative role. Although a few of these measures actually were passed by both the House and Senate, no final regulatory legislation ever emerged from Congress—an outcome largely due to the lobbying of scientific groups and a few genetic-engineering companies. Despite the brief flurry of legislative activity during 1977 and 1978, the NIH guidelines remained the sole mechanism for regulating genetic research.

As they first emerged under the growing public outcry for safety—with the added threat of federal legislation in the wings—the NIH guidelines were fairly stringent, but that soon gave way to industrial and scientific accommodation. In 1978, for example, Eli Lilly and Genentech, then involved in a joint venture to commercially produce genetically engineered insulin, objected to an NIH rule limiting laboratory batches of genetically produced substances to ten liters, a volume far too small for commercial production. Subsequently, the Pharmaceutical Manufacturers Association began pressuring the NIH's parent department—then the Department of Health, Education and Welfare (HEW)*—to propose new recombinant DNA guidelines that would be voluntary for industry.† This proposal also

*Now the Department of Health and Human Services.

†In 1979, the pharmaceutical industry also blocked an FDA proposal to make the sale of any genetically engineered drug contingent on mandatory compliance with the NIH guidelines.

included the creation of "biosafety committees"—local committees for each company or research institution—that would register with the NIH and approve research projects rather than having NIH do the supervision directly.

Joseph Califano, then HEW secretary, pressured the NIH committee to adopt the new voluntary regime, exempted the Genentech-Lilly venture from the ten-liter requirement, and called for the adoption of new guidelines to accommodate large-scale commercial production of recombinant DNA bacteria. Some of the NIH committee members balked at the new voluntary scheme, and a few walked out of a March 1980 committee meeting protesting the HEW-backed revisions. Nevertheless, the new rules then went into effect. By September 1981, the NIH had dropped its plans to have facilities doing large-scale fermentation with genetically altered bacteria comply with its guidelines.

Ironically, the early NIH guidelines had prohibited the "deliberate release" of organisms containing recombinant DNA into the environment, but by 1978 "waivers" of this prohibition were provided, and by 1982 "review and approval" procedures were substituted for the waiver proviso. The NIH guidelines were revised and amended to specifically allow for the environmental release of genetically altered crops, livestock, and microbes. Moreover, none of the NIH field-test approvals—even those approved for federally funded research*—complied with the National Environmental Policy Act.

The NIH guidelines—as they initially emerged and evolved over an eight-year period—were primarily oriented toward the biomedical side of biotechnology. The NIH committee focused almost exclusively on containing the microbes of DNA research within the laboratory, and its experts came predominately from the medical and pharmaceutical sides of microbiology. In fact, the NIH com-

*In addition to Steven Lindow's experiment, NIH had also approved two other field tests; however, Lindow's was the first to reach the actual testing stage. In August 1981, the NIH committee approved a field test of corn plants modified by recombinant DNA techniques for Dr. Ronald W. Davis, a professor of biochemistry at Stanford University. In April 1983, NIH approved an experiment to field test genetically modified tomato and tobacco plants (transformed with *E. coli* and yeast DNA, using pollen as a vector) for Dr. John C. Stanford, a professor of plant genetics at Cornell. As of late 1984, neither of these experiments had been field tested.

mittee was never designed to be a regulatory body* and was not scientifically equipped to deal with the explosive parade of genetically engineered products that would affect the larger environment outside the laboratory.

In July 1983, the Environmental Protection Agency (EPA) moved to claim jurisdiction over the regulation of certain genetically engineered products. The EPA was claiming that it had the jurisdiction to regulate genetically engineered substances under the Toxic Substances Control Act, but expected that industry would challenge that interpretation in court.† Others thought that in some cases pesticide laws would apply. Still others wondered where genetically altered plants‡ and livestock might fit in. Would the USDA and FDA have a role?

Some industry spokesmen worried about EPA's projected new role. "I have a feeling," said Winston Brill of the Cetus Corporation, who had left the NIH committee only two months earlier, "that EPA won't be so adaptable and could stifle the movement, both in research and commercial-wise." The NIH "already regulates the sorts of functions EPA is starting to look at," said Harvey Price, executive director of the Industrial Biotechnology Association, a thirty-member trade organization of companies such as Genentech, Du Pont, and Eli Lilly. "It isn't clear to a lot of people," added

*Anthony Robbins, a former director of the National Institute for Occupational Safety and Health and now with the House of Representatives Committee on Energy and Commerce, has noted, for example: "As a research institution, NIH has always rejected the idea that it might perform a regulatory role. It has even been queasy about advising regulatory agencies on purely scientific matters. It was ironic that, faced with moderate pressure to regulate, and strong pressure to protect universities from regulation, NIH chose to set up its own 'regulatory' program to assure proper containment of genetically engineered organisms. It chose to regulate only its own constituency and only to the extent that was tolerable to the regulated. It created a voluntary program of regulation."

†The Toxic Substances Control Act governs all new chemical substances, and EPA's Director of the Office of Toxic Substances, Donald R. Clay, argued that certain genetically engineered substances would qualify as chemicals. "If you look at the definition of 'chemical' under Section 3 of the [Act]," Clay told a House subcommittee, "I think DNA meets it."

‡Jeremy Rifkin would later argue, for example, that a plant genetically engineered to resist pests should be treated like a pesticide. "In the past, a pesticide was a spray," he said. "Now a pesticide can be a piece of genetic information imprinted on a plant."

Price, "that there's any need for the EPA to look over NIH's shoulder on these kinds of issues."

Some major corporations and biotechnology companies began maneuvering to keep the NIH in the driver's seat. Eli Lilly, Chevron, Schering-Plough, Genentech, Genex, and the Industrial Biotechnology Association all urged the NIH to expand its regulatory role rather than diminish it, and to continue reviewing private-sector proposals. Industry feared that if the NIH withdrew from its quasi-regulating of genetic engineering over the "environmental release" question, commercial progress in biotechnology might be thwarted for years while Congress, the federal agencies, and the courts decided exactly what the proper regulatory course should be. Moreover, without the appearance of some kind of governmental sanctioning, every commercial release of a genetically altered substance could become a potential legal liability.

By February 1984, Representative Albert Gore's subcommittee issued a staff report entitled "Environmental Implications of Genetic Engineering." Among other things, the report called for the creation of a federal interagency task force to review environmental release proposals, establish a risk assessment program, and develop guidelines for the environmental release of genetically engineered substances. "No deliberate release should be permitted by EPA, NIH, USDA or any other federal agency," said the subcommittee's report, "until the potential environmental effects of the particular release have been considered by the interagency review panel." Moreover, the report recommended that the NIH "cease its practice of evaluating and approving proposals for deliberate releases from commercial biotechnology companies," and that the EPA's regulatory role be expanded.*

These issues came into sharper focus when Steven Lindow and the University of California tried to go forward a second time with

*In an October 1984 speech before the Animal Health Institute, Robert B. Nicholas, Chief Counsel and Staff Director of the Gore subcommittee, and a key player in the debate, explained the report's three chief conclusions: (1) that there are many potential benefits to be derived from the release of genetically engineered substances into the environment, but a "small number of such releases may present a small risk of significant adverse consequences"; (2) that the ability to do a risk assessment—to describe and evaluate the likelihood of any particular risk—is not well developed; and (3) that no federal agency has both the expertise and legal authority to make decisions concerning the release of a genetically altered organism into the environment on a commercial scale.

their field test of Lindow's microbes in May 1984, but were enjoined from doing so by Judge John J. Sirica, responding to a request for a restraining order by Jeremy Rifkin and his co-plaintiffs.

In his review of the Lindow experiment and NIH's review of that proposed field test under the National Environmental Policy Act, Sirica found that NIH's approval was "without benefit of a specific or a general investigation into the environmental hazards of deliberate release experiments." While Sirica's decision was based on the larger question of whether NIH should do environmental impact statements for the Lindow and other out-of-the-lab experiments, he ruled that NIH should not approve any other federally funded experiments until such procedure was adopted. But Sirica's ruling did not prohibit NIH from approving similar experiments by private industry.

Subsequently, on June 1st, the NIH advisory committee recommended the approval of an experiment identical to Lindow's for Advanced Genetic Sciences (AGS)—the company that had financed part of Lindow's research. Companies such as AGS are technically not required to have their experiments cleared by the NIH before conducting them, but industry has used the NIH imprimatur as a kind of legal shield to protect itself against liability (in addition to AGS, the Cetus Corporation and Biotechnica International had also brought proposed field tests to NIH). Inadvertently, a double standard was created, with industry facing less stringent standards than federally funded projects.

By this time, however, the Reagan White House had become involved in the regulatory tussle. Christopher DeMuth, an Office of Management and Budget (OMB) official—with help from the Departments of State and Commerce—began to challenge EPA's claim of regulatory authority. DeMuth, described by *Biotechnology* magazine as a person who "has made a career in Washington working to deregulate industries or to create incentives for voluntary compliance with such goals as clean air and water"—was sympathetic to the view that regulation hampers innovation. In a March 12, 1984, memo to some cabinet officials, DeMuth noted that heavy-handed regulation might hurt the biotechnology companies, which he called "extraordinarily innovative" and "unusually sensitive to regulatory delays and costs."

DeMuth and Secretary of Commerce Malcolm Baldrige succeeded in wresting control over the biotechnology issue from the

EPA. The high-level Cabinet Council on Economic Affairs, which Baldrige then chaired, assumed control of the regulatory question in May 1984. Prior to that shift, the issue had been dealt with in the Cabinet Council on Natural Resources and the Environment, chaired by William Ruckelshaus, EPA Administrator. Baldrige, for one, was concerned that EPA regulation would hinder American biotechnology companies in the economic race with Japan to commercialize certain products. George A. Keyworth, Reagan's science adviser and head of the White House Office of Science and Technology Policy (OSTP), was also of the view that the administration should not do anything that might stifle the technology. Others in the White House were also concerned with sending the wrong signals to Wall Street.

Meanwhile, some interests in the biotechnology industry were already saying that environmental interests were threatening the American lead in biotechnology with regulatory overkill. "U.S. scientists have held a commanding lead in the exciting field of gene splicing," said *Chemical Week* in a February 1984 editorial. "But just as the technique promises to pay off in the form of a host of commercial products, a coalition of environmentalists is demanding new restrictive studies that would seriously delay the introduction of such products. The environmentalists' demands are ill-advised; their fears, ill-founded." (The magazine admitted, however, that recombinant-DNA technology had "grown faster than has the necessary regulatory policy," and that few federal regulatory agencies were "equipped to cope with genetically engineered products," but felt that the NIH could handle the situation.)

"Any move now to curtail [biotechnology] would erode U.S. lead," added the magazine, arguing that the United States had the "toughest laws in the world" on drugs and food additives, and was being "even more conservative" about recombinant-DNA and its products. *Chemical Week*'s concern was that "other, less scrupulous nations might release some really bad actors into the environment," noting that any dangerous products of recombinant-DNA would have "no respect for national borders."

Subsequently, what began to emerge from the White House was a proposal for a "super-RAC"—a larger NIH review panel that would comprise a body of biotechnology experts to provide advice to appropriate federal agencies. The White House and industry were eager to avoid at all costs any new law for regulating biotechnology.

Bernadine Healy Bulkley of the White House Office of Science and Technology Policy argued at the time that it would be "premature to jump into new law." In Congress, some members, such as Representative James Florio of New Jersey, introduced legislation to define genetically altered microbes as chemicals under TSCA, which could throw much of the jurisdiction over genetic engineering to EPA. Other congressmen and senators were just beginning to discover the issue.

However, some major corporations with biotechnology products nearing the market were becoming impatient. "It is unacceptable to leave biotechnology in a regulatory limbo," said Du Pont chairman Edward G. Jefferson at a September 1984 dinner preceding the dedication of the company's $85 million life sciences complex in Wilmington, a portion of which would house biotechnology laboratories. While recognizing the need for "the appropriate oversight" of industrial biotechnology by government, Jefferson noted that "no major agency has yet established guidelines or protocols to deal with the special concerns that surround the movement of a genetically engineered product from the laboratory, through development, to the marketplace." Jefferson exhorted the executive branch to "coordinate regulation of biotechnology among appropriate existing agencies."

Monsanto, also supporting the need for regulation of genetically engineered products outside of the laboratory "under existing statutes," acknowledged that the situation was a bit confusing and wondered which agencies would regulate certain products. "Should the EPA or the USDA oversee experiments involving seeds, living plants, and microorganisms?" the company asked in its 1984 brochure entitled "Genetic Engineering: A Natural Science." "Confusion about government agency authority could delay the commercialization of beneficial new products," explained the company, "leaving an opening for other nations to take the lead. . . . It could also deprive the United States of significant social benefit and the opportunity to preserve and create thousands of jobs." Monsanto, a company used to working with EPA, had endorsed a plan by the agency to coordinate biotechnology regulation among federal agencies, and was ready to submit some of its genetically engineered microbial pesticides to EPA for review under existing pesticide law.

By October 1984, the White House; fifteen federal agencies including the Departments of State, Commerce, Health and Human

Services, EPA, FDA, and USDA; the National Academy of Sciences; and at least four Congressional Committees were looking into the issue of regulating biotechnology or related questions concerning its environmental impact. Long political and legal battles loomed ahead on the governmental and special-interest playing field that is Washington, D.C.

Whatever finally happens on the regulatory front, one thing is clear: the race is on among a myriad of corporate interests and scientists to capture the billion-dollar markets and scientific kudos promised by genetic engineering. For corporations there is great pressure to be the first in the market, to show profitability, and to impress the investment community; for scientists there is the possibility of a Nobel prize or a lucrative discovery; and for the federal government there is great political pressure to be the world leader in biotechnology. But as it now stands, there is virtually no governmental apparatus in place, nor any clear legal authority, for reviewing or regulating the commercially produced substances of genetic engineering that will soon be introduced by the thousands into the environment.

"BUT A FEW WILL"

The conventional wisdom among scientists and businessmen—especially those now on the cutting edge of commercial biotechnology—is that there are more benefits to come with biotechnology than there are risks. That may well be true. Yet there is a long list of once highly vaunted new technologies—including nuclear power, sulfa drugs, antibiotics such as penicillin and streptomycin, synthetic urea, DDT, and the internal combustion engine—that proved to hold undesirable and negative consequences for society once they were developed and put into widespread use.

"During the initial development of the chemical industry, or at the time when the use of pesticides was just beginning," explains Cornell's Martin Alexander, "little or no hazard existed for society . . . and no threat was posed to major natural ecosystems. But as those technologies became more widely used and moved in new directions, the environmental and health problems became quite apparent."

In addition to society's record of later-discovered side effects for once highly-vaunted technologies, there is also a well-documented history in pharmacology and medicine of harmless organisms that convert to pathogenic or harmful forms after undergoing simple genetic changes. In the cases of *Mycobacterium tuberculosis, Staphylococcus aureus,* and *Neisseria meningitidis,* for example, each organism can undergo simple genetic changes which make them more of a disease problem for humans. Moreover, the recent evidence on the genetic exchange (via plasmids) of antibiotic resistance among bacteria common to both man and livestock, poses even larger questions for genetically engineered changes to these and other populations of bacteria. And with genetic engineering, as in any other technology, there is the human factor to consider as well as the prospect for accidental creations.*

In the realm of ecology, history shows that when exotic species of plants and other organisms are introduced into new environments, the results, like that of the gypsy moth and chestnut blight, can sometimes be disastrous, both ecologically and economically. While many introductions of such species do not create a problem, or in fact prove beneficial to society, the point is that the results are unforeseeable. "The consensus amongst ecologists," says Oak Ridge ecologist Frances Sharples in her 1981 study, *Spread of Organisms with Novel Genotypes,* "is that the outcome of a species introduction is not predictable. Rather, the introduction of exotics

*In the summer of 1980, Ian H. Kennedy, a scientist at the University of California at San Diego, thought he had cloned copies of a virus called *Sindbis,* a fairly harmless bug that lives inside mosquitoes. In fact, Kennedy had unknowingly duplicated another virus named *Semliki* forest virus, a virus infectious enough to be among those agents banned by the NIH in federally funded cloning experiments. According to Kennedy, a laboratory vial containing the infectious *Semliki* virus being shipped from England, had broken in transit and, unbeknownst to him, contaminated his stock of *Sindbis* virus. Since the two viruses appear somewhat similar under a microscope, initial detection of the mistake was not made until a few students noticed some odd behavioral patterns in the virus.

There are also suspected instances of accidental creations resulting from scientists using cloning and cell fusion techniques. For example, scientists working in the laboratory with cell cultures may have created a canine *parvovirus*—which became a deadly disease among puppies in recent years. It turns out that the canine *parvovirus* is a close relative of a cat virus, feline *leukopenia,* a common blood disease. It is suspected that scientists working to produce a vaccine for the cat virus, but who were growing the *leukopenia* virus in dog cells, accidentally "invented" the canine *parvovirus* in the process.

is viewed as a game of chance with the possibility of both high risk and rewards." And while some scientists argue that the experiences with exploding populations of introduced exotic organisms cannot be used to foretell how "more exactly" (and therefore, they say, more predictable) genetically engineered organisms will behave, others think it provides a reasonable scientific record of how ecological systems respond to new genetic introductions generally.

Yet in the tiny world of microorganisms, we don't know much about "ecology," or what even *non-engineered* organisms already do in the environment, let alone fortified, genetically altered ones bred for persistence and use in large numbers. Inside the organism, throughout the plant and animal kingdoms, we know even less about the cellular goings-on of genes, their movements among organelles, their expression, and their evolutionary history as interrelated entities. All of this leaves some uncertainty about predicting what might happen with genetically altered organisms ecologically, their secondary demands on the environment and society, and what the full range of economic costs as well as benefits might be for a particular introduction.

"In general," says Cornell's Martin Alexander, "whenever a new technology is introduced, there is a possible hazard. My belief is that the same likely is true of genetically engineered organisms. Most of them introduced into the environment won't survive. But a few will. Most that survive won't have an effect. But a few will. And most of those that have an effect won't do any damage—but a few will."

Similarly, the 1984 report issued by Representative Albert Gore's Science and Technology Subcommittee on Investigations and Oversight characterized the potential dangers of genetically altered organisms as being "low probability, high consequence" events. In the subcommittee's words, "while there is only a small possibility that damage could occur, the damage that could occur is great." And the subcommittee report warned that "Assessing the risks presented by deliberate releases should not be simply a game of biological roulette. If it is, many benefits of this new technology will be lost, and disastrous environmental consequences could well be permitted to occur."

CHAPTER 13

GREEN REVOLUTION II

> The revolution is green only because it is being viewed through green-colored glasses.
> —*William C. Paddock (1970)*

In the summer of 1981, agents from Prime Minister Indira Gandhi's government were quietly dispatched to the United States to buy $260 million worth of wheat to shore-up India's dwindling grain reserves. Two poor harvests—the result of erratic monsoons—had dramatically reduced that nation's grain reserves from 20 million tons in 1979 to only 3 million tons. Gandhi, however, was not anxious to have the world know she was in the market for grain; only a few years earlier her government and international organizations such as the World Bank had proclaimed India to be "self-sufficient" in agriculture.

Indeed, for four years before this purchase, India had been self-sufficient in food grains, due in part to something called the "Green Revolution"; a revolution in plant breeding that produced new high-yielding wheat and rice varieties that helped India and other countries feed millions of people—but not without a price. Critics said that the Green Revolution was in fact part of the problem, making India and other countries like it more, not less vulnerable. For the high-yielding seed of the Green Revolution was only high yielding when plied with the ingredients of modern agriculture; ingredients which worked as a system of technology sustained by hefty infusions of energy and capital.

Today, the biogenetic revolution in agriculture is being touted as a second Green Revolution. And once again, some critics are wondering how this new Green Revolution will differ from the last. They wonder if it will bear truly "appropriate technologies" for nations in need, or if instead it will bear the seeds of hidden costs that create more problems and greater vulnerability for those nations' agricultural systems.

The crucible of the first Green Revolution, however, and the necessary precursor to understanding its further extension through biotechnology, is the Mexican countryside of the 1950s where one plant breeder practiced his science.

THE APOSTLE OF WHEAT

In October of 1970 Norman Borlaug was standing in the middle of a Mexican wheat field—a field of semi-dwarf wheat born of his own hand to be exact—when his wife Margaret brought him the news that he had just won the 1970 Nobel Peace Prize. According to the official announcement, the Nobel Committee had decided to award the Prize to Borlaug "for his great contribution toward creating a new world situation with regard to nutrition . . . The kinds of grain which are the result of Dr. Borlaug's work speed economic growth in general in the developing countries."

Borlaug, the tough-minded son of Norwegian immigrant farmers from Cresco, Iowa (a town called "little Norway"), was the first agricultural scientist ever to be selected for the prestigious international award. After more than a quarter-century of hard, grinding agricultural work at the Mexican research center, and with billions of his wheat plants being used around the world to help keep famine at bay, Borlaug had made a contribution to humanity and world peace that few ever dream possible. At the official Nobel ceremony in Stockholm in December 1970, Madame Aase Lionaes, head of the awarding Nobel committee, praised Borlaug as an "indomitable man who fought rust and red tape . . . who, more than any other single man of our age, has provided bread for the hungry world . . . [and] who has changed our perspective."

Norman Borlaug came to Mexico in 1944, spirited away from a classified wartime laboratory job he held at Du Pont by Dr. J. George Harrar, head of a new Mexican research program, and Dr.

Frank Hanson, an official of the Rockefeller Foundation in New York City. Borlaug was initially recruited as a plant pathologist to serve as one member of a Mexican/American research team, but he soon became a star performer as a wheat breeder.

At that time, Mexico was importing half the wheat it consumed. Facing a deteriorating economy and rapid population growth, the Mexican government had asked the United States in 1940 for assistance in developing a comprehensive agricultural program. However, the United States was very nearly at war, and the Roosevelt Administration asked the Rockefeller Foundation to prepare a study on the Mexican situation. The resulting report led to the creation in 1943 of a cooperative agricultural improvement program established by the Rockefeller Foundation and the Mexican Ministry of Agriculture. The initial goal of this program was to increase the country's production of basic food crops. By 1953, an interdisciplinary team of agronomists—including most notably Borlaug—had succeeded in vastly improving Mexican wheat yields.

In his Mexican research, Borlaug worked as a classical plant breeder, conducting selections and crosses with different kinds of wheat plants, attempting to move the right genes into the right plant. After years of experimentation and endless cross-breeding, Borlaug developed high-yielding, semi-dwarf wheat varieties; plants that would produce abundant grain heads and strong, short stalks, and which would not "lodge" or fall over when fertilized heavily with nitrogen. These "short-statured wheats," as they are sometimes called, are highly productive because they channel photosynthetic activity into grain production rather than stem growth. Borlaug's wheats made Mexico self-sufficient by 1956.* Their wide adaptability later

*After the success of Borlaug's wheats in the 1950s, the Mexican government began to press the Rockefeller Foundation for control of the research center. By 1960, Mexico assumed control of the facility under the name of the National Institute for Agricultural Research. At the same time, however, a new Mexican-based Rockefeller organization, the Inter-American Food Crop Improvement Program, was established, with Borlaug heading up the wheat program—one of three crop-research programs. In 1959, anticipating the changes that were to come in Mexico, the Rockefeller foundation sent Borlaug on a fact-finding tour of Brazil, Argentina, Bolivia, Peru, Ecuador, and Chile to survey the prospects for extending the Foundation's agricultural work in those countries. By this time, the United Nations had also tapped the services of Borlaug as its representative for another program of agricultural research around the world.

made them successful in countries from the Equator to the fortieth parallel.

In 1960, Borlaug, addressing a group of scientists and United Nations officials gathered in Rome to consider the world food problem, proposed that a special program be established in Mexico for training young agronomists from nations all over the world. He called his proposal a "practical school for wheat apostles." Citing the severe shortage of educated agronomists in underdeveloped countries, Borlaug saw this training program as one of the best ways such nations could begin to meet their mounting food problems. Borlaug's proposal became part of the FAO's international wheat program, with the Rockefeller Foundation providing the financing, and the Mexican government the training facilities.

Borlaug's apostles—who initially came from countries such as Afghanistan, Cyprus, Egypt, Ethiopia, Iran, Iraq, Jordan, Libya, Pakistan, Syria, Saudi Arabia, and more than ten countries in South America—were trained by CIMMYT* scientists in genetics, agronomy, soils, and plant breeding for one year, and then sent back to their native lands to preach the new agricultural gospel.

For Borlaug, the apostles also played a key role in testing CIMMYT's new wheat varieties throughout the world. One former apostle, Ignacio Narváez, recalls Borlaug's method: "As well as training . . . the young men from different countries, [Borlaug] placed high importance on testing . . . seeds in the countries from which the apostles had come," says Narváez. "With myself and other colleagues, he sat for hours assembling collections of little packets of different seeds from the latest advanced varieties. These were for the apostles to take across the world with them to plant in the soils of their own lands. From that . . . Borlaug would know how they responded to different environments."

When Borlaug's new wheat varieties began to be released from Mexico in the mid 1960s, they spread quickly throughout India, Pakistan, Turkey, and other nations. By 1963, semi-dwarf wheats comprised 95 percent of Mexico's 1.5 million acres of wheat, and yields in some places were up to 100 bushels per acre. Mexico's spring wheat harvest that year was six times what it was in 1944.

*"CIMMYT" is the Mexican acronym for Centro Internacional de Mejoramiento de Maiz y Trigo, the International Maize and Wheat Improvement Center.

The success of CIMMYT prompted the Rockefeller and Ford foundations to establish other such centers around the world. In 1962, the International Rice Research Institute (IRRI) was established in the Philippines, and by 1966 had created "miracle" rice varieties that joined CIMMYT's wheats in developing countries throughout the world. In 1967, the Ford and Rockefeller foundations established two more international research centers, one in Nigeria and another in Colombia. By the early 1970s, however, the two foundations had reached their financial limit, and the auspices for what were then being officially called International Agricultural Research Centers (IARCs) shifted to the United Nations.* Yet Norman Borlaug was still the man most responsible for moving the idea of the Green Revolution into reality, and one of his greatest challenges still remained.

BORLAUG'S WHEATS IN INDIA

The groundwork for the Green Revolution in India was laid in 1952 when the Indian government asked the Rockefeller Foundation for help in overcoming its perennial and overwhelming grain production problem. Almost in exact duplication of its first involvement in Mexico, the Foundation dispatched a study team to India, and shortly created the Indian Agricultural Research Institute, headed by a Rockefeller appointee, Dr. Ralph Cummings, to help train agricultural scientists in India. After nearly ten years, however, neither the Institute nor India's fifty agricultural colleges seemed to be making much headway in improving Indian food production.

Then, in 1962, a few grains of Borlaug's miracle wheat seed made their way to India for testing and observation. An Indian agricultural official, Dr. M. S. Swaminathan, who would later direct the International Rice Research Institute, observed that these were indeed very special seeds. Through the Rockefeller Foundation, Swami-

*Robert McNamara, a former Ford Foundation board member, was then president of the World Bank and was intrumental in shifting the primary responsibility for the IARCs away fom the foundations and toward a United Nations–type arrangement. The IARCs are now governed by an umbrella organization called CGIAR—the Consultive Group on International Agricultural Research. CGIAR, in turn, is governed by the United Nations and the World Bank, and is financially supported by a consortium of thirty-four donors including foundations and national governments.

nathan arranged for Borlaug to visit India in 1963 and offer his advice for improving that nation's agricultural situation. Borlaug, of course, was mindful of the new variables that faced him in India. He knew, for example, that his Mexican-bred wheats, though widely-adopted, might not stand up to Indian diseases, and that Indian agriculture, still in the wooden-plow-and-water-buffalo stage, might not be able to handle the equivalent of the Mexican grains. Nevertheless, he decided to go for broke.

"The evidence at this time is circumstantial," wrote Borlaug in his 1963 report to Swaminathan, "but fragmentary as it is, there is the prospect of a spectacular breakthrough in grain production for India. The new Mexican varieties that I have seen growing incline me to say that they will do well in India and grow beautifully." But while such opportunities did exist, said Borlaug, "there are conditions that must be met with fact and action. . . . If the disease resistance holds up in its present spectrum . . . and if the quantities of fertilizer and other vital necessities are made available, then there is a chance that something big could happen."

By the summer of 1965, India and Pakistan together had ordered 600 tons of wheat seed from Mexico. While this was only enough seed for a sprinkling of plantings throughout the two countries, the idea was to dispense the high-yielding wheat seed strategically in selected areas. A vivid demonstration of its power would create a "hunger" among farmers for the new seed and the other Green Revolution ingredients. Despite some difficulties in this first year, Borlaug's semi-dwarfs performed well: India's 1965 wheat harvest was up 1 million tons, a new record.

Drought hit India hard in 1966 and 1967, yet Borlaug had convinced the Indian government to invest in the future and to buy more semi-dwarf wheat seed. In the fall of 1966, India spent $2.5 million for 18,000 tons of Mexican wheat seed. With it, the government planted 700,000 acres of wheat which was harvested not for food, but for seed to be used in the 1968 planting. In 1968, Indian wheat production increased by five million tons, and grain production overall soared 28 percent.

By this time, Borlaug's dwarfs accounted for 42 percent of India's wheat yield, and he was now advising Indira Gandhi that her country could be the world's third leading wheat producer by 1970 if she could get her government totally behind the new program. The following year the Indian government launched the largest national wheat research program in the world, and Indira Gandhi told her nation: "We are on the threshold of self-sufficiency in grain food production. The technical advances in agriculture should enable us to meet our rising demands. This year our total grain production will cross the 100-million-ton mark for the first time."

Over the next six years, India's wheat production doubled, rising at an average rate of about 10 percent a year, with overall grain production surpassing the 100 million ton mark every year until a severe drought hit in 1972. With the help of the Green Revolution, India was feeding her population with home-grown grain. By the early 1970s, the Philippines joined India and Mexico in making the claim of self-sufficiency in grain production, and Green Revolution wheat or rice varieties (or both) were growing in Iran, Algeria, Morocco, Tunisia, Iraq, Saudi Arabia, Turkey, Kenya, Egypt, Pakistan, Brazil, Indonesia, and other countries.

The successes of the Green Revolution were remarkable, yet accompanying them were some very painful lessons and a few very serious failings. The most alarming of these—exposing the vulnerable, high-tech underbelly of Green Revolution agriculture and modern agriculture worldwide—came with the energy crisis of 1973–74.

ENERGY: PAYING THE PIPER

The crucial ingredient for the success of the high-yielding Green Revolution wheat and rice varieties was fertilizer, especially nitrogen fertilizer, and no one knew this better than Norman Borlaug. When Borlaug spoke at a promotional luncheon at one of India's new tractor-assembly plants near New Delhi in 1967, he was emphatic about the role of fertilizer in the new revolution: "If I were a member of your parliament," he told an audience that included politicians and diplomats, "I would leap from my seat every fifteen minutes and yell at the top of my voice, 'Fertilizers! . . . Give the farmers more fertilizers!' There is no more vital message for India than this. Fertilizers will give India more food," said Borlaug. "And if there is no more food, there will be an exploding volcano beneath the feet of (the) political leaders of this land. If you feed these soils of India, there will be no more crowds of tractors standing around this factory, or any other tractor factory that may be built. . . . Only with fertilizer can you dare to hope."

In 1971, nitrogen fertilizer was cheap, selling for $50 a ton, the result, in part, of industry-wide overcapacity during the 1960s, that contributed to abundant supplies.* By late 1973, however, when OPEC began raising the price of oil, fertilizer costs shot up to $225 a ton in Southeast Asia (it takes about a ton of oil to make a ton of ammonia, which is then converted into two to three tons of fertilizer). In short order, the energy crisis became a crisis for agriculture, particularly agriculture that depended on petroleum and petroleum-based products. Caught in the resulting economic cross-fire was the Green Revolution and those nations that adopted it.

According to United Nations consultant John Mellor, as much as 60 percent of the increase in grain production during some of India's very best agricultural years was "attributable to the complex of factors associated with the intensified use of fertilizer." Between 1964 and 1974, fertilizer use in India rose at a rate of more than 20 percent annually, and fertilizer imports poured into the country at a rate of 500,000 to 1 million tons a year.†

India's bill for fertilizer imports between the late 1960s and 1980 rose by more than 600 percent, an amount greater than what the country spent for food imports in the worst years of famine. And fertilizer wasn't the only Green Revolution ingredient with energy costs that India was importing—there were also tractors, combines, irrigation pumps, and pesticides. In 1969, more than 200,000 "tube wells" had been punched in the Indian landscape for irrigation, and some 35,000 small tractors were purchased abroad. Moreover, many

*At the outset of the Green Revolution there was an excess of fertilizer capacity worldwide. During the 1960s major oil companies and fertilizer interests had gone on a building spree. Between 1963 and 1968, for example, more than $4 billion had been spent by American fertilizer companies in opening mines, building new plants, and broadening distribution networks. The prospects for cheap and abundant fertilizer were bright, and American companies were anxious to move their inventories and recoup their investment.

†During the late 1960s, the United States and World Bank began applying pressure to India and Pakistan to encourage western chemical companies—by offering them certain inducements—to build fertilizer plants in their countries. Subsequently, Standard Oil of California, International Minerals & Chemicals, and other companies built a few new plants on the Indian subcontinent. Despite these investments, however, India was still dependent on outside sources for as much as 40 percent of its fertilizer needs.

of the tube wells were run by diesel or electric power. With the rise in energy prices, some of these wells were removed from production.

The rising energy costs did not stop the Green Revolution, but they did make it more expensive, particularly for those farmers who could least afford it. Many farmers went into debt trying to keep up with the costs of energy-intensive Green Revolution inputs; others went out of business. However, large farmers and wealthy landowners who could afford expensive inputs and expand their operations profited from the Revolution. Very shortly, because of the millions of dollars flowing out of countries like India to pay for expensive agricultural supplies, questions began to be raised about whether this new agricultural revolution might be helping the developed world more than it was helping developing countries.

THE BOOM FOR AGRIBUSINESS

The first recognition that there was money to be made in the Green Revolution came from the value placed on the seed itself. Early signs that this seed would become hot property came directly from Norman Borlaug's test plots in Mexico. During the 1950s and 1960s, Mexican farmers would occasionally steal handfuls of Borlaug's wheat seed from these fields. However, other, more substantial thefts were in the making.

By 1965, samples of Borlaug's wheats in Pakistani test plots had proven themselves especially good for making chapatis, the traditional Pakistani wheat pancakes. But when West Pakistan's Governor tried to order fifty tons of "Mexipak" seed from CIMMYT in 1966, the order could not be filled because of an inadequate supply of seed. At this point, Ignacio Narváez, who had worked with Borlaug in the wheat fields at CIMMYT, but who was then the Rockefeller appointee in Pakistan, stepped in. Narváez paid $25,000 to a farmer in Mexico's Yaqui Valley to grow the fifty tons of Mexipak seed, then arranged to have it illegally smuggled into Arizona and shipped to Pakistan as "animal feed." Borlaug's wheat was becoming more than just a highly valued agricultural commodity. By 1968, some of his "three-gene" dwarf wheat seed was selling for fourteen dollars a gram in Pakistan's black market—the equivalent of about $6,000 a pound in commercial trading.

Major corporations were not far behind the black market in

noticing the increasing economic value of "miracle" wheat seed. Two of the world's largest grain corporations, Cargill and Continental, had discovered that Mexican seed was a lucrative enterprise, especially since demand was outstripping supply and prices for the new "miracle" seeds were running two to three times above those of regular wheat grain. Both companies began "forward contracting" with Mexican farmers and Mexican cooperatives. In some cases the companies would cover the Mexican farmers' extra costs for raising the wheat needed to produce the seed. When independent seed buyers in Europe tried to purchase "miracle" wheat seed from Mexico in the fall of 1977, they discovered that Cargill and Continental had a corner on the market.

Other western corporations had discovered the Green Revolution even before the 1970s. For example, as a result of "John Deere field days" held in India during the late 1960s—when as many as 10,000 farmers would attend a combine demonstration put on by the giant farm equipment manufacturer—the government-owned Punjab Agro-Industries Corporation ordered sixty Deere combines to help bring in the wheat and rice harvest. Hundreds of Deere combines and other farm products—as well as occasional custom harvesting services—were also sold to Indian farmers.

Pesticides were part of the sales picture, too. President Lyndon Johnson's Science Advisory Committee estimated in 1967 that if the Third World was to double its food production by 1985 through modern agricultural techniques, a 600 percent increase in the rate of pesticide application would be required. Exxon, for one, opened hundreds of "agro-service" centers in the Philippines and other developing nations—often in conjunction with existing local filling stations—where farmers could buy everything from tractor fuel to fertilizer.

In 1967, the Indonesian government contracted Ciba-Geigy to provide the technical apparatus for an experimental Green Revolution rice production project. Following this contract, companies such as Hoechst, AHT, Mitsubishi, Coopa, and Ciba-Geigy all worked with the Indonesian government in dispensing the ingredients of the Green Revolution—including fertilizer, pesticides, management services, and the miracle seeds themselves. During 1969 and 1970, these companies brought Green Revolution technology to more than 20 percent—roughly 2.5 million acres—of Indonesia's wet-rice land.

PUSHING OUT THE OLD

The Green Revolution meant a virtual transformation of agricultural customs, practices, and tradition in developing countries. With the high-yielding miracle wheat and rice seed, came the ingredients of modern, high-tech agriculture. But in this meeting of the modern and the traditional, the differences were often as great as those between a five-acre subsistence farm in India tended by a peasant family and a modern 1,000-acre wheat farm in Kansas equipped with the latest farm machinery.

Unlike "modern" agriculture, traditional agriculture is less concerned with high yield than it is with regularity. Subsistence farmers' main goal is that there be at least some crop each year, rather than a huge harvest every year, or that next year's crop be larger than that of this year. Consequently, the kinds of crops that subsistence farmers have selected for hundreds of years are those crops that have done well and have "co-evolved" within environments of limited natural resources.

Rice breeder and Rockefeller Foundation associate Peter Jennings, explaining the ways in which traditional crops "fit" their environments, says that "In traditional agriculture the fertility of the soil is often the factor that limits growth. The native crop varieties extract nitrogen and other nutrients from the soil with great efficiency. . . . develop extensive root systems, drawing on a large area of soil, and . . . exhibit vigorous growth, which suppresses weeds that compete for the available nutrients. Having been bred by traditional methods of selection for thousands of years," says Jennings, these crops "have acquired a precise, although narrow, adaptation to the local conditions, including peculiarities of the soil, the water supply, the length of the growing season, average and extreme temperatures and number of daylight hours."

Moreover, in a population of traditional wheat plants, there is greater genetic variability than there is in modern commercial wheat varieties, since the individual plants are not all purebreds, even though they may all *look* alike. With regard to this, Jennings says that the traditional varieties, called "land races," have helped the traditional farmer by having "at least partial resistance to insect predation and disease, and partial tolerance (to) environmental stresses such as drought." Thus, he notes, if a crop becomes infected with

…articular disease, some of the strains in the land race are likely …o be susceptible, while others may well be resistant and will survive. Moreover, the nonuniformity of the plant population tends to limit the maximum number of pests and disease organisms that can invade, thereby preventing disastrous crop failures. "The net effect of this agricultural system," says Jennings, "is to give the farmer a measure of security."

With the Green Revolution, the security inherent in subsistence agriculture gave way to monoculture and mass production. In these practices, traditional crop varieties were often abandoned in favor of high-yielding "modern" varieties. Traditional wheat varieties, for example, are taller, produce little grain, and require more field space to forage for nutrients. Green Revolution varieties, by contrast, are shorter (typically semi-dwarf), have less-developed root systems, mature more quickly, and produce much more grain. They are built for speed of growth and allow, in some cases, three crops a year.*

From the perspective of Green Revolution agriculture, the "problem" with traditional varieties is that they tend to convert a higher proportion of fertilizer into stem growth than for grain production. As they grow taller, traditional varieties tend to "lodge," or fall over, making mechanized harvesting difficult if not impossible. Yet in the absence of any nitrogen fertilizer, a traditional rice variety such as Peta will do just as well as if not better than a Green Revolution variety.

Today, new crop varieties are being developed by some of the IARCs that are less pesticide- and fertilizer-dependent. However, the introduction of earlier Green Revolution varieties displaced the cultivation of traditional varieties in many areas of the Third World. Today, many of these traditional varieties—and their priceless genetic variation—are disappearing, forcing some Third World agricultural economies increasingly in the direction of Green Revolution monoculture.

*One interesting side-effect of Green Revolution agriculture, with practices such as multiple cropping, is its potential impact on local custom. Clifton R. Wharton, Jr., writing in an April 1969 edition of *Foreign Affairs,* warned Green Revolution proponents that "there may be resistance [to the use of new varieties] if the new harvest pattern conflicts with religious or traditional holidays which have grown up around the customary agricultural cycles." Yet in some cases, the modern methods have won out over local custom.

Ironically, in displacing traditional agriculture, the Green Revolution also displaced the very genetic variability that once worked to resist crop disease and insect infestation. Without pesticides, the early Green Revolution crop varieties—although high-yielding under favorable conditions—were often vulnerable to considerable insect damage. This initial shortcoming sent Green Revolution plant breeders on a catch-up program of plant breeding for disease resistance that is still going on today.

THE GREEN REVOLUTION TREADMILL

One early "miracle" rice variety developed by the International Rice Research Institute (IRRI) in 1966—"IR-8"—was quickly adopted for use throughout the Philippines and Southeast Asia. However, the new variety was bred for quantity, not quality, and its success was short lived. IR-8 was particularly susceptible to a wide range of diseases and pests; in 1968 and 1969 it was hit hard by bacterial blight and in 1970 and 1971 it was ravaged by another tropical disease called tungro.

IRRI officials were not totally unprepared for the onslaught of problems that beset IR-8. Dr. Nyle Brady, former head of the IRRI, recalls that his Institute "knew IR-8 had weaknesses when it was introduced. But those early rice strains gave us a breathing spell. They were trying to keep food production ahead of population growth. IR-8 showed what could be done." Other officials, pointing to the big picture of overall harvests, argued that the increased yields compensated for the insect attacks. Despite this, many farmers were taking a beating. In 1975, for example, Indonesian farmers lost a half million acres of Green Revolution rice varieties—IR-8 among them—to leafhoppers.

By the early and mid-1970s, IRRI scientists were back in their test plots developing stronger successors to IR-8. The goal was to retain the good features of IR-8, while cross-breeding it with other varieties that were found to be resistant to the problem diseases. In 1977, the first rice variety with resistance to more than one disease—IR-36—was released for use. This new variety, its developers claimed, was resistant to at least eight major known diseases and pests, including bacterial blight and tungro. It's parentage spanned six countries—India, Indonesia, China, Vietnam, the Philippines, and the United States—and involved thirteen different lines. By 1982, the popular new variety covered some 11 million hectares in Asia, and in some areas, IR-36 was the only variety used. Yet, when a new virus disease called "ragged stunt" and another unknown virus labeled "wilted stunt" began to ravage the IR-36 rice crop, IRRI scientists were already generating replacements. In 1980, three new varieties were released, followed by two others in 1982.

Watching this treadmill of new crop-variety development, some farmers and scientists have wondered if incorporating some aspects of traditional agriculture back into the "modern" system might not be a better strategy. In the mid 1970s, an attempt was made in the Philippines to remedy the vulnerabilities of Green Revolution rice varieties by returning to a native strain. "A few years ago," recalls Dr. Norman Meyers, recounting the events that led up to the Philippine attempt, "one of the prized developments of the Green Revolution (was) a strain of rice known as IR-8." When that variety was hit by tungro disease in the Philippines, says Meyers, rice-growers then switched to another Green Revolution hybrid which soon proved vulnerable to grassy stunt virus and brown hopper insects. Philippine farmers then moved on to yet another super-hybrid that was exceptionally resistant to almost all Philippine diseases and insect pests. But this variety proved too fragile to the islands' strong winds. Thereupon, recounts Meyers, IRRI plant breeders decided to try an original Taiwanese rice strain that had shown an unusual capacity to stand up to winds —only to find that it had been all but eliminated by Taiwanese farmers, who by then had planted virtually all of their rice lands with IR-8.

A GRAND EXPERIMENT

When initially brought to the fields of various Third World countries, the first generation of Green Revolution wheat and rice varieties—though highly productive—came up short on performance in a number of other important areas, including disease and insect resistance. These new "pedigreed" varieties produced high yields, but only under the most favorable of circumstances. Their most successful use was on good alluvial soils with plenty of fertilizer and irrigation, and some amount of chemical pesticides to keep insects and diseases at bay.* When stripped of favorable circumstances and accompanying supports, these varieties were often highly vulnerable to stress and predation.

The second and third generations of Green Revolution crop varieties—coming six to eight years after the first varieties—were geared to account for the shortcomings of their forerunners. In many ways, at least in its early years, the Green Revolution was a huge and costly learning experiment—especially in regard to the disease and insect susceptibility of its crops; a process of releasing a new variety, observing its frailties, improving it through cross-breeding, and releasing another new variety. Yet in this grand process of scientific trial and error, the agricultural systems of entire nations were altered, trade balances were affected, and thousands of farmers were ruined.

Because the governments in many developing countries were at first reluctant or slow to provide incentives for small farmers and tenants to use the new varieties and accompanying technology, they were first adopted by the wealthier and more advanced farmers. These farmers, wrote Clifton R. Wharton, Jr. in an April 1969 article in *Foreign Affairs,* found it is "easier to adopt the new higher-yield varieties since the financial risk (was) less and they already (had) better managerial skills." Moreover, the doubling and trebling of yields meant a corresponding increase in the wealthy farmers' incomes, said Wharton. And one indication of this wealth at the top was the large number of new farm-management consulting firms in the Philippines that were advising large landlords on the use of the new seed varieties and making handsome profits out of their share of the increased output. "As a result of different rates in the diffusion of the new technology," he concluded, "the richer farmers will become richer."†

Consequently, in many countries adopting Green Revolution

*"Virtually all the new wheats in Mexico, India, Pakistan, and Turkey," said Green Revolution critic William C. Paddock in 1970, "are grown under artificial irrigation . . . on the very best land in the nation, the most expensive land, the land which receives the largest capital investment, and the land with the best farmers." The new rice varieties, said Paddock, "also require carefully controlled irrigation. However, on nonirrigated land (they) do no better than the standard ones."

†In fact, Wharton perceptively warned in 1969 that if more of the progressive farmers captured markets previously served by the smaller semi-subsistance farmers, additional problems of welfare, equity, and unrest could occur. "If only a

strategies, land values and rents soared, farm consolidations accelerated, and small farmers were displaced by the millions, often exacerbating unemployment and worsening conditions in crowded urban centers. While overall grain yields rose in the developing world, so did financial, economic, and political dependencies.

Despite these unhappy side effects, the Green Revolution is here to stay, institutionalized in the United Nations, the IARCs, and the national governments of many developing countries. And while its advocates have slowly learned from past mistakes, the Green Revolution is now about to become a part of an even more potent revolution: the worldwide biotechnology revolution.

BIOTECHNOLOGY: THE NEW GREAT HOPE

As the debate over the effectiveness of the Green Revolution continues, biotechnology and genetic engineering are now being touted for their agricultural potential in developing countries. And once again, the conventional wisdom among officials in international agencies, such as the United States Agency for International Development (AID), is that what is good for American farmers and industrialized nations will be even better for the farmers of developing countries. Says former IRRI director Nyle Brady, now a senior administrator in science and technology at AID: "In the long run, techniques such as recombinant DNA technology, protoplast fusion, and the . . . development of truly pest-resistant plants, drought- and salt-tolerant varieties, and energy-efficient nitrogen-fixing bacteria may be achieved." And although such developments will be of "revolutionary value" to American farmers, says Brady, they "will be even more helpful to the resource-poor farmers of the developing countries."

Brady sees the genetically improved seed becoming "the nucleus

small fraction of the rural population moves into the modern century," he wrote, "while the bulk remains behind, or perhaps even goes backward, the situation will be highly explosive. For example," noted Wharton, "Tanjore district in Madras, India, has been one of the prize areas where the new high-yield varieties have been successfully promoted. Yet one day last December, forty-three persons were killed in a clash there between the landlords and their landless workers, who felt that they were not receiving their proper share of the increased prosperity brought by the Green Revolution."

of technological packages that the Third World farmer will accept," as illustrated by "the ready acceptance in the 1960s of new high-yielding wheat and rice varieties, along with the chemical packages that made them productive."

Some Third World scientists and governments see biotechnology as a way out of the fertilizer and pesticide dependencies that came with the first Green Revolution. According to former Indian agricultural official Dr. M. S. Swaminathan, now director of IRRI, "nearly every developing country has plans or programs for harnessing the tools of biotechnology for national development."

In the Philippines, for example, the National Institute of Biotechnology and Applied Microbiology has set a high research priority for nitrogen fixation and microbial insecticides. In India, the National Biotechnology Board has targeted genetic engineering, photosynthesis, and tissue-culture work among its priority research areas. But these countries are competing with wealthy multinational corporations and the governments of developed nations looking to sell their genetic products or reap international prestige from the new Green Revolution. And in that race, developing countries face tremendous odds in determining their own fate.*

In September 1983, the United Nations Industrial Development Organization (UNIDO) advanced a proposal for an International Center for Genetic Engineering and Biotechnology aimed specifically at the research and training needs of Third World nations. However, that idea has run into opposition from some advanced countries, including the United States, France, Great Britain, and West Ger-

*Biotechnology programs, with some emphasis on agriculture, have been established in Japan, Canada, England, France, West Germany, Australia, the United States, and other developed nations. Very often, these programs involve the very largest corporations in those countries—such as Elf Acquitaine in France, ICI and Shell in England, and Mitsubishi in Japan—often with funding and political support from the national government. In the race to capture biotechnology patents, developed countries such as the United States, Japan, and the Soviet Union are far and away the dominant players. One survey of 2,400 patents issued between 1977 and 1981 found that the member countries of the Organization for Economic Cooperation and Development (OECD), comprised predominantly of developed nations, had captured over 80 percent of the patents. Meanwhile, there is a worldwide shortage—even in the United States—of trained scientists in molecular biology and microbiology, with corporations and national governments in the developed countries paying huge salaries and getting the pick of the lot.

many. According to one report on the proposed International Center in *Science* magazine, "U.S. officials apparently made it clear that they were unlikely to support an initiative which would not only mean funding a new U.N. project, but might also boost foreign competition in a field in which the United States is striving to maintain economic leadership." In addition, a draft report from the White House Office of Science and Technology Policy claims that the UNIDO Center's work "could prove overly ambitious in light of (its) operating budget," but that the center "is unlikely to produce world class research."

Developed nations, which supply millions of dollars worth of agricultural commodities to the less developed countries, have a considerable vested interest in maintaining such trading patterns. And while biotechnology could radically alter global agricultural trading relationships, experts say, it will most likely do so in ways that will benefit developed nations.

"Trade in agriculture," says a 1982 paper prepared by the Organization for Economic Cooperation and Development (OECD), which represents industrialized nations, "is particularly susceptible to changes initiated by biotechnology." Traditional agricultural trading patterns between the industrialized and developing countries, says the OECD paper, "may become progressively dislocated," when industrialized trading partners suddenly become self-sufficient in "colonial crops." For example, the need for a crop such as sugar cane could be made obsolete as a result of adopting cheaper biotechnology processes which make sugar substitutes such as high-fructose syrup. In addition, says the OECD, because of the "keenness of industrialized countries to export biotechnology to the Third World," other trading patterns could change "without either trading partner being fully apprised of the result."

American-based agricultural genetics firms, some backed and financed by multinational corporations, are now gearing up to move into the developing world. In 1982, the International Plant Research Institute (IPRI), which has been supported by Eli Lilly and ARCO, entered two joint ventures with the Malaysian Sime Darby Group aimed at the agricultural market in Southeast Asia. Sime Darby, the largest and most profitable corporation in Malaysia with annual revenues of more than $1.1 billion, is a major plantation owner and major agricultural influence throughout Southeast Asia.

According to IPRI's Executive Vice President Robert Abrams, one of the Institute's joint ventures with Sime Darby, known as the Asean Biotechnology Corporation, "will begin operations with more than 150 employees in Malaysia" and will serve as "the focus for the introduction and expansion of new technologies developed by IPRI at its headquarters in California." The new venture, says Abrams, "will represent one of the first efforts to apply genetic engineering and recombinant DNA technology to a broad range of tropical agricultural crops in a practical way." The second of the two joint ventures, called the ASEAN Agri-Industrial Corporation, will essentially sell agricultural biotechnology services to other companies and governments in Southeast Asia.

Norman Goldfarb, chief executive officer and chairman of Calgene, a company now working with a Nestlé subsidiary and Finland's largest chemical company, Kemira Oy, to develop genetically engineered herbicide-resistant crop strains, believes that biotechnology will help the world's unsophisticated farmers. "In Africa," he says, "there are a lot of unsophisticated farmers." In that country, he explains, "you can't even expect (farmers) to drive a tractor straight; you might ask them to put the seed on the field evenly." The new products of biotechnology, in Goldfarb's view, will bring 'built-in farm management' to such farmers, and products such as herbicide-resistant crop varieties "will be good for the Third World."

Unilever, the world's fourth largest corporation outside the United States, is using biotechnology to genetically alter the tropical oil palm, which accounts for about 15 percent of all vegetable oil traded on world markets. Unilever's new oil palms now produce 30 percent higher yields than most existing varieties, and at one of the company's molecular biology laboratories in Vlaardingess, Holland, scientists are working to isolate and control the genes in the oil palm that regulate other traits of the plant as well. If successful, Unilever will be able to produce palms under contract that yield specified kinds of vegetable oil for food processors throughout the world. The company is already growing cloned palm trees at its Unipalm Plantation in Jahore, Malaysia, and its researchers are working to develop hybrids of both normal and dwarf coconut palms.

American and European seed companies—as well as chemical and pharmaceutical corporations—have a strong interest in selling new crop varieties and other biotechnology products globally. Pioneer, DeKalb-Pfizer, Sandoz, Ciba-Geigy, and Cargill are all looking to the export market to continue selling hybrid corn seed. Lubrizol, Upjohn, Sandoz, Celanese, ARCO, and Limagrain all export vegetable seed. Today, 50 percent of American seed exports go to developed nations, but there is a growing market for American seed in developing countries as well. American seed exports to South American countries increased by 47 percent in 1979, and sales to Mexico have quadrupled since 1976 and now account for 21 percent of all American seed sold abroad. Other countries, including Libya, South Africa, Iran, Iraq, Jordan, Saudi Arabia, and Syria all purchase more than $1 million worth of seed annually from the United States.*

ALTERNATIVES ARE POSSIBLE

Despite the emerging pressures from corporations and industrialized nations to push the new ingredients of biotechnology into the agricultural systems of less developed lands, there are signs that some of these countries' governments want to turn away from the high costs and treadmill nature of "modern" agriculture. In the Philippines, for example, a seed-collecting and -banking program has been initiated under the order of Prime Minister Cesar Virata to find traditional varieties of rice, corn, and vegetables that have become scarce because of the Green Revolution. Green Revolution rice varieties now cover 6.4 million Philippine acres, 75 percent of the country's rice land. As a result, Philippine farmers now spend about forty-two dollars on chemicals for every acre planted to such

*Complementing the prowess of American and European seed companies in moving their new varieties into developing countries are several international organizations and trade associations working to advance patenting laws for seed and biotechnology products. One intergovernmental organization in particular, the International Union for the Protection of New Varieties of Plants (UPOV), is working to advance the adoption of plant patenting laws around the world. Created in December 1961 with the blessing of the United Nations' World Intellectual Property Organization, UPOV has twelve member nations from the developed world, mostly European, and is now trying to add the United States, Japan, Canada, Mexico, New Zealand, and Australia to its roster of member countries. The worldwide seed- and plant-patenting laws it advocates will facilitate the marketing of popular crop varieties all over the world. Helping UPOV achieve these goals are organizations such as the International Association of Plant Breeders (ASSINSEL).

newer varieties, and many have gone into debt to keep up with their costs.

"Productivity gains in cereal production," says one Asian Development Bank study on the social and economic impact of the Green Revolution, "are coming more slowly and more expensively than they did during the past decade." But in the Philippines, it has been found that the use of local varieties—including those of rice and corn—in place of Green Revolution varieties can cut the high cost of farming by as much as 66 percent, creating a substantial economic impetus for collecting and saving traditional seed. "The seed-retrieval program," says Philippine program director Domingo Panganiban, "seeks to rebuild the national genetic pool of plants geared to small farmers."

In Tanzania, a new national agricultural policy now emphasizes crop rotation, composting, and village-based agriculture over the high-tech practices of the Green Revolution. Recalls Tanzanian President Julius K. Nyerere of his country's earlier involvement with "modern" agriculture: "When (we) became independent [in 1961], our ambition was to 'modernize'. . . . It appeared to us that if you wanted a productive agriculture, you had a mechanized agriculture, and you used chemical fertilizers, chemical insecticides, and—to be completely up to date—even herbicides. That, at least, was our vision of American and Canadian agriculture." As a result, says Nyerere, "We urged the use of chemical fertilizers; we established a fertilizer factory—heavily dependent on imported components—and we ensured that chemical fertilizers were used in growing certain of our export crops, especially cotton and tobacco. We also adopted a World Bank–assisted maize program, which depended upon using chemical fertilizers. . . . We even stopped teaching compost making in our schools, regarding this as a discredited and old-fashioned technique, which was irrelevant to our future."

Nyerere now speaks of the new ways of agriculture* with some emphasis on a new kind of productivity. He points out that Tanzania's recently published Agricultural Policy Statement places great

*For an excellent account of how Tanzania arrived at its new agricultural policy see Robert Rodale's "The Road to Morogoro" and "Tanzania Self-Reliance: A New National Policy" in *The New Farm*, September/October, 1983.

emphasis on improving the nation's peasant agriculture—increasing its per-acre output through the use of compost, manure, crop rotation, and intercropping. These methods, says Nyerere, provide the answer to the peasants' questions about how they can increase their income without becoming dependent upon an unreliable supply of expensive fertilizers and other inputs. "When we talk about self-reliance as being the foundation of national self-reliance," he says, "this is the kind of thing we should be talking about."

However, countries like Tanzania, the Philippines, and others wishing to free themselves from the dependencies of Green Revolution I or II will still need some help from beyond their borders, and most logically from the United Nations and the International Agricultural Research Centers (IARCs).

NEW ROLE FOR THE IARCS

Although the IARCs may have fueled some of the problems brought on by the Green Revolution, in that their research and crop varieties often had devastating effects on local agricultural systems, today there are many scientists in the IARC system who are sensitive to these problems and whose work and perspective are important as counterpoints to more purely commercial interests. As a publicly-responsive system of research and scientific opinion, the IARCs can be used as a check on the real worth of new agricultural products introduced for use in less developed countries. Yet there is always the danger—through funding sources and political pressure—that the IARCs will become captive to the interests of industrialized nations, or simply become the research servants of political interests in Third World governments. Clearly then, appropriate agricultural priorities must be established, but these must come from the governments of individual nations themselves.

Some IARCs have begun to integrate new genetic techniques into their research on crops and livestock. Scientists at IRRI in the Philippines, for example, are using cell- and tissue-culture methods to develop rice varieties that are tolerant to the high levels of salt and aluminum found in many Southeast Asian soils. They are also working to produce varieties with an increased lysine and protein content. Beyond this, IRRI scientists are studying nitrogen-fixing agents appropriate for the Third World, such as wetland algae and a "green manure" plant, *Azolla,* that can be grown with rice to help fertilize it. And with the University of Nottingham in England, the IRRI is working on protoplast fusion techniques in an attempt to produce an *Azolla*/rice hybrid. Some of this work has the potential of helping small-scale agriculture, reducing production costs, and decreasing agrichemical use. And designed to meet these goals, more of it should be funded.

Besides their new research in biotechnology and with farm production alternatives, many IARCs have also developed extensive collections of seed and plant materials, some of which are quite rare. These genetic resources will be extremely important in developing appropriate biotechnological options for Third World agriculture, as well as in providing the option for reinstating some of the more valuable and less technology-dependent features of traditional agricultural systems. The IRRI, for example, has collected samples of 60,000 of the world's 120,000 rice strains in its "seed bank," and the International Board for Plant Genetic Resources is now collecting and storing seeds for thirty-two of the world's major food crops.* More of this kind of work also needs to be undertaken.

However, the IARCs are beginning to run into funding problems. In 1983, funding for the thirteen IARCs fell short of projected needs by 6 percent. The total IARC budget of $162 million is about $100 million less than what Du Pont now spends annually on life-sciences research. "We are now at a point," said Lloyd Evans, a member of CGIAR's Technical Advisory Committee, in November 1982, "where we might have to contemplate closing a center." While no IARCs have been closed, a planned center for water-conservation research was dropped, and training programs have also been cut back. This means that in a time of potentially great agricultural change fueled by genetics research, the IARCs may be standing still rather than moving forward.

In addition to the activities of the IARCs, a number of United Nations and other international research programs also deal with biotechnology and microbiology, or with research training in these

*It is important to note, however, that many seed collections have already "missed the boat" in terms of containing some of the most important strains that are already extinct, and that some seemingly large and impressive collections are "genetically redundant" in terms of what they have in storage. See Chapter 10 for other views on this subject.

areas.* Yet the extent to which these various programs can help developing nations improve their foothold in food production with appropriate biotechnology—and offset the political interests of these nations' governments, the economic interests of multinational corporations, and existing trading dependencies with developed-nations—is unclear. Key variables in this struggle will be political pressure and the driving forces of food demand and population growth.

POLITICS, CHEAP FOOD, & THE FUTURE

Despite the progress of the Green Revolution in lifting grain yields with high-tech agriculture, and the newer promises of agricultural abundance through biotechnology, a substantial food problem still exists in the less-developed nations of the world. Speaking of the gap between grain production and demand in ninety developing countries, IRRI's Dr. M. S. Swaminathan explains that "the net cereals deficit of 36 million tons in 1978–1979 will have doubled by 1990 and again doubled by 2000." This means that the grain demand in the developing world will exceed the supply by 144 million tons within only fifteen years.

Politicians in these ninety developing nations will be looking for food and food-production technology anywhere they can find it. Consider for example, how some governments embraced and supported the Green Revolution technologies of the past.

The guarantee of farm-price supports was an interesting political development that accompanied the early adoption of Green Revolution wheat and rice varieties in some countries. By 1970, for

*Among them are UNESCO (United Nations Educational, Scientific and Cultural Organization); UNEP (United Nations Environmental Programme); UNIDO (United Nations Industrial Development Organization); UNITAR (United Nations Institute for Training and Research); UNCSTD (United Nations Conference on Science and Technology for Development); UNERG (United Nations Conference on New and Renewable Sources of Energy); ICRO (International Cell Research Organization); IOBB (International Organization for Biotechnology and Bioengineering); WFCC/WDC (World Federation for Culture Collections/World Data Centre); CSC (Commonwealth Science Council); IDRC (International Development Research Centre); IDB (Inter-American Development Bank); EFB (European Federation of Biotechnology); CEC (Commission of the European Communities); IUPAC (International Union of Pure and Applied Chemistry); and IUM (International Unions of Biology and Microbiology).

example, Mexico, Turkey, India, and Pakistan were supporting the price of wheat at levels 30 to 100 percent above its world price. Some Green Revolution price-support programs also subsidized the use of fertilizer and pesticides. According to the International Association of Plant Breeders (ASSINSEL), in one irrigated area of India "farmers received subsidies of 10 percent to 20 percent on fertilizers and 25 percent on pesticides, plus government-backed loans to enable them to pay for these inputs and tide them over to harvest." Not surprisingly, fertilizer consumption in this same area of India between 1969 and 1979 rose from 3.6 to 50.1 kilograms per hectare. "Without government intervention," says ASSINSEL, "it is unlikely that progress could have been so rapid."

National governments in developing countries moved into the business of providing price-supports and loans for Green Revolution farmers primarily to cover the costs of high-technology farming. "The new varieties require irrigation water, fertilizer, and additional labor," said William C. Paddock in 1970, explaining that the financial risk this entails for the farmer "is justified because of the support price. (But) take away that crutch and fewer would take the risk," said Paddock, noting that price supports therefore became necessary to get more farmers to participate in the Green Revolution.

Yet these price-support programs became a huge and mounting expense for governments with large numbers of farmers and few financial reserves. In fact, there was some speculation that the United States and other industrialized nations' credit programs were both indirectly and directly helping to prop up the Green Revolution price-support system.

After the energy crisis of 1973–74, price supports for the Green Revolution escalated even more dramatically as farmers' costs soared. And as the costs of such programs rose, an incentive developed to reduce the number of farmers by embracing even more high-yielding agricultural technology.

In the meantime, urban population growth, and the political need to feed restless millions, also continues to pressure Third World governments to seek high-yield agriculture. By the end of this century, twelve of the world's largest fifteen cities will be in developing countries. And by the year 2050, according to a 1984 World Bank report, there are expected to be at least twenty Third World cities with populations of *30 million each*. Mexico City is already swollen with 14 million people, and Calcutta and Sao Paulo have 11 million each. But with governments embracing more high-technology agriculture, more farmers will be driven off their land, and more of them will come to the cities. However, in the Hobbesian tradeoff of a million more farmers leaving the countryside, versus the need to feed 12 million people already in a huge city, the "city vote" will win every time. In fact, the mounting need to produce cheap food means that national governments will be more interested in getting new crop varieties and new biotechnology products to highly productive farmers than in keeping more small farmers on the land.

The IARCs, for example, have little power over how national governments in the Third World use the products of their research. Says John L. Nickel, director of the International Center for Tropical Research (CIAT) in Cali, Colombia: "For political reasons we can't bypass the national agricultural programs. . . . We are pressured in different directions. Our donors (i.e., foundations) are all interested in the rural poor, and they want measurable results in a fairly short time. On the other hand, many of the national governments we are working with argue that if we focus on small farmers' needs we are just institutionalizing poverty."

Interestingly, some national governments are also more interested in attaining high yields in new crop varieties than in disease or pest resistance, even if such qualities are genetically "built-into" plants. They would rather have chemical-dependent crops that guarantee massive amounts of food than pesticide-free crops that produce lower volumes of food. And if that means eliminating even more small farmers, so be it. Says Nickel: "Some governments argue that, if we try to develop plants that don't need inputs, we are robbing the world of the maximum agricultural yields. The fear is that, if we work too hard on developing resistances, we will produce plants that don't have the potential for high yields. The world is going to have to produce food at maximum efficiency, it is argued, and if small farmers can't afford the inputs that are necessary they will have to be replaced by huge producers who can, whether they . . . be state-owned or capitalist."

CIAT rice breeder Peter Jennings, who has developed semi-dwarf varieties suitable for some South American countries, says that the focus at CIAT "is much more on the consumer than on the farmer, and I'm not convinced we should focus on the marginal producer."

The farmer with only one or two hectares cannot be helped, observes Jennings, and the farmer with three or four hectares can at best expect an income that will only enable him to compete with factory wages. And Douglas R. Laing, CIAT's director for crop research, adds that "These countries are pushing all out for cheap food. They know that, by the year 2000, 70 to 80 percent of the population of Latin America will be in the cities. If we develop anything that can be grown on a large scale, many of these governments would grab it and kill the small farmer."

Cuba is already growing CIAT bean and cassava varieties on some 100,000 acres of large-scale, state-owned plantations. Brazil is using CIAT cassava varieties to develop huge alcohol fuel projects, and Mexico is eyeing the cassava as a feed source for a huge national chicken industry. Biotechnology will undoubtedly be used to fuel more large-scale projects such as these and, in the process, give national governments and the corporations that attend their needs considerable control over what is produced and how it is produced.

Yet by embracing biotechnology to help build large-scale, high-technology food-production systems, the less developed nations of the world could in fact be sowing the seeds of long-term dependency, and therefore, some measure of vulnerability and potential political instability.

CHAPTER
14

PROTEIN FOR BUSINESS

If people didn't have taste buds,
it would be easy to feed the world.
—*Dennis Steadman,*
Chase Econometrics

The name of the company is Genetic Engineering, Incorporated. The large black and white sign hanging from a Western-styled branch frame at the entrance to a long gravel driveway tells you that. At the center of the metal sign, in the middle of the word "genetic," there are the tell-tale dents of some recent shotgun shelling, probably the handiwork of rural pranksters. From the roadside, this Northglenn, Colorado, company located about 45 minutes north of Denver just off Interstate-25, looks like just another farm, with Holstein and Simmental cows grazing on green pastures. It's not until you've made your way down the long gravel driveway that you learn what Genetic Engineering is all about. Under a microscope there, with the aid of an enthusiastic lab technician, you can see a half-dozen live, seven-day-old Holstein embryos the size of pinheads just "flushed" from a "donor" cow.

Bovine embryos—selling them, splitting them, transferring them into other cows, identifying their sex, and genetically engineering them—is what Genetic Engineering, Incorporated, is all about. Embryo transfer and other livestock bioengineering are being practiced in the United States to upgrade breeding stock, perpetuate certain favored pedigrees, and increase beef production and dairy yields.

However, the real boom growth for the embryo engineering business is expected to come from foreign markets, and especially from nations interested in building instant herds of dairy cows or beef cattle to help feed their people.*

Even amid the recent worldwide economic recession, some experts have been predicting that per capita income and the demand for meat- and dairy-based protein will rise in the years ahead, not only in developing nations, but also in the Soviet bloc, and that this demand will translate into a need to build and expand dairy and cattle herds, as well as poultry, swine, and other livestock operations. Bioengineering companies with experience in the fields of embryo transfer, sex determination, and embryo transport—such as Genetic Engineering—will be ideally situated to capture such markets.

Yet others have some doubts about exporting the genes of protein development to foreign countries, whether in the form of Holstein embryos or protein-making microbes. They worry that such genes may spawn entire new systems of "protein businesses" that generate capital spending and related economic activities, but which also inflate the steps involved in producing protein rather than providing it more directly to people through the simplest and most cost-effective manner.

EXPORTING EMBRYOS

"We're not going to solve the world's nutrition problem," says Bob Chapman, Genetic Engineering's co-founder and senior vice president, "but we're going to get a start on it." Chapman appears to be in his late thirties, is dressed comfortably in a green, Ralph Lauren polo shirt, tan jeans, and cowboy boots. With sandy blond hair, a ready smile, and an easy-going western style, Chapman has a Robert Redford look about him. He talks openly and confidently

*In 1982 some 35,000 calves were born as a result of embryo transfer techniques. Today, there are approximately 140 companies in the United States performing embryo transfers, doing about $25 million worth of business annually. Although only about 1% of all American cattle are involved in embryo transfers, and most of these currently involve top-of-the-line pedigrees, the number of transfers is expected to increase dramatically as the cost of providing the service comes down and both domestic and foreign demand rise.

about his new business. "We're interested primarily in the international market for dairy and beef embryos," he says, "but we'll sell anything: pregnancies, herd management advice, or training programs."

In 1980, while working at the College of Veterinary Medicine and Biomedical Science at Colorado State University in Fort Collins, Chapman met Dr. Ed Adair, a urologist who was then developing a method of separating sperm cells into those that carried the chromosomes for determining male and female sex. On the basis of Adair's research and Chapman's background in animal reproduction, the two men decided to form a livestock genetics company. The third principal in the company is Charles Srebnik, a former Wall Street investment banker and part-time cattle breeder, who helped arrange some early financing and underwriting for the company.* Today, Chapman, Srebnik, Adair, and another Genetic Engineering director, 54-year-old biochemist and businessman Dr. Arthur J. Rosenburg, own about 45 percent of the new company.

Initially, Genetic Engineering developed around Ed Adair's semen-separation technique and embryo-transfer technology. But the new company really hit pay dirt in 1982 when its scientists discovered and patented a technique for identifying the sex of six-day-old cattle embryos—a dramatic breakthrough that will give the new company an edge in embryo sales both in the United States and internationally. As Genetic Engineering's President, Dr. Jan Clee, said after the new technique proved successful: "This procedure will permit cattle breeders for the first time to pre-select male or female bovine embryos according to whether they breed for milk or beef. From now on they can enlarge their herds with unprecedented efficiency." According to Clee, the new breakthrough would also "hasten the production of quality herds throughout the world, and contribute to the solution of the enormous problem of food shortages in many areas."

The embryo transfer method of building livestock operations is attractive because it can be done relatively quickly and in large

*Chapman recalls that he and his partners once "visited with Armand Hammer," the chairman of Occidental Petroleum, to discuss financing. (Hammer is also a part-time cattle breeder.) They also held some discussions with Monsanto, says Chapman, "but they wanted too much of the rock."

numbers. Dr. Autar Karihaloo, director of embryo transfer technology for the Carnation Company, which is also involved in the embryo trade, states flatly, "You can ship 2,000 cows under your airplane seat."

In addition, embryos do not carry the diseases and parasites that fully grown animals might spread. Furthermore, implanting a Holstein embryo conceived in the United States into a native African cow, to take but one example, insures that the resulting calf will have the surrogate mother's immunities to native African diseases. However, genetic traits of a negative variety introduced through such techniques would be difficult and expensive to recall. Yet the economical advantages of shipping thousands of tiny embryos anywhere in the world may facilitate the spread of popular traits and/or breeds worldwide.

In the near future, says Bob Chapman, his company and others will be "exporting wholesale" the genetic expertise of the American cattle industry to developing nations." And, he adds, "we'll be saving them one hundred years of work, bringing them a long way forward in terms of what they have now."

"They're really hurting out there," he says, referring to the food situation in developing countries. "They're very poor, with practically no money to pay for anything." Asked then how he expects to sell embryos and training programs to such foreign nations, Chapman replies that one approach is "through third-party bartering." He explains that Genetic Engineering has a contract with the Cyrus Eaton World Trade Group to arrange bartering agreements with foreign governments interested in buying embryos, with the embryos being paid for with goods those nations have in abundance. "We sell them embryos for rice, shoes, oil, or whatever they have," says Chapman. "Eaton does the bartering, and we get the cash." In fact, former U.S. Secretary of Agriculture Bob Bergland, who worked with Cyrus Eaton in 1982–83, helped negotiate a few such deals for Genetic Engineering. A Cyrus Eaton representative sets up a briefing time for an interested buyer, and Chapman or another Genetic Engineering official explains the program. "It's not hard to sell the program," says Chapman. "An Eaton representative sets it up, we explain it, and Eaton does the haranguing."

Genetic Engineering has already sold frozen cattle embryos to Romania and several other countries, and is now negotiating with further prospective customers, including Italy and South Africa. Chapman explains that selling the training programs that go along with embryo transfers will also be an important part of his business. "Selling this new technology to farmers in developing countries is like giving a teenage kid a Ferrari. If he doesn't know what to do with it, he might wind up putting regular gas in the thing when it really needs hi-test," says Chapman. "Train, train, train will be the name of the game," he explains.

Today, Genetic Engineering is not alone in the embryo export business. The Carnation Company, better known for such products as its evaporated milk, Slender Diet Foods, and Instant Breakfast, is also exporting frozen cattle embyos to developing nations, and is also involved in other livestock- and agriculture-related businesses, both in the United States and abroad.* Carnation negotiated embryo sales with several African nations in 1982, and plans to intensify its sales effort in the future. Beyond this, there are more than one hundred other companies involved in the embryo business as well as trade associations such as the Holstein Friesian Association, which has sold cattle embryos to nations such as Hungary, Spain, and Italy. But the embryo business itself is only part of the equation.

Certainly, genetic technologies such as embryo transfer, sexing

*For example, in its 1980 annual report, Carnation details some of its involvement in the feed and livestock genetics business: "Carnation feeds—sold by the sack or by the truck load—are consumed by millions of four-footed or feathered farm creatures from New York to California and in more than 20 foreign countries. We also sell frozen semen from top-rated bulls to those interested in improving beef or dairy stock anywhere in the world; we sell Holstein heifers to countries wanting to upgrade their dairy industries; and we provide technology related to animal husbandry and crop improvement." In addition, Carnation is also involved in agricultural development and training programs in several foreign countries. "In demonstration programs in Mexico, Peru and the Philippines, with the endorsement of each of the governments, Carnation is sharing its technical and management skills in farming and dairying," explains the company. "Animal husbandry, how to grow good forage and preserve it as silage, how to prevent erosion and conserve water are a few of the subjects taught at Carnation-operated farm schools. In Mexico and Peru, Carnation successfully introduced dairying into new areas, and in the Philippines, Carnation's fieldmen have worked with villagers to discover cash crops and forage crops that can be grown on the hitherto unplanted land under coconut trees. They have also assisted in the genetic improvement of native cattle through artificial insemination, which has resulted in increased milk production."

techniques, twinning, and embryo engineering will make it attractive for foreign nations to rapidly develop their livestock industries. But after the instant herds materialize in those countries, new feeding requirements will also emerge. New livestock industries will require huge amounts of feedstuffs—either in the form of imported grain, crops grown locally, or other substances. And here is where the process of making protein can become inflated beyond the needs of people as well as available local resources. It is here also that questions of efficiency begin to arise.

PROTEIN FOR LIVESTOCK

In the United States and other advanced nations, huge amounts of cereal grains—including nearly 90 percent of all American corn, sorghum, oats, and barley—is fed to livestock for the production of meat, milk, and eggs. In 1980, for example, 4.1 billion bushels of American corn production—fully 62 percent of the entire crop—was fed to domestic livestock, and much of the rest (2.5 million bushels) was exported to feed livestock in other countries.

Some critics observing the increasing use of grain crops as livestock feed have underscored the nutritional inefficiency of this process. In her book *Diet for a Small Planet,* Frances Moore Lappé explains that beef cattle are the most inefficient protein converters of all livestock, requiring sixteen pounds of grain and soybean meal for every pound of meat they produce.* To produce the same pound of meat, hogs require six pounds of grain and soybean meal, turkeys four, and chickens three. On the basis of these conversion factors, Lappé calculates that for every 140 million tons of grain and soybean meal fed to cattle, poultry, and hogs, only 20 million tons—or one-seventh of what is used as feed—is returned in the form of meat. She calls the rest "protein loss."

Lappé argues that grain protein used as feed could be better used

*"Another way of assessing the relative inefficiency of livestock as protein converters," explains Lappé, "is by comparison with plants. An acre of cereals can produce five times more protein than an acre devoted to meat production; legumes (beans, peas, lentils) can produce ten times more; and leafy vegetables fifteen times more." These figures are averages, says Lappé; "some plants in each category actually produce even more. Spinach, for example, can produce up to twenty-six times more protein per acre than can beef."

as food. Moreover, she explains, "animals like cattle, sheep and goats don't need to *eat* protein to *produce* protein," since such animals have a second stomach, called the rumen, that can convert a wide variety of waste products such as orange rinds, cocoa residue, and even bacteria into protein.

However, the economic interest in continuing to expand feed-grain-based agricultural systems (part of which flows from surplus crop production in developed nations)—from feedlots to large hog-confinement operations—is considerable, and explains in part why such systems are now being exported to developing nations. But this export opportunity also raises possibilities for seed and biotechnology companies. Aside from the genetics business found directly in the expansion of livestock herds worldwide, there is also genetics business that will flow from the feed-grain and other supply activities connected with livestock agriculture. For example, much of the genetic work now being undertaken in the name of "protein improvement" of cereal grains is being done at the behest of the feed-grain establishment and the livestock industry, including feed producers, animal drug companies, and others. But what of protein for people?

FEEDING THE WORLD'S HUNGRY?

Approximately 50 percent of the world's protein needs and 70 percent of its calories are satisfied through the direct consumption of cereal grains. In some parts of Asia, the Near East, Africa, and Latin America, cereal grains such as wheat, corn, rice, and barley account for as much as two-thirds of all dietary protein.

Protein is essential for human growth and the maintenance of all bodily functions—the hair, eyes, skin, and muscles are all primarily composed of protein. The "building blocks" of protein are the amino acids. Twenty of these amino acids are needed to manufacture animal protein, and ten are considered essential because they cannot be made internally by humans or animals; they must be derived from food or feed.* Yet most of the world's important cereal crops

*These "essential amino acids," as they are called, include lysine, leucine, methionine, phenylalanine, threonine, tryptophan, arginine, isoleucine, histidine, and valine.

are very low in at least one of the essential amino acids, most often lysine or tryptophan.

For a period during the 1950s—when the "protein gap" theory of malnutrition was in the ascendancy among nutritionists and world food planners—some plant breeders turned their skills to increasing the protein content of cereal grains such as wheat, barley, and rice. Others began scouring seed collections all over the world for high protein-variants, and during the 1960s and 1970s, high-protein varieties of corn, barley, sorghum, and wheat were discovered.

Finding these high-protein grain varieties in seed collections was one thing, but using them in breeding programs and commercial agriculture was quite another. It was soon learned that in practically all of the cereal varieties with a higher-than-normal protein content there was a catch—an undesirable genetic linkage. The genes for yield and those for protein content were connected to one another, but not in a complementary way. In fact, the more breeders selected and cross-bred for an increased protein content, the lower the yields of a cereal variety would be. Conversely, breeding for yield often lowered the protein levels of a crop.

"In wheat generally," says Robert Romig, research director for the Northrup King Company, "protein and yield are inversely correlated." But Romig is quick to add that there are instances in which an increased yield, or some improved harvesting characteristic such as early maturity, actually increases the net amount of protein per acre. As an example, Romig points to "Era," a University of Minnesota semi-dwarf wheat variety that produces a 20 percent increase in yield but a 2 percent decrease in that variety's protein content. Yet this variety, explains Romig, because of its higher yielding capacity, will give the farmer more protein per acre.

Charles Krull, director of Temperate Corn Research for DeKalb AgResearch, agrees with Romig. "In terms of pounds of protein per acre, you're always better off with (a variety that has) higher yield," he says. But while DeKalb is now working on high-protein corn, Krull explains, "we haven't been too successful in marketing them. It's kind of dangerous to advertise that you have higher protein varieties, given the inverse relationship [known by farmers] between protein content and yield."

For other scientists, however, there is more to crop productivity than yield. Yield per acre, they say, may not be the best way to measure productivity when it comes to feeding people efficiently, from either a land-use or a nutritional perspective. For example, Dr. R. Bressani has suggested that using high-protein corn varieties in some countries could save both land and crop volume, while also improving nutritional yield. Bressani calculates that using high-protein corn may make it possible to supply the yearly protein needs of an average adult from half the crop and half the land area normally required with regular corn varieties.

If higher protein cereal crops were grown in countries where cereals constituted an important part of the diet, nutritional efficiency and agricultural productivity might both increase. However, much of the work on high-protein cereals seems destined to have more of an impact on the livestock industry than on feeding the world's hungry.

"Nutritionally superior cereals may have a more immediate and greater impact on animal feeding and the feed industry in the industrialized regions than on the diets of the poor populations (of) developing countries," say biologists C. R. Bhatia of India's Atomic Research Center and R. Rabson of the United States Department of Energy. One of the many reasons for this, say Bhatia and Rabson, is that "There is much greater control over animal diets than is possible with human diets. The formulation of feed is based on nutritional needs and the cost of each component of the formula is adjusted to keep the production costs low in a competitive market."

But if the genetics behind improving the nutritional profile of corn and other cereal grains is aimed at reducing livestock feeding costs, that will probably facilitate the worldwide spread of the feed-grain/meat system of food production worldwide—a system that critics such as Frances Moore Lappé and others charge is protein-inefficient. Consider, for example, the history and present use of high-lysine corn.

HIGH-LYSINE CORN

In 1963 at Purdue University, Dr. Edwin Mertz, a corn breeder, and Dr. Oliver E. Nelson, a geneticist, discovered a corn variety whose kernels were high in the amino acid lysine. Lysine is one of the essential amino acids that humans, pigs, and chickens cannot synthesize for themselves and must obtain from outside food sources.

Mertz and Nelson later identified the gene in the new corn variety that was responsible for its high lysine content—called the "opaque" gene because it produced non-translucent kernels. When bred into an ordinary variety of corn, the opaque gene would nearly double the level of lysine in the kernels.

This was an important discovery. In a number of tests and trials, high-lysine corn was fed to hogs, poultry, and children with positive nutritional results. Hogs fed the new corn gained five times more weight in a forty-five-day period than others from the same litter that were fed regular corn. There was also great hope that in countries in which the diet consisted primarily of corn, high-lysine varieties might help eliminate the disease called "kwashiorkor," that in children causes an enlarged liver, a bloated stomach, and, in some cases, death. "High-lysine corn can have miraculous effects on these malnourished children," says *Saturday Evening Post* science writer Cory SerVaas, M.D. Pointing out that when corn high in lysine and tryptophan was substituted in the diet of an emaciated boy brought to a Colombian hospital in 1967 who had been existing on a typical poverty-level diet of corn gruel and corn soup, Dr. SerVaas noted that his condition improved immediately. "Not only did he gain weight," she explains, "but his bones began growing again as well. In three months, he had recovered completely."

In the early 1970s, when the tragedy of famine hit Biafra and other African nations, high-lysine corn began to be touted as "a complete protein for people." Unfortunately, its yields, like those of other high-protein cereals, were lower by 10 percent or more. And there were other problems too. The kernels of the new varieties were soft, which meant more insect problems in the field—and, in industrialized countries, less-than-enthusiastic acceptance among corn millers and food processors.

"Del Monte uses a hard, endosperm dent corn," explains Purdue University's Edwin Mertz. "They want grit to make their corn flakes." Mertz also notes that high-lysine corn has a lower starch content than ordinary corn, for which reason dry millers and wet millers haven't touched it. "If it's got less starch," Mertz explains, "they say, 'forget it.' "

Nevertheless, during the late 1960s and early 1970s, a number of the major seed companies—including Pioneer, DeKalb, Funk Brothers, and others—all tried to develop commercially acceptable high-lysine corn varieties. But by the mid-1970s almost every seed company that tried to overcome the genetic riddle of combining a high-protein content with a high-yield capability and good grain quality gave up. Furthermore, when farmers "penciled out" the advantages and disadvantages of the new corn varieties, few would take the chance of trying them because of their publicized yield and insect problems.*

One seed company, however, stuck with it. The Crow's Hybrid Corn Company of Milford, Illinois has been working on high-lysine corn for more than fifteen years and now sells a number of high-lysine varieties that are comparable in yield to regular corn hybrids. Crow's overcame the ear-rot problem in high-lysine corn and raised its yield levels. Today, some dairy and hog farmers are becoming increasingly interested in the newer high-protein corn varieties for feeding purposes—one reason for which is that dairy cows produce more milk when fed high-lysine corn. But while acceptable varieties of high-lysine corn are beginning to be used on a limited scale, largely because of the persistence of Crow's, and more recently that of CIMMYT in Mexico, this nutritionally improved corn, at least in the United States, will be used primarily to feed livestock.

High-lysine corn, if developed extensively, will probably follow the expansion of livestock industries around the world. In South Africa, for example, where it was originally tried as a nutritionally-improved food source for some of that country's native tribal people, high-lysine corn is now grown primarily to feed a developing livestock industry, with farmers being paid a 10 to 12 percent "premium" or bonus for growing it instead of regular corn. Nevertheless, in some parts of the developing world, high-lysine corn is also beginning to be used as a nutritionally superior subsistence crop, although it is not being sold by major commercial seed companies. Government and cooperative international agencies like CIMMYT are helping to develop and distribute high-lysine corn seed—still on a very limited scale.

While most American seed companies remain skeptical about the

*Farmers who feed livestock on their farms might use high-lysine corn if it could displace the added ration and added expense of soybean meal in hog feed—soybeans being a good source of lysine for hogs. However, soybean meal is also a very cheap source of lysine, and the high-lysine corn varieties were expensive.

future business opportunities for high-lysine corn, a few biotechnology companies have taken a special interest in the high-protein corn. Sungene Technologies Corporation of Palo Alto, California, entered a joint agreement with Crow's Hybrid Corn Company in 1983 to screen Crow's corn stocks for high-protein mutants with the goal of using protoplast fusion or gene splicing to develop new varieties. Molecular Genetics, Incorporated (MGI), is pursuing high-lysine corn varieties with an eye toward selling them to hog and poultry farmers in the United States. While MGI's high-lysine project was begun under its own initiative, the new genetics firm is partly owned by American Cyanamid, a major chemical company. MGI notes, however, that there may be some future compatibility between animal health products, such as vaccines and growth hormones, and improved feed grains such as high-lysine corn. The biotech company is currently conducting vaccine research for Cyanamid and SmithKline Beckman, both of which are heavily involved in the livestock drug business.

As plant breeders and genetic engineers now working on the genetics of protein content in grain crops seem to be doing so primarily in terms of developing varieties suitable for livestock nutrition,* biotechnology is also being used to develop a "new kind" of

*Similar attention to the "performance characteristics" of wheat varieties suitable for the baking industry and barley varieties for the brewing industry are also being pursued by plant breeders and genetic engineers. For example, a December 1982 press release from the USDA explained:

"Wheat can be made an even more nutritious food for the world's hungry if a genetic puzzle about this food staple can be solved." The genetic puzzle to which USDA is here referring is the wheat's DNA and how it affects the crop's protein level. The press release was issued to explain one of the Department's research projects employing genetic engineering techniques to unlock the secrets of wheat protein.

"We are attempting to identify genes which will help us improve wheat protein," noted Dr. Frank C. Greene, the research chemist heading up the project at USDA's Western Regional Research Center in Albany, California. "We also want to know how these genes are 'turned on and off' in plants," he said. Greene explained he was particularly interested in the part of the DNA responsible for gluten, a storage protein found in the wheat kernel's endosperm.

Milling & Baking News, reporting on Greene's research in 1983 said that despite the fundamental nature of the project, the research is focused "on a part of the wheat plant used by millers and bakers, and thus ultimately, of great practical significance." But of great practical significance for whom? For the "world's hungry," as USDA's press release said, or for the milling and baking industry?

protein that may one day be used to feed both people and livestock, displacing the need for high-protein corn and soybeans, and building a whole new industrial "protein system" in the process.

PROTEIN FROM FACTORIES

In the early 1970s, several of the world's major oil and chemical companies, including Exxon, Du Pont, British Petroleum, and Standard Oil of Indiana, were at work on what they thought would be a new cheap source of animal and human protein that most people have never heard of; a substance called "single-cell protein." This protein is made by microorganisms, and is extracted from their dried cells. The microbes used in this process—fungi, bacteria, or algae—are given a food source (also called a feedstock) and are "grown up" in large batches in fermentation tanks. The multiplying microbes then serve as tiny "protein factories" as they reproduce, and the protein extracted from them can be fabricated into livestock feeds, used as food additives, or even turned into products that look like conventional foods.

Why did the energy companies take such an interest in SCP, as it is called? For one thing, they had surplus oil that might serve as a feed source for the microbes, and for another, given the world food situation, the protein so derived might prove a profitable venture.

Exxon, in fact, embarked on its program with the idea of producing a human-grade food product that would be the nutritional equivalent of an egg, and at one point entered into a joint venture with Nestlé, the giant Swiss food conglomerate, to produce and market such new high-protein foods in Third World nations. That partnership, and Exxon's venture in single-cell protein, ended when it ran into difficulty proving to regulatory authorities that the new foods would be free of hydrocarbon residues and would contain only those amino acids that were specified.

In fact, the oil industry's enthusiasm for single-cell protein—which was founded on the continuing prospect of cheap energy—cooled significantly when energy prices soared after the 1973 OPEC embargo. Because petroleum costs then constituted about 50 percent of the operating costs for making single-cell protein, many of the oil companies simply dropped their ventures. Single-cell protein, produced in a petroleum-based fermentation process, could not compete with conventionally made food and feed. Today, however, with the application of biotechnology and genetic engineering, that may be changing.

"With the advent of new biotechnology and the threat of potential world food shortages," says the Congressional Office of Technology Assessment, "interest in SCP may once again return. . . . The high protein content, good storage properties in dry form, texture, and bland odor and taste of SCP suggest real potential in feed and food markets." Using biotechnology and genetic engineering, for example, and with the aid of a non-petroleum-based energy source to "feed" the microbes, it is possible to drive down the costs of producing single-cell protein. Microbes used in the process could be genetically altered to be more efficient in their consumption of energy, or engineered to produce proteins with a more complete range of amino acids. They could also be genetically instructed to excrete protein, making its harvesting easier and more economical.

Not all of the energy and chemical companies involved with SCP in the early 1970s abandoned their investments; those that stayed with it are now using biotechnology to improve their processes, and several have very big plans for selling their protein systems and products all over the world.

One of the most aggressive corporations now producing and pushing SCP is England's largest company, Imperial Chemical Industries (ICI), a company also involved in fertilizers, agrichemicals, and pharmaceuticals. ICI clearly sees the world's protein needs expanding dramatically, assumes that a livestock-based "meat-milk-and-eggs" expansion will occur in many countries, and believes that its new SCP product, called Pruteen, is the feed ingredient that will make it all possible.

"As populations increase in affluence, they desire to eat an increasing proportion of their protein as meat, milk or eggs," explains ICI in its brochure, *Pruteen*. "The animals producing such foods require protein, most of which is obtained from soyabean meal and fishmeal.

"Many countries don't have the land area or suitable climates to produce sufficient quantities of protein. They have to import to allow their people to achieve the desired levels of nutrition. With the world population forecast to increase by 40% to 6 billion by the year 2000, it will be difficult to meet protein requirements from existing sources." It is against this background that ICI has developed "Pruteen," a 70% protein product which the company sees as a replacement for traditional protein sources. ICI's method of making SCP utilizes strains of bacteria which obtain their food energy from a methanol feedstock (methanol is an alcohol that can be made from oil or natural gas).

In ICI's brochure, the company juxtaposes two world maps, one depicting world protein demand which highlights the major protein-importing and -exporting regions of the world. The second map identifies over seventy locations throughout the world the company designates as "regions of surplus (natural) gas." ICI's brochure explains that there is an imbalance in the world protein trade, particularly because traditional protein production is limited by climate and land availability. "In essence," says ICI, "a few nations in the Western Hemisphere provide the bulk of high quality proteins to feed the animals and people in the rest of the world. As a consequence, most of the world suffers from a protein trade deficit and their populations have a lower than acceptable plane of nutrition."

ICI's plan to remedy this situation is to match its SCP process with those protein deficient areas of the world where a surplus of natural gas exists. Says the company in its brochure: "The production of SCP from natural gas through methanol is a unique opportunity for certain countries to convert an abundant, often wasted resource—flared gas—to a scarce resource, food. Such a transfer of resources increases a nation's independence and security and avoids the vulnerability of trading in two world markets—selling hydrocarbons to buy food. Such security, along with savings in imports and foreign exchange, make the adoption of SCP production irresistible in many countries."

Indeed, ICI claims that its brand of SCP has already been approved in thirteen countries for feeding farm animals and fish, and according to the company the process of adoption is just beginning. "The age of SCP has arrived and can add a new dimension to world food production patterns," says ICI. "It will establish a better balance in the regional supply and demand for protein. . . . It will offer new trade and employment opportunities and provide the base for a high technology industry—all of which will make a positive contribution to increased prosperity."

In the United States, the Phillips Petroleum Company, the nation's eighth largest oil corporation, has produced a SCP product similar to ICI's using its own patented strain of yeast. Like ICI, Phillips agrees that "there is a food shortage in many parts of the world" and that its product, Provesteen, provides a solution. "The Provesteen process was developed by Phillips to meet the demands of many of the world's countries unable to supply their own protein needs because of climate or geographic conditions," says the company in its booklet —*Provesteen.*" Phillips' brand of SCP is now sold as an animal feed ingredient, but eventually it may become suitable for sale in the human food market, according to Phillips. ICI, too, sees its product moving into the human food market. "Laboratory developments have confirmed that 'Pruteen' has the potential to become a highly acceptable ingredient of human foods with many of the desirable functional characteristics required by food manufacturers," says the company in its brochure.

The United States' Food and Drug Administration has approved Phillips' SCP that is made from a molasses feedstock for use as a food additive. However, Dr. Nevin Scrimshaw, an expert in single-cell protein and the director of the Massachusetts Institute of Technology's Clinical Research Center, says that the nucleic acid content of this product will have to be reduced before it could be approved as a food product. On the other hand, FDA has not been asked to approve Phillips' SCP made from methanol because it is not expected to be used in the United States.

Meanwhile, Phillips sees a booming world market for its SCP within the next two decades, and foresees a day when Provesteen will produce annual revenues in excess of $1 billion annually. "Most of the population growth will take place on land that is presently devoted to agriculture," says David Dreisher, vice president at Phillips' subsidiary, the Provesta Corporation. As the amount of arable land shrinks," he explains, "production of single-cell protein will play a greater role because it can be produced year-round in a controlled environment undisturbed by weather, blight, or pests."

In addition to single-cell protein systems developed by Phillips and ICI using genetically instructed microbes, there have also been other kinds of cellular-based protein systems envisioned utilizing plant tissue.

"There is the prospect that someday people will see huge factories pouring out soybean endosperm . . . at the end of a long, slow-moving conveyor belt," writes *Boston Globe* science writer Robert Cooke in his 1979 book, *Improving on Nature*. "In such a factory . . . small snips of soybean tissue would be prepared and loaded onto the belt (and) as they moved toward the far end of the factory—would be doused with nutrients, hormones, and light," explains Cooke. "By the time they reached the end of the trail they would have multiplied many times to produce large amounts of protein-rich soybean cells. No plants need to be involved, except for supplying the first few bits of starting tissue."

Cooke points out that soybeans are already widely used throughout the world as a source of protein, and that in countries such as the United States, the beans are now processed into substitutes for meat and other staples. "If soybean meal could be produced in even more massive quantities, cheaply enough and with scant use of energy," Cooke offers, "the results could make a big difference in the food picture everywhere. . . . A factory rolling out tons of soybean meal . . . would also be isolated from the vagaries of weather, from the problems of crop diseases, poor soil quality, and insect attack." Thus, concludes Cook, "by eliminating the need for the plant, it's also possible to eliminate the need for sunshine and soil."

PROTEIN INFLATION, PROTEIN POWER

The protein-factory model of food production—whether single-cell protein, soybean endosperm, or huge livestock feeding operations—has the potential to consolidate the control of food production while inflating the number of steps involved in protein manufacture. Single-cell protein systems, for example, assume a dairy, poultry, hog, fish, egg, or cattle operation as the automatic "next step." The related "protein steps" in such a system include producing the SCP in the factory, feeding it to fish or livestock, then feeding the fish, eggs, meat, or dairy products to people.

There is, of course, nothing "wrong" with people eating meat or dairy products, or with companies aiding in the production of these products, but there are questions about cost and efficiency—questions that relate to how protein makes its way through the food system, and how much it will cost a nation and its people to produce and consume. The key question, then, is protein for whom and for what purpose—to feed people or to inflate the "value-added" function of agribusiness; to serve food-processing needs; or to expand the business bases of certain kinds of agriculture?

Soybeans and corn or other crops fed directly to people in some countries would seem to be a more cost- and resource-efficient way of producing and supplying protein than the feedlot system—and certainly more efficient than the use of "surplus" natural gas or petroleum to build a "single-cell protein-with-livestock" system.

However, for some foreign governments, the high-tech protein systems hold the initial promise of mass-producing "food" very cheaply and in huge quantities, and without the anxiety of traditional resource requirements or the vagaries of weather. Yet the installation of these systems in foreign nations will not come without costs and potential vulnerability. New dairy, poultry, swine, and cattle operations throughout the world are likely to use the same "best breeds." Genetic vulnerability will certainly follow. Diets, the price of food, and farming systems will be altered too.

The ability to instantly produce huge herds of livestock in foreign countries through embryo engineering, coupled with the prospect of genetically developed high-lysine corn or high-methionine soybeans to feed them with, could dramatically transform cereal-based diets and dietary habits in developing countries. A single-cell protein system based on petroleum or natural gas could become very expensive should energy costs suddenly rise. Moreover, such systems, depending on how rapidly developed, might also displace local farmers growing protein crops and/or feed grains. Alternatively, single-cell protein systems based on crop feedstocks would likely shift local cropping patterns and may invite plantation-styled and only-for-contract kinds of agriculture using specified, uniform varieties to supply the factories.

CHAPTER 15

THE POWER OF PATENTS

If I had a child headed into a career now,
I'd want him to be a patent lawyer—
preferably a biotechnology patent lawyer.
—*Orrie M. Friedman, Chairman*
Collaborative Research, Inc.

In 1981, S. Leslie Misrock, a senior partner with the New York law firm of Pennie & Edmonds, hired three Ph.D.'s—one in biochemistry, one in cell biology, and another in chemical engineering—and sent them off to law school. Misrock, and others like him in the influential world of corporate patent law, are anticipating a new kind of lawyering. They are preparing for the myriad lawsuits expected to come with commercial biotechnology. And only lawyers grounded in science will do, for only they will be able to cross-examine a microbiologist in a patent fraud case or deal adroitly with the nuances of federal regulations in environmental litigation. The "big ticket" cases for law firms like Pennie & Edmonds and their new stables of high-tech lawyers will be the patent cases.

Patents are government grants of limited monopoly awarded to individuals and institutions to create an incentive for research and innovation by giving inventors an exclusive marketing opportunity for a specified period of time.* The inventor makes money and

*According to University of California professor of law Lawrence A. Sullivan, author of *Handbook of the Law of Antitrust,* "the patent system presupposes that we cannot rely on the market to determine the socially optimal investment in innovation."

society gains useful products—or at least that's the theory of how the system should work.

The first patent rights were granted in Venice in 1474, and since then "intellectual property law" has been evolving in various forms, and occasionally with some controversy. Over the centuries, kings and emperors, popes and statesmen—and today, economists, lawyers, and politicians—have argued about the social utility of patents, torn between providing incentive for inventiveness and innovation, and the social and economic consequences of giving monopoly powers to obtain those benefits.

The social benefits generally attributed to a patent system, in addition to those of spurring innovation and new knowledge, include distributing the inventions and knowledge throughout society and increasing social welfare through the widespread use of beneficial products. Social costs, on the other hand, can include monopoly pricing, market barriers to new competitors, redundant research activity, and the downplaying of what economists call "externalities" or third-person effects—such as the hidden, long-term costs of developing new chemicals and disposing of their waste.

Since the inception of the nation's first patent law,* written by Thomas Jefferson in 1790, much has changed. For one thing, "authors and inventors"—in the sense of individuals as the terms no doubt meant in 1790—have today become corporations, governments, and universities. Inventiveness is no longer the individual inventor's flash of genius, but now an institutional process. The scope of the patent system has also expanded greatly since 1790, covering a range of living things such as fruit trees and strawberries (since the 1930 Plant Patent Act); seed-bred plants such as wheat and soybeans (since the 1970 Plant Variety Protection Act); and genetically modified microorganisms (since the 1980 Supreme Court decision in *Diamond v. Chakrabarty*).

In highly technical societies, patents have become a source of power. They are the legal basis for making a commercial leap forward with a new product, and increasingly, the means for staying

*The basis for the patent system in the United States rests in Article 1, Section 8 of the U.S. Constitution, inserted there in 1787, and reads: "The Congress shall have the power to promote the progress of science and useful arts, by securing for a limited time to authors and inventors the exclusive right to their respective writings and discoveries."

on top of a market and subduing competitors. Corporations fight long and hard with one another to defend their patents, with the loser often bowed and bloodied, sometimes driven out of business or severely weakened as a result. The winners, on the other hand, can emerge in an even more powerful position.

In 1983, in one of the largest patent cases in American history, Pfizer was awarded $55.8 million in compensation from the International Rectifier Corporation after Rectifier infringed on Pfizer's patent for the animal antibiotic doxycycline. In place of an actual cash award, Pfizer received the animal health and feed-additive businesses of Rectifier's subsidiary, Rochelle Laboratories. Prior to this award, Pfizer was already a major player in the animal health business.

For many corporations, the principal purpose of a patent is to serve notice of ownership. It is not unlike the action of a Texas rancher putting his "T-Bar" brand on the hide of a Longhorn steer. The patent, in this sense, is more like a warning flag that says "litigation is expensive"; a signal to existing and would-be competitors that the owner is prepared to fight long and hard to maintain his patent position—and the market that goes with it. Among business interests, the patent is more a statement of territory than it is necessarily a measure of innovation. But when those "territories" are food and the substances behind the production of food, there is an order of protected power that is broad and far reaching.

Today, with the arrival of bioengineering in the food and agriculture system, there is a new round of patent gathering and patent lawyering, for much is at stake. New companies such as the Cetus Corporation were budgeting $1 million a year by 1983 to file for and defend new biotechnology patents; established corporations spend even more. But as both new and old companies begin maneuvering in the biotechnology patent sweepstakes, there are still questions about patenting life forms that have not been adequately debated.

PATENTING GENES

Since the *Diamond v. Chakrabarty* decision, allowing the patenting of man-altered microbes, science and commerce have assumed that all of biology is now eligible for patenting and private ownership. Yet one of the issues left hanging by the Supreme Court is where to draw the line in patenting life forms—including those used in agriculture and food production. If a genetically altered microbe or seeds can be patented, then why not new breeds of cattle, chickens, or hogs?

In fact, in 1974, one farmer who discovered a hen with the genetic trait for dwarfness applied for a United States patent on the hen. The farmer discovered that the dwarf hen, when mated with normal cocks, produced regular-sized offspring, but—because she was smaller—did so on less feed, thus reducing costs. Dwarf hens of this kind could save poultry producers a bundle of money.

However, the patent examiner reviewing Merat's application rejected it, in part because he believed the chicken was not a "manufacture" under the terms of the patent law, and that the farmer did not make a distinct enough claim on his "invention," using the term "subnormal" to describe the dwarf hen. On the farmer's appeal to the U.S. Patent Office Board of Appeals, the board agreed with the patent examiner, adding that breeding chickens was not a "process" either, which might make it eligible for another kind of patent. A second appeal, this time to the Court of Customs and Patent Appeals, also resulted in the denial of the patent, so farmer Merat never patented his dwarf hen. But all of that was before the 1980 *Chakrabarty* decision. Now genetically altered microbes, and all other living organisms as well, provided they were changed in a novel way and had no twin in nature, might be eligible for patenting—or at least that is the current interpretation of the Supreme Court's decision by industry and many lawyers. Before long, there may well be patents issued on dwarf hens and other kinds of livestock.

In fact, some argue that it is the genes of a microbe that are being patented. "A given microorganism," says David Jackson of the Genex Corporation, "is ultimately defined by the sequence of base pairs in its DNA molecule or molecules." Jackson says that the nature and the identity of a microorganism are defined by its set of genes. "This is surely how the most precise definition of a mircroorganism will ultimately come to be specified," he says. But Jackson takes this argument one step further by suggesting that there are "many pieces of a DNA molecule smaller than the entire genome of a microorganism" that are, in his view, "eminently patentable." In other words, Jackson believes that individual genes* are patentable.

Given the individual gene theory of patenting, all manner of specific genetic traits in the plant and animal world—traits that control and determine food production—might soon be eligible for patenting. This means, for example, that a unique corn gene for yield, or those for disease resistance, high-protein content, or any one of a thousand other traits, might be routinely patented. It also presumably means that such genes could be owned and marketed as exclusive property for a seventeen-year period, and with recent changes in Federal law, as long as twenty-two years for some pharmaceutical products.

However, some companies engaged in agribusiness haven't always upheld the right to patent substances in the realm of nature.

THE JONES HYBRID CORN PATENT & THE CORPORATE DEFENSE

In the early days of producing hybrid corn, one of the difficulties encountered by seed companies and researchers in cross-breeding was the need to control pollination among corn plants to insure that the right genetic materials combined in timely fashion to produce a hybrid. This problem was overcome by detasseling some corn plants so they would then serve as females in the hybrid cross. The removal of the male, pollen-bearing tassels was done by hand—over hundreds of acres on the commercial level—and for years provided gainful summer employment for hundreds of high school students in the Midwest. Eventually, however, hand detasseling was eliminated in commercial seed production, thanks to some fortunate discoveries of genetic traits in corn plants.

In 1931, a scientist named Rhoades discovered a corn plant that produced sterile tassels, and for a while there was great excitement among plant breeders who believed the discovery would lead to the end of hand detasseling. And while this male-sterile trait was in-

*The patenting of individual genes has also been discussed by the International Union for the Protection of New Varieties of Plants (UPOV). In 1976, the French government recommended at a UPOV meeting that a study of the question be undertaken, reasoning that changes in new plant varieties by way of plant breeding were actually changes incorporating one or more useful genes into the new variety.

herited in a way that produced sterile tassels, it also resulted in hybrid seed that produced cobs without kernels. Something else was needed to restore fertility in the final cross.

In Texas meanwhile, two other scientists, Paul Mangelsdorf and John S. Rogers discovered another male-sterile mutant, one that carried the infamous T-cytoplasm that would later facilitate the spread of the 1970 Southern Corn Leaf Blight. As a graduate student, Mangelsdorf gave some of the Texas strain to Donald Jones who thought there might be a way to restore its fertility in the making of hybrids. In 1948, Jones discovered a fertility-restoring gene that could be used in combination with the new male-sterile cytoplasm. This meant that the hybrid seed the farmer used would carry the fertility-restoring trait, thus insuring a crop.

When Mangelsdorf and Jones applied for a patent in 1948 on only the Texas male-sterile cytoplasm, they were turned down. Yet in 1956, Jones and Mangelsdorf were awarded a patent for the fertility-restoring system; the latter Jones-Mangelsdorf patent was awarded because of the *method* Jones had devised. Although this patent gave Jones and Mangelsdorf the right to collect royalties from seed companies that used the scientists' male-sterile system in producing hybrid corn seed, no such royalties were paid until 1970, when a court action required the companies to make the payments.

After Jones and Mangelsdorf were awarded their patent, they turned it over to the Research Corporation, a New York firm which specialized in representing university interests. In 1963, after the Research Corporation was unsuccessful in warning the seed companies that they were infringing on the patent, the Research Corporation filed a class-action suit against several hundred seed companies then using the Jones-Mangelsdorf technique. By that time, nearly all the hybrid corn seed grown in the United States was produced from inbred lines containing the Jones/Mangelsdorf system.

"The seed companies fought the Jones patent," recalls Dr. James Horsfall, a plant pathologist who worked at the Connecticut Agricultural Experiment station, and who chaired the National Academy of Sciences committee that wrote the 1972 report on the Corn Blight. "They objected (to the patent) on the basis that you can't patent living things, that you can't patent nature and that sort of thing," he said. After seven years of legal maneuvering by both sides, the issue was finally settled out of court in 1970. The seed companies agreed to pay five and a quarter cents royalty on every bushel of hybrid corn seed produced using the Jones-Mangelsdorf method during the years 1961 through 1973, which came to about $2.5 million.

Since 1970, the American seed industry has been changed by its own patenting law, and many seed companies, and especially some of their parent corporations, have become vigorous champions of the plant patenting laws as well as the patenting of other genetically altered substances in agriculture. In fact, some seed companies are now suing farmers and seed dealers who do not abide by the provisions of the 1970 Plant Variety Protection Act.

PATENT LAW INTIMIDATION

When patenting came to the American seed industry in 1970 by way of the Plant Variety Protection Act (PVPA), it changed the way business was done in that industry—but not overnight. The concept of property rights in the seed business was alien to many who worked in the front lines of the business—seed dealers, farmers, and plant breeders at land-grant colleges who worked with public domain seed. Farmers were accustomed to trading and selling seed with each other and obtaining it cheaply from local seed companies or cooperatives that used university-developed strains. Seed dealers would occasionally mix or blend public-domain seed varieties to suit local needs or to experiment with their own mixtures. Public plant breeders were used to exchanging seed freely with breeders in industry and with their colleagues at other universities for the purpose of further research. But within ten years of the PVPA, those practices began to diminish dramatically. The first lawsuits came in the late 1970s.

In a 1978 case of patent infringement, seed mixing, and improper advertising, the North American Plant Breeders Company (NAPB) of Mission, Kansas, then owned by the Olin Corporation, sued an Oregon seed dealer named Dick Haynes and his business, Farmterials, for $150,000 in actual damages plus $300,000 for treble damages. Although Haynes was protected by his business insurance, and the final court award to NAPB was only about $7,000, Haynes says NAPB's action "scared the hell out of me—to have that much liability."

Haynes believes that the NAPB suit, which resulted in a permanent injunction of Farmterials from ever selling NAPB's "Lud" barley—the variety that Haynes sold under another name—was aimed at Farmterials as an exemplary action to show other seed dealers in the region that NAPB meant business.

For Haynes, the NAPB lawsuit hurt his business relationships with other seed companies. He notes, for example, that Germain's Seed Company, one of the largest companies handling barley and alfalfa varieties in the Northwest, would not permit Farmterials to be an exclusive handler of their varieties. According to Haynes, "they weren't satisfied with the way I was doing things . . . and they heard some of my radio commentaries* taking issue with the Plant Variety Protection Act."

Farmers and farm co-ops have also been sued under the PVPA and the Federal Seed Act—a law that was tied to the PVPA in 1970.* Technically, under the PVPA farmers are allowed to save

*One of Haynes' radio ads went like this: "Hello, this is Dick Haynes at Farmterials in Baker. When buying alfalfa and grass seed this Spring, be careful of some of the new patented varieties. The PVPA has enabled large companies, many of international origin, to patent new varieties. These varieties to be patented only have to be different, not better. So many times farmers are falling victim to expensive national advertising campaigns or glossy brochures and paying high prices for seed that is actually inferior in performance to many of the old standard varieties. Alfalfa like Ladak, Vernal, and Ranger (Public varieties) are proven high producers and long lived, while many of these new patented varieties are short lived and grown in southern climates. If you want long lived winter hardy alfalfa that is drought resistant, buy our Ladak. . . . The price is only $1.50 per pound. Deal with a seed company you can trust . . . that's Farmterials in Baker."

*In 1970, it became a violation of federal law to *sell* uncertified seed of certain plant varieties registered under the PVPA. With this legislative action, the commercial aspects of *plant patenting* were now tied to the regulatory aspects of *seed certification*—a governmental function designed to protect farmers from false labeling and misrepresentation in seed quality. Yet, according to one Congressional Research Service review of the connection between the PVPA and the Federal Seed Act:

> . . . Persons in the Seed Regulatory Branch of AMS [Agricultural Marketing Service, USDA], who asked to remain anonymous, complain that the use of section 501 [section that cross-references PVPA and Federal Seed Act], to protect the interests of plant variety certificate holders is a misuse of the Federal Seed Act, the purpose of which is to protect the rights of seed purchasers by assuring accurate labeling.

the seed from their own crop, use it themselves the following year, and even sell it to their neighbors. But after an 1981 amendment to the farm bill, the "farmer exemption" in the PVPA for sale to neighbors was effectively eliminated for any variety that was classified as "certified seed" under the Federal Seed Act, of which most patented varieties are. Farmers selling such seed to neighbors, "over the fence" as it were, would then be selling certified seed, and so would be in violation of the law.

During the late 1970s and early 1980s, when the first legal actions were taken to enforce the PVPA, many farmers who received letters of notification from company lawyers simply didn't know the seed patent law existed. Others, learning of the law's power, became resentful of its reach. "In this area," explained Falls County, Texas agricultural agent Ronald Leps in 1982, "farmers simply aren't buying certain varieties. They don't like being threatened by seed companies."

Yet, after a rash of lawsuits, many brought by a seed-company subsidiary of a major corporation, farmers and seed dealers got the message.* The old days of freely trading and selling seed under the casual order of verbal promises and handshake transactions had given way to the certainty of court-supported patent laws and proprietary interests. A way of doing business for over 200 years had changed quite abruptly; farmers' access to the supply of seed had changed; and ultimately, the way crops get into the ground had changed.

THE EFFECT ON UNIVERSITIES

At some land-grant universities, secrecy in plant breeding and plant-genetics research, and restrictions on the exchange of seeds

*Enforcement of the PVPA/Federal Seed Act provisions so far has come about through several court actions in addition to the NAPB/Farmterial case, including: a case brought by Ring-Around Products of Alabama (owned by Occidental Petroleum) against two certified seed growers—one in Missouri and another in Kentucky—for unauthorized sale of "Mitchell" soybean seed; a suit by Dixie Portland flour mill against some Kansas farmers for selling "Plainsman V" wheat to their neighbors; a suit brought by Coker Seed Company (KWS AG) of Hartsville, South Carolina, against a Texas farmer and two local cooperatives for selling "Coker 68-15"; and a suit brought by the Delta and Pine Land Co., of Scott, Mississippi (owned by Southwide Corp.), against two cotton gin co-ops for selling the seed company's "Deltapine 41" cotton.

are becoming commonplace. Critics charge that this began in the early 1970s with the enactment of the PVPA, and say that lately, with the growth of commercial biotechnology, university plant breeders have become even more reluctant to exchange scientific information. Moreover, some universities have resorted to the practice of "exclusive release" of new crop varieties to seed and biotechnology companies in return for funding,* leading to concern that secrecy for patents and exclusivity together will "choke off" all scientific exchange in plant genetics.

One barley breeder at Arizona State University, attempting to obtain some plant germ-plasm (seeds and other plant materials) from other breeders at Washington State University for his own breeding program, was told that he couldn't have it "because it may be patented." Virtually the same response followed a similar request to Montana State University. Scientific communication between breeders at some universities has also stopped. "Montana doesn't talk to Idaho," he explains, and Purdue has become especially covetous of its disease-resistant plant materials. "You can't get anything from Purdue," he says, "and you can't get anything out of (the University of) California." Iowa State University and the University of Illinois, he says, are also on the list of land-grant universities where barley breeders are reluctant to share their plant-breeding stocks.

However, this Arizona barley breeder has a stern warning for those universities now restricting the exchange of their seeds: "Once you close your [breeding] program in," he says, "you're not going to produce good varieties. Those that take this path," he says, "will self destruct because they won't have anything worth selling." Eventually, he claims, the "seed companies will take over; they have more money and more land. Public varieties won't be worth growing (and) farmers won't buy them."

*Cornell's Department of Plant Breeding and Biometry, stung by state and federal budget cuts, has entered into an agreement with several seed companies in which the companies get exclusive access to new lines of vegetable seed in return for financial support. According to one vegetable plant breeder working for the Agrigenetics Corporation in Minnesota, the Cornell arrangement is essentially a *quid pro quo:* "If you (the seed company) provide us with x amount of dollars," he says of the arrangement, "we will provide you with first shot at our releases." As this industry plant breeder points out, a "new release" in a vegetable line can amount to "a pretty big jump" on another competitor.

WILL PHOTOSYNTHESIS BE PATENTED?

Today, with biotechnology and the ability to patent genes, the legal maneuvering of business and government has gone beyond plants and seeds. Major corporations and biotechnology companies are now patenting very broad biological and genetic processes, some of which will be key to food production in the years ahead.

According to E.F. Hutton's genetics expert Zsolt Harsanyi, the trend among companies involved in biotechnology research is "to play it safe and patent everything." Genentech's Vice President for Legal Affairs, Thomas Kiley, adds, "When you have a chance to write a clean slate, you can make some very basic claims, because the standard you're compared to is the state of prior art, and in biotechnology there just isn't very much." By 1984, Genentech alone had 1,400 pending patent applications.

A few of the biotechnology patents granted so far have been so broad as to constitute some of the very basic biological processes necessary to conduct even rudimentary genetic engineering. For example, in December 1980, Stanford University was awarded a patent for the Cohen-Boyer method of gene-splicing, a basic recombinant DNA process. Because Stanford holds this patent, it now sells licenses on the process to those who want to use it at a cost of $10,000 each (since obtaining its patent, Stanford has issued more than 75 licenses, mostly to companies such as Upjohn, Eli Lilly, Monsanto, and Atlantic Richfield).

"I'm shocked at the [broadness of the] patents people are seeking," says Boston Biomedical Consultants' biotechnology analyst Barbara Lindheim. "It goes against the give-and-take and sharing that has been the major impetus for major biological breakthroughs." In agricultural biotechnology, basic methods of cell culturing, new techniques for hybridizing plants, and genetic methods for sexing seven-day-old livestock embryos have been patented.

The Agrigenetics Corporation, now owned by Lubrizol, has patented a biotechnology process that allows for the rapid development and commercial production of hybrid seeds without the normal limitations imposed by traditional hybridization techniques. The Agrigenetics technique has the potential of reducing the number of generations needed to develop hybrid seed, and is described as having "broad commercial ramifications" because it can be applied to numerous crop species.* In addition to this patent, Agrigenetics also has an exclusive license arrangement with Oregon State University for pending patent rights relating to a biotechnology technique for improving nitrogen fixation in certain plants.

Atlantic Richfield has applied for a patent on a new method of regenerating whole plants from protoplasts (plant cells which have had their cell walls removed). ARCO's new method has been used to regenerate whole tomato plants from tomato protoplasts with a high degree of success—a process that was previously rare and unpredictable, and one that will now speed the screening of tomato plants for desired genetic traits.

It is only a matter of time before other companies and universities begin to patent other important genetic techniques, or perhaps even entire packages of genes, in key biological processes such as nitrogen fixation and photosynthesis—processes that are central to food production as well as essential to broader ecological functions.

PATENTS & POWER

Patenting has always been important to industrial development, technological progress, and useful social innovation. But now, patentable science has crossed an important biological and genetic threshold. "Bio-patenting"—the patenting of plants, seeds, genes, and microbes—is contributing to a consolidation of corporate power in the ability to produce food; a consolidation in the ownership and use of key ingredients in agriculture.

Today, for example, more than 1,200 seed patents have been issued by the United States Office of Plant Variety Protection. However, more than half of those seed patents are held by the subsidiaries

*The Agrigenetics patent, however, was not well received by seedsmen in the United States, and some English scientists were quite upset with it as well. Professor Neil Innes, of UK's National Vegetable Research Station and also chairman of the British Association of Plant Breeders, claimed that many elements of the patented Agrigenetics procedure were both well known and in frequent use. In fact, Innes was so concerned that he petitioned the U.S. Patent Office to withdraw the Agrigenetics patent on the grounds that it was "obvious" and based on already published materials. Part of the procedure, involving the micropropagation of plants in tissue culture, had been discussed in the scientific literature as early as 1978, according to Innes. The petition is still pending.

of fifteen corporations. Five corporations—Upjohn, Sandoz, Lubrizol, Royal Dutch-Shell, and ITT—account for 30 percent of all PVPA patents issued.* In some individual crop species, five or fewer companies hold most of the patents.

Agrigenetics, Upjohn, and Sandoz are among the leading patent holders in soybeans, wheat, and vegetables. Pioneer Hi-Bred International—the dominant firm in hybrid corn seed—is also gathering patents in soybeans and wheat. Cargill has patents on safflower, ARCO leads in onions, and Occidental Petroleum tops the list in soybeans. Sandoz wheat varieties now grow on thousands of acres of farmland in Minnesota, Montana, and the Dakotas, while one Adolph Coors barley variety is used extensively in Colorado.

As for biotechnology patents†—or what the U.S. Patent Office calls "patent subclass mutation or genetic engineering"—a similar pattern exists. As of year's end 1982, the top ten patenting institutions—Upjohn, Ortho Pharmaceutical, Ajinomoto, Harvard College, the University of California, Research Corporation, Cetus, Genentech, the Noda Institute, and Agroferm AG—accounted for nearly half of all biotechnology patents. Upjohn and Ortho alone held nearly 30 percent of those patents.

Patenting is also helping to shift capital investment in agricultural research, molecular biology and plant genetics from the public domain to the private sector, and primarily to major corporations. All of the nation's land-grant universities put together, for example, have obtained about 12 percent of the PVPA seed patents, about equal to the number acquired by one major pharmaceutical corporation, Upjohn. And while it is true that Harvard and the University of California now number among the top ten recipients of biotechnology patents, this is, to some extent, a function of early discoveries by scientists who have now left those universities. Moreover, as "patent fever" has gripped the universities in their search

*It is also interesting to note that five foreign corporations—Sandoz (Swiss), Royal Dutch-Shell (Britain), KWS AG (W. German), Limagrain (French), and Royal Sluis (Dutch)—account for 20 percent of all PVPA patents.

†Prior to the outcome of the *Chakrabarty* case, the U.S. Patent Office had suspended action on some 130 applications for other microorganisms, ranging from genetically altered microbes proposed for use in mining to those capable of fermenting corn starch into ethanol. After the *Chakrabarty* decision, most of these patents were granted, and by 1982 more than 200 applications a year were being filed with the U.S. Patent Office for biotechnology-related "inventions."

for new sources of revenue, their research mission has become more commercial and less academic.

Corporations, meanwhile, want more. Some American corporations, for example, want uniformity in patent laws around the world, and they want foreign countries—particularly Third World countries—that don't have such laws now to adopt them and make them consistent with those in the developed world.

"The competitive position of U.S. industry (in biotechnology) would be improved if there were international conventions that would provide greater uniformity with respect to patentability and (property rights)," said Du Pont's Ralph Hardy in a May 1984 speech at the National Academy of Sciences. "There are some countries that do not recognize (property rights)," said Hardy, "and this will significantly retard the development and early commercialization of products that would improve the health and food supply of these countries." Yet those speaking out most often in favor of worldwide property rights are not the developing countries, but multinational corporations.

"The major challenge to genetic engineering scientists and companies, as well as national governments," says Monsanto's executive vice president Nicholas L. Reding, "is to support strong uniform worldwide property rights." Reding told an international gathering of agrichemical specialists in 1983 that patents were "under attack," and that some governments did not keep certain kinds of scientific data secret from competitors, and were lenient about pirates who counterfeited products. "No rational manager," said Reding, "is going to introduce products developed at great expense into a nation that does not protect the exclusive rights to their manufacture and sale." Without the assurance of adequate legal protection for their investment in foreign nations, said Reding, corporate patent owners "will avoid these countries."

In the United States, some corporations and biotechnology companies want the federal government to reinstate a 1977 "fast track" patent procedure that was instituted by the Commerce Department to give priority to genetic engineering. At the time, however, Joseph Califano, then Secretary of Health, Education and Welfare, asked that the procedure be suspended while a review took place. In February 1984, the Association of Biotechnology Companies (ABC) petitioned the U.S. Commission of Patents and Trademarks to reinstate the 1977 procedure. ABC believes that the fast track allowance—which might mean that biotechnology patents could be approved one year sooner—would help spur the development of the new industry.

THE PUSH FOR LONGER PATENTS

In its 1980 annual report, the Monsanto Company expressed concern about the climate for technological innovation in the United States, noting a decline in industrial research and development. Over the last several decades, the chemical company explained, certain developments "have eroded the traditional rights of patent holders." The registration of pesticides, said the company, involved lengthy pre-market testing and regulatory review which "effectively reduced the patent term well below the seventeen years intended."* In fact, for some of Monsanto's pesticides, the "usable" patent term after testing was as short as eight to ten years. Drug companies also complained that the regulatory process was cutting into the patent life of their products, barely allowing enough time in the market to recover research costs. This, they said, discouraged further research and innovation.

"As one step to return the patent system to its position as a major incentive for innovation," said Monsanto in 1980, "the term of patent protection for regulated products and their uses should be extended to restore the years of patent life lost in the review process." Before long, the National Agricultural Chemicals Association agreed, and so did the Pharmaceutical Manufacturers Association. A bill called the Patent Term Restoration Act was introduced in Congress—a bill that would extend the patent life on products for the years "lost" during pre-market testing and regulatory review. This bill, however, eventually grew to include drugs, pesticides,

*The United States patent statute was derived from earlier English and American colonial laws. The seventeen-year term of the patent is a product of both historical precedent and political compromise. The fourteen-year patents under early English law were based on the idea that two sets of apprentices could, in the time of seven years each, be trained in the new techniques. Occasionally, a third seven-year term was granted in exceptional cases. The seventeen-year term now used in the American system was the "middle ground" between the historical fourteen-year term and the exceptional twenty-one years of protection. But today, major corporations are pushing for at least twenty-two years.

animal health products, new industrial chemicals, and genetically-engineered products.

The push for longer patents in Congress initially came from the pharmaceutical industry with a bill introduced in 1980 by Indiana Senator Birch Bayh. By 1982, however, this bill and a companion measure in the House of Representatives ran into a maelstrom of opposition. Editorials in a number of newspapers around the country opposed the bill; some of the bill's opponents argued that there wasn't much evidence that drug research had been stifled. In fact, drug industry research investment had increased every year between 1965 and 1978, and the eight best-selling drugs in 1980 enjoyed an average patent life of 15.1 years.

In Congress, Senators Edward Kennedy and Howard Metzenbaum charged that the bill was "not in the public interest," noting that in 1980, "the drug industry earned 20.5 percent on equity, the nation's fourth most profitable industry . . . far above the 14.5 percent American average."* Nevertheless, the bill cleared the Senate, and with the backing of the Reagan Administration, the American Bar Association, the U.S. Chamber of Commerce, the American Heart Association, numerous medical societies, and several universities, was headed for sure approval in the House. There, however, it met the opposition of consumer groups, generic drug makers, and sympathetic Congressmen who defeated the bill temporarily.

Meanwhile, pesticide, animal health, and biotechnology interests had been clarifying and expanding sections of the bill that covered agricultural and biotechnology products. In April 1981, for example, Genentech's general counsel, Thomas Kiley, proposed that the Patent Term Restoration Act be broadened to include "process patents" such as those used in genetic engineering and the biosynthetic production of substances such as insulin, interferon, and animal growth hormones. By the time the bill was introduced in 1983, it contained a provision for including process patents. Yet the consumer opposition to the drug patent bill remained, and was in fact

*In the U.S. pharmaceutical industry, there are about twelve American-based multinational corporations—including Eli Lilly, Merck & Co., SmithKline, Upjohn, and Pfizer—that account for about half of all U.S. drug sales and well over two-thirds of all private drug research. In addition, foreign corporations such as Hoffman-La Roche and Ciba-Geigy and are among the top twenty firms in U.S. drug sales, with $100 to $200 million in drug sales each year.

strengthened in 1983 when the American Association of Retired Persons and the AFL-CIO joined the fight. Then a bargain was struck.

Democratic Congressman Henry Waxman of California, who had opposed the bill for consumer reasons, but who also had a bill pending to speed approval of generic drugs, allowed a deal to be fashioned. The Pharmaceutical Manufacturers Association—the chief proponent of the patent bill—and the Generic Pharmaceutical Industry Association—the chief proponent of the generic drug bill—agreed on a combined new bill that would allow drug patents to be extended by as much as five years *and* generic drugs to be approved more quickly. That package was approved by Congress in 1984.

Agrichemical interests, not part of the drug deal, were advised to pursue their own bill, parts of which were taken from the original drug patent bill and fashioned into "the Agricultural Patent Reform Act of 1984." This "pesticide patent bill," as it was nicknamed, allowed for extended patent terms of up to five years for pesticides, animal health products, new chemicals, and biotechnology products, and had a long list of supporters, including, most prominently, the National Agricultural Chemicals Association, the Animal Health Institute and the Chemical Manufacturers Association. Other organizations, such as the American Seed Trade Association, the National Farm Bureau, the National Cattlemen's Association and the National Association of Wheat Growers also supported the bill.

Industry supporters of this bill thought they could easily move it through the Congress in the wake of the drug bill without much trouble. But some opponents of the bill, including the National Audubon Society, Congress Watch, the Environmental Policy Institute, and the National Coalition Against the Misuse of Pesticides, pointed out the fact that one pesticide reform bill and another pesticide food safety bill were both stalled in Congress, and that those bills should move before consideration of any pesticide patent bill. Despite some last-minute discussions by proponents and opponents to fashion a compromise package around the patent bill and the pesticide food safety bill, no deal was agreed to, and the legislation died as Congress adjourned for the 1984 elections.

However, during the four years of Congressional fighting and dealing on patent-term extension bills, there was little convincing evidence substantiating a real decline in research investment or innovation in the chemical and pharmaceutical industries.

BEYOND PATENTS: SECRET HEALTH & SAFETY DATA

Monsanto, one of the corporations in the vanguard of the patent-term extension push, has gone to some unusual lengths to try to protect its products from competitors and, in that process, keep certain kinds of information secret and out of public view.

In 1981, Monsanto sought to include a special section in the nation's pesticide law—the Federal Insecticide, Fungicide and Rodenticide Act (FIFRA)—by adding a new category called "innovative methods or technology" which would protect indefinitely, and exempt from public disclosure, certain pesticide testing technologies used by the chemical industry. Monsanto, it seems, had invented a novel method of tracking the environmental effects and breakdown of its herbicide Roundup, using a technique called radiosynthesis. Although Monsanto could have patented radiosynthesis under the patent statutes, such a patent would then be made public, and Monsanto did not want its competitors to have this information, even under a patented situation. What Monsanto was attempting to do by way of amendment to FIFRA was to create a special property right provision in the pesticide law. But there was more to it than that.

In the new section proposed for FIFRA, the language also exempted metabolites of pesticides—the secondary chemical reactions and substances of a pesticide which begin to form in the soil, air and/or water after the pesticide is released into the environment. Under the proposed amendment, these metabolites—sometimes potential carcinogens and sources of toxic pollution—and the testing data on their impacts, would be treated confidentially as "innovative composition of matter." Environmental organizations, the American Association for the Advancement of Science, and a multitude of independent scientists all objected to the provision, and coalesced to oppose the creation of this special category of protection.

Because of the controversy, the "innovative technology" amendment to FIFRA was never passed. But Monsanto and the National Agricultural Chemicals Association (NACA) didn't give up. Monsanto was already in court challenging a 1978 FIFRA provision for public disclosure of pesticide safety data, and NACA was pushing hard for the patent-term extension bill in Congress.

Then on June 26th, 1984, the U.S. Supreme Court ruled 8–0 in the *Ruckelshaus v. Monsanto* case, that chemical companies could not keep secret health and safety data used in developing new pesticides. That was essentially what the 1978 amendments to the pesticide law required—the provisions Monsanto challenged as an unconstitutional "taking" of private property under the Fifth Amendment. However, the Supreme Court found otherwise. "A voluntary submission of data by an applicant in exchange for the economic advantages of [EPA approval]," wrote Justice Blackmun, "can hardly be called a taking."

Less than five weeks after the Supreme Court's decision, new statutory language that would allow pesticide manufacturers to decide what information could be withheld as a "trade secret or confidential or financial information" mysteriously made its way into the House version of the Agricultural Patent Reform Act of 1984, which was then making its way through a House Judiciary subcommittee. Neither the bill's Congressional sponsors, EPA, nor NACA could explain how the language got into the bill *after* it was approved in a Judiciary subcommittee mark-up. The official "explanation" was that legislative counsel for the committee, seeking the advice of industry lobbyists in an effort to "simplify" the legislation, drafted the objectionable language. Although the "simplification" was subsequently removed by the full Judiciary Committee in an effort to appease environmental and consumer organizations who objected to it, the incident was sufficiently embarrassing to kill the controversial provision and slow the consideration of the pesticide patent bill.

PATENTS, INNOVATION, & THE SOCIAL GOOD

It is often argued that the nation's patent system contributes to technological innovation, continuing investment in research, and the development of useful products. Yet in the newest areas of agricultural technology, there is some concern about the kind of innovation now taking place; the lifetime of agricultural products produced; and the "capital momentum" behind certain kinds of research fostered and protected by patents.

In both the chemical and pharmaceutical industries, for example, it is not uncommon for a company to try to spread its research and development costs for new pesticide products over as many patentable years as possible, particularly through the use of "variations-on-a-theme" chemistry. There is some question, then, about the extent to which all new pesticides and animal drugs qualify as genuine innovations as opposed to "copycat chemistry," "freshened-up" products, and/or trivial improvements.

Because of the huge revenue possibilities*—combined with patent protection—some pesticide makers become very committed to one class of chemistry, and devise product spin-offs which often derive from one kind of chemical molecule. Says *Chemical Week:* "what passes for new herbicides are often analogues of previous successes." For example, Ciba-Geigy's atrazine herbicide Aatrex, started a whole line of high-performance herbicides very much like Aatrex. Monsanto's herbicide Bronco is a chemical relative of Roundup. Monsanto has also devised Super-Lasso, a successor to Lasso. Rohm & Haas' new herbicide Goal is, in fact, a derivative of an existing herbicide named Blazer.

It appears that even the United States Patent Office can be fooled by some molecular twists, issuing patents for pesticides that are less than unique. For example, Ciba-Geigy's director of biological research, John F. Ellis, referring to the similarity between Ciba-Geigy's Dual and Monsanto's Lasso, admits to "copy-cat chemistry." "I don't know how we ever got a patent," says Ellis, "but we found a loophole."

*Although a chemical or pharmaceutical corporation may spend as much as $50 million (lower figures of $20 to $30 million have also appeared in the literature, and for chemical analogs, costs would be considerably lower) developing a new pesticide or livestock drug, a popular patented substance, in a few years time, will produce annual revenues that may run as high as $500 million to $1 billion. Eli Lilly's herbicide Treflan, reaping $350 to $400 million annually between 1979 and 1981, has accounted for at least 10 percent of the company's total corporate income since 1978. Registered for use on 50 crops in the United States, Treflan has held as much as 70 percent of the dinitroaniline herbicide market in recent years. Similar herbicides from American Cyanimid and BASF hold the other 30 percent. At its peak in 1975—sixteen years after it was patented—Ciby-Geigy's herbicide Aatrex accounted for one-quarter of the corporation's total sales. In 1982, Monsanto's herbicide Lasso produced sales of nearly $400 million. Another, named Avadex, had $125 million in sales. The company's popular herbicide Roundup is expected to reap $1 billion annually by 1985. "Products like Aatrex, Treflan and Roundup," says *Chemical Week,* "guarantee years of high earnings."

In the seed industry, too, there have been charges that the Plant Variety Protection Act fosters mostly "cosmetic" changes in new plant varieties contributing to what some call "chrome- and tail-fin" plant breeding. Other critics see the plant patenting system as nothing more than a glorified registration process, where coming up with a name for a new variety is all that really matters.

In both the chemical and pharmaceutical industries—particularly among the largest companies with the largest research labs—there appears to be a patent-on-patent building process where one patented product that does well begets a new and improved version of the same product, another patent, more profitability, more research, and still more patents. After crossing a certain threshold, and after hitting it big with a $100 million-a-year product or two, the momentum for research and patenting appears to favor the largest companies.

In the long run, at least for some chemical and pharmaceutical companies, patents appear to encourage rear-guard actions rather than forward-looking research and risk-taking innovation. Patents, in other words, do not always mean that improvements will be made with the "innovation" of new products, but that existing products—with a few minor changes that make them "distinct" and/or "novel"—will be perpetuated. The guiding rule for issuing patents is that the "invention" has to be different, not better. And in some cases, they may be a good deal worse. At the U.S. Office of Plant Variety Protection, for example, one soybean patent was awarded to a company for developing a new variety of soybean whose "novel and distinguishing feature" was that it was *susceptible* to a certain plant disease.*

One of the most powerful biotechnological developments coming to agriculture is the potential manufacture of products that will link synthetic herbicides, insecticides, plant and animal growth hormones, and the like with specific varieties of crops and/or breeds of livestock—chemical molecules matched to genes. It is also conceivable that microbes or insects found commercially useful for

*It is argued by industry, however, that what is being protected by this patent are unexplored genes for future research rather than a viable commercial product. Inbred corn lines have also been patented under the PVPA for legal protection. Yet in both cases there is some question whether the law was intended to be used for such purposes.

performing one agricultural task or another will be matched genetically with plants or livestock, or to kill other microbes or insects. These products, in becoming commercial, will also be patented and/or trademarked, and their marketing heavily capitalized for the life of the patent.

Such new patenting arrangements in agriculture, and the research investments behind them—as in the case of herbicide-resistant crop varieties, for example—could contribute to a continuation of some synthetic chemical products which are dangerous to the environment and public health. Coupled with the "incentive" to milk one product line and one patent term for all their worth, it is likely that a "pesticide momentum" will continue to skew agricultural research in one direction, "locking out" research in other directions, such as minor crop development or non-chemical pest management strategies. Under such weighty legal and economic conditions favoring established products, truly beneficial agricultural innovations will continue to be discouraged.

Alternatively, coupled with "variations-on-a-theme" chemistry, the new biotechnological and genetic changes now possible in agriculture may present farmers, consumers, and patent examiners with an ever-increasing catalog of products that trivialize and inflate the food production process with more technical steps rather than making it more productive. It is possible, therefore, that future product lines for agriculture could include new generations of "novel" crops or chemicals unique and different enough to qualify for new patents, but not truly beneficial for society.

Finally, even if the new products of agricultural biotechnology eventually become chemical-free and wonderfully productive, there is still the issue of granting legal control over the substances of food production to private interests for extended periods of time. The question of patenting the biological and genetic basis of food-production has never received the attention and serious debate it deserves. With mounting population, and the increasing political tension certain to come with the inequitable allocation and use of food-making technologies worldwide, such a debate cannot wait much longer.

CHAPTER

16

AGRIGENETIC POLITICS

Agripower should not be a political tool. Feeding people . . . is too serious a matter to be left to political manipulation.

—*Richard E. Bell,*
former Assistant Secretary
USDA, 1976

In the closing days of the 1984 Democratic Presidential Primary, when U.S. Senator Gary Hart and former Vice President Walter Mondale were fighting for the crucial states of New Jersey and California, Hart made a short campaign stop at a new high-technology company in southern New Jersey—a company named DNA Plant Technology Corporation (DNAP). Donning a long white lab coat, and using a revolving stand of red-capped test tubes and a nearby, recently plowed farm field as props, Hart took one of the company's test-tube tomato seedlings and planted it. "This shows that New Jersey is on the cutting edge of change," said Hart, referring to both the tomato and the company that produced it. High-tech companies like DNAP were springing up in New Jersey and other states, and the Colorado senator was making the most of that trend. Hart was after votes, of course, trying to identify with the changing economic order, something American politicians have been doing ever since the Louisiana Purchase, the Erie Canal, and the first transcontinental railroad.

But Hart's brief rendezvous with test-tube tomatoes and a high-tech company in New Jersey was also a signal that the political landscape of agriculture was changing, and that Hart and DNAP were part of it. And there were other indications as well, such as in the American seed industry, where political loyalties and coalitions were already changing. Consider, for example, the story of one South Dakota seed company.

GRASS ROOTS CLOUT

In early October 1981, Keith Price, the vice president of the Gurney Seed & Nursery Company of Yankton, South Dakota, wrote a letter to his U.S. Congressman, Democrat Tom Daschle, then a member of the House Agricultural Committee. "Dear Mr. Daschle," Price began, "I am writing to solicit your support of the 1981 Farm Bill with its Modified Sugar Loan Program of 18 cents per pound."

Sugar? Why would a top officer of a seed company in the middle of South Dakota, which had nothing to do with the production or sale of sugar, be writing his congressman about the sugar program in the 1981 farm bill? The answer to that question came in the next paragraph of Price's letter. "Gurney's and its 800 employees are . . . interested in the Sugar Loan Program," wrote Price, "as our parent company, Amfac, Inc., is one of the major domestic producers of sugar. This year Amfac will suffer huge losses in its sugar operation due to depressed prices, and these losses have to adversely affect Gurney's and its employees."

In another letter, Gurney's merchandising manager Jim Waltrip put the request to Daschle a bit more directly: "The survival of the domestic sugar industry," he wrote, "is very important to me personally as well as (to the) Gurney Seed & Nursery Company and its employees. It is an agricultural industry that needs help." Pointing out that Amfac losses in its sugar operations could "adversely affect the ability of Gurney Seed & Nursery Company to grow and expand in South Dakota," meaning "fewer jobs and less compensation for our employees," Waltrip concluded that the answer was for Amfac "to get some relief."

"I think you will agree, Mr. Daschle," wrote Waltrip, "that Gurney's progress is very important to the economy of South Dakota. We've been here for 115 years."

While Gurney's had indeed been operating for more than 115 years, the difference in 1981 was that the South Dakota company was now owned by one of the largest corporations in the sugarcane business headquartered thousands of miles away in Honolulu, Hawaii.

Amfac is a $2-billion-a-year sugar, food, and resort conglomerate that in recent years has made a strong bid to dominate the $200 million mail-order horticultural business in the United States. After a series of acquisitions between 1968 and 1980,* Amfac told its shareholders it had a 15 percent share of the mail-order horticulture business, and boasted that it was "the largest wholesale ornamental horticulture company in the U.S." Amfac reported that it moved into the seed and horiculture industry because it offered "a great opportunity to achieve a national perspective in an agricultural related business" which also had "excellent potential" for growth and profit. One reason for Amfac's venture into the seed and nursery business was to compensate for lower earnings from its sugar business, which was facing a decreasing world sugar demand. But in 1981, Amfac would use its newfound political leverage gained through its seed-company acquisitions to try to improve its position in the sugar market.

In Washington, Amfac's chief lobbyist in 1981 was Ray Pope, who was involved in the Congressional drafting of the sugar portion of the farm bill. On several occasions during the bill's consideration he visited the offices of many congressmen. On one of his visits to Daschle's office, Pope reminded the congressman, "You know, we bought the Gurney Seed Company . . ."

Before the fight on the 1981 farm bill was over, Gurney officials wrote Daschle again by Western Union Mailgram. The occasion was the final House vote on the farm bill in mid-December 1981. The farm bill "package" that was coming before the House at that

*In 1968, Amfac acquired the $22 million-a-year South Dakota company, beginning a twelve year spree of seed company and nursery acquisitions. In addition to Gurney, Amfac acquired at least six other seed, nursery and garden products businesses during the 1970s and early 1980s, including the Glenn Walters Nursery in Oregon (1977); Select Nurseries of southern California (1977); the Jenco Nurseries of Dallas and Austin, Texas (1979); American Garden Products and its two California nursery subsidiaries (1980); and the Henry Field Seed & Nursery Co., of Shenandoah, Iowa (1980).

time included the sugar program that the industry wanted, but it also included some unfavorable features for grain farmers, a major constituency in South Dakota. Nevertheless, Gurney's Keith Price urged Daschle to support final passage of the bill, underlining its "critical importance to my parent company, Amfac, Inc." Price noted that "the vote within the next 24 hours is expected to be extremely close. Your vote could make the difference." Indeed, the vote was as tight as could be, with the bill passing by a one-vote margin. Despite the pressure from Amfac and its South Dakota friends, however, Tom Daschle did not vote for the bill.

In a letter to Price explaining his position, Daschle said: "I have, as a member of the House Agriculture Committee, consistently supported a price-support system for sugar. However, the Farm Bill which we voted on yesterday encompassed much more than just sugar and I could not, in good conscience, vote for a bill which will probably lock grain farmers into prices below their costs of production." Tom Daschle, at least, knew where the best interests of South Dakota lay.

The lobbying activities of Amfac-through-Gurney on the 1981 farm bill illustrates one of the handiest kinds of political windfalls that can come to corporations when they acquire smaller companies: grassroots clout. And while Amfac's interests were not advanced in Tom Daschle's vote, it is the rare congressman or senator who will act contrary to pressure from business in his own home state. By picking up smaller companies and their employees through an acquisition, or even a friendly investment or research contract, larger companies also pick up local credibility and grassroots lobbyists. And this can have an important impact in Washington.

The Amfac/Gurney example also illustrates how the process of political representation in agriculture is changing; how it is becoming more centralized and more subject to the influence of major corporations. Chemical, pharmaceutical, and agribusiness companies are now pouring millions of dollars into the seed industry, biotechnology, and agricultural genetics research. These investments are helping to transform the politics of agriculture, as can be seen in the changing power base of the American seed industry.

GOOD OLE BOYS & THE BIOTECHNOLOGY BLOC

Although the seed industry is represented in Washington by the American Seed Trade Association (ASTA), which claims more than 600 member companies, ASTA is becoming increasingly dominated by non-seed corporations. Since the late 1960s, more than 120 ASTA member companies have become subsidiaries of major corporations, and some speculate that at least half of the industry is now up for sale. ASTA—like the rest of agriculture—has generally moved toward a more "big business" style of politics.

A second organization—the National Council of Commercial Plant Breeders, established in 1954 as an organization of major seed companies interested in international trade and plant patenting laws—is also becoming dominated by corporate interests. Of the fifty-six companies that comprise the Council, at least twenty are subsidiaries of companies such as Monsanto, Ciba-Geigy, and Upjohn, while another seven are major companies with plant-breeding and/or biotechnology programs such as Pioneer, DeKalb, Gold Kist, Land O'Lakes, Cargill, Del Monte, and Heinz. Many of these same companies are also found in key positions in state and regional seed-trade organizations, or on ASTA committees which formulate industry policy.

In recent years, ASTA and the seed trade have lent their name and political weight to legislative fights concerning export policy, patenting, pesticides, and other agribusiness issues. The seed industry, in fact, played an instrumental role in creating the U.S. Feed Grains Council, a Washington-based commodity group active in pushing for expanded farm exports. ASTA funnels more than $100,000 annually into the Feed Grains Council for overseas market development.*

Besides this, ASTA has supported the positions of the National Agricultural Chemicals Association on most recent pesticide legis-

*In addition to ASTA funds, some of the larger individual seed companies have also put sizeable amounts of money into export promotion and market development organizations. In December 1982, for example, Pioneer Hi-Bred International pledged a three-year fund of $630,000 to support the export and market development activities of three grain organizations: $160,000 annually to the U.S. Feed Grains Council, and $25,000 annually each to the American Soybean Export Foundation and U.S. Wheat Associates.

lation—including patent term extension for pesticides and veterinary products—and has occasionally lobbied to bring about the approved use of specific pesticides. The Association worked for nearly eight years, for example, to lift the EPA restriction on Captan, a fungicide used as a seed treatment for corn and other seed. In 1981, when the EPA lifted its long-standing ban and allowed the use of Captan-treated corn seed for hog and cattle feed, the seed industry was ebullient. "The announcement (of EPA's Captan decision) during the [ASTA] convention," reported *Seedmen's Digest* in August 1981, "was greeted with broad smiles from seedsmen." (Prior to EPA's lifting of the Captan ban, seedsmen with surplusses of Captan-treated seed were forced to discard it.)

For years the politics of the seed trade worked pretty much like the rest of the farm lobby, more or less depending upon the skills of a Washington lobbyist who knew how the agricultural committees worked, fit in well with the rest of the farm groups, and got to know congressmen and senators personally. For the past ten years in Washington, the man who did that best for the seed industry was Harold Loden, a cigar-smoking Georgian known for his striped seersucker suits and southern gentleman's style.

One long-time U.S. Senate staffer who had watched Loden work the agriculture committee during the 1970s, said of him: "Harold Loden is a nice guy; a lot of people know him and he's been around for a long time. Loden generally gets what he wants. . . . Seed is not a major commodity like corn or wheat, and the senators generally do what Harold Loden tells them to do."

In 1983, Loden retired from ASTA and was replaced by William Schapaugh, a twenty-three-year veteran of the Upjohn-owned Asgrow Seed Company, a company which presently holds more seed patents than any other in the United States.

Companies like Upjohn represent a new wave and a new kind of power in the seed industry. They are multinational companies building a science and commercial base in the agricultural genetics industry. But more importantly, they are also positioned to build a new kind of political base to go along with their new commercial interest. Upjohn, for example, owns seed companies in Kalamazoo, Michigan; Spokane, Washington; and three locations in Iowa—places where Upjohn is now a political constituent.

Similarly, Pfizer, a New York City–based corporation with headquarters on East 42nd Street in Manhattan's 15th U.S. Congressional District, is also represented in the 14th Congressional district of Illinois (Delkalb-Pfizer Genetics); the 3rd district of Iowa (Clemens Seed Company); the 2nd district of Minnesota (Trojan Seeds); and the 4th district of Missouri (Jordan Wholesale Company). Other Pfizer research and agricultural facilities put the company in the 7th district of Indiana (Terre Haute Animal Research Center); the 2nd district of Connecticut (Pfizer Central Research at Groton); and through its hybrid corn seed research stations, in at least 12 other congressional districts in Georgia, Indiana, Iowa, Illinois, Wisconsin, Ohio, North Carolina, Kansas, Nebraska, Minnesota, Missouri, and Hawaii.

During the last five years, Monsanto has acquired two seed companies, has made substantial investment or research arrangements with at least five biotechnology companies, and has research contracts with both Washington University and Harvard. Not only is Monsanto a corporate citizen of the 2nd Congressional district of St. Louis, Missouri, where its corporate headquarters are found—as well as of eighteen other Congressional districts in 12 states where its various plants and production facilities are located—but it is also "represented" in Massachusetts, Arkansas, California, and Maryland where it has acquired seed companies, made biotechnology investments, or contracted university research.

Upjohn, Pfizer, and Monsanto, of course, are not alone in this regard. A number of other chemical, energy, pharmaceutical, and agribusiness giants have similarly broadened their political bases by way of seed-company and biotechnology investments. Even foreign corporations such as Ciba-Geigy, Sandoz, and Hoffman-La Roche are now represented in the American political process through their seed companies or their equity positions in biotechnology companies. In all, during the last ten years or so, more than 85 Congressional districts in at least 34 states—many of them agricultural—have been "picked up" by major corporations through their acquisitions or new agrigenetic investments.

In acquiring seed companies, major corporations automatically broaden their political influence by moving into a more traditional sphere of "good-ole-boy" agriculture. Seed-industry representatives can lobby in places where chemical and drug companies might have a hard time. And by moving into equity relationships with new biotechnology companies or contractual research with prestigious universities, major corporations bring the voices of cutting-edge science and university respectability to their cause. But these corporations also bring something else to the political process—money.

In recent years, most major American corporations involved in politics have created huge Political Action Committees (PACs), which contribute millions of dollars to candidates running for public office. Today, there are roughly 1,450 corporate PACs in operation; in 1980 they poured more than $55 million into congressional elections, and in 1982, more than $100 million.* "The first thing we do," says International Minerals & Chemicals' vice president Nicolaus Bruns, Jr., of his company's PAC, "is look at candidates where we have facilities or people living." In the 1983–84 political season, for example, Hercules Chemical distributed PAC funds in some 30 congressional districts where it has facilities. Celanese's employee PAC contributed to thirty-two candidates in districts where that company operates.

Even the American Seed Trade Association has formed its own political action committee, called SEEDPAC. In an August 1982 letter soliciting contributons to SEEDPAC, an ASTA spokesman explained, "The name of the game in the political arena is personal contact and financial support, and ASTA needs to support the maximum number of those candidates who have helped us accomplish seed industry objectives." Furthermore, said the letter, "individuals in the seed industry working together through SEEDPAC can accomplish more in Washington than can any one individual or firm," and cited as an example the changes in the Plant Variety Protection Act that had been supported by ASTA. "If you agree with me," concluded the letter, "please make your personal check out to 'SEEDPAC' and send it to me."

THE NEW FARM LOBBY

Just as the new genetically based ingredients of agriculture are shifting control of agricultural production from farmers to corpo-

*In the 1980 elections, for example, Dow Chemical's PAC dispersed $350,000; Occidental Petroleum's $134,000; Monsanto's $105,000; American Cyanamid's $60,000; and Stauffer Chemicals's $34,000.

rate suppliers, so too is the locus of farm politics moving increasingly in the direction of major corporations. More and more, the most powerful players in the "farm lobby" of the future will be companies such as Monsanto, Du Pont, Pfizer, and Cargill. While it is true that some companies such as these are already political powers in agriculture, and that some of them have always had agricultural interests ranging from exports to pesticides, their political influence is broadening because of the kinds of products they are now bringing to agriculture and the kinds of policies they must pursue to protect their new interests.

As always, such corporations are concerned with maximum agricultural production at home or abroad, since "scientific farming" worldwide will mean booming sales of seed, feed, pesticides, fertilizer, and animal drugs—whether made conventionally or through genetic engineering. But because of their new-found interest in the science and commerce of agricultural genetics, many major corporations are also concerned with federal research spending in basic biology, the availability of trained research scientists, federal agency regulations, international patent law, and export control laws. Politically, this means new kinds of "agricultural" coalitions.

It is conceivable, for example, that on some issues of common concern in the biotechnology area, very broad and influential industrial/agricultural coalitions will be put together to lobby Congress on bills of specific interest to the chemical, pharmaceutical, or agricultural industries. As a result, one may find organizations such as the newly formed Industrial Biotechnology Association, (IBA) teaming up with the American Seed Trade Association, the National Agricultural Chemicals Association, the Pharmaceutical Manufacturers Association, or any one of the "big farmer" trade groups, such as the National Association of Wheat Growers or the U.S. Feed Grains Council.

The Industrial Biotechnology Association was formed in July 1981 around a nucleus of seven new genetics firms, and chiefly J. Leslie Glick of Genex, who was concerned about the public perception of genetic engineering, calling it "very confused." In part, the IBA was formed to clarify that muddled public perception on behalf of big companies as well as small ones. "We could not have gone on without an association," said John Donalds, then Dow Chemical's director of biotechnology and an officer of the IBA;

"there are government and public image issues that need to be addressed, and a trade group can handle things that a company cannot."

Today, the IBA's roster includes some thirty-two companies,* ranging from such industrial giants as Du Pont, Monsanto, and Exxon to new biotechnology companies such as Biogen, Genentech, and Agrigenetics, each of which pays $10,000 a year to belong to the organization. The IBA's executive director is Harvey Price, formerly vice president and general counsel for the Atomic Industrial Forum. Price is working to build the IBA into a larger organization.

"Our intent," says Price, "is to become an international association. We are trying to recruit members in Canada, Western Europe, and Japan as well as in the U.S." However, he adds, "the focus will be strictly on things that affect the biotechnology climate in the U.S. . . . The overseas firms will be involved primarily because of their interest in the U.S. market and their interest in our role as a leader in biotechnology."

"With time," says Price, IBA wants to establish itself as "a credible source of information on biotech and its products, both for the media and the general public," using pamphlets, filmstrips, movies, and other materials "that explain biotechnology in a sense that can be used with community groups and school children. Basically," he concludes, "we want to make people more comfortable and knowledgeable about the things biotechnology does."

THE HIGH-TECH BANDWAGON

In the United States Congress and all across America, "high technology" is being welcomed with opened arms. At a time when many

*As of February 1984, the IBA's member companies included the Agrigenetics Corporation; Allelix; Amgen; Amicon Corporation; the Bendix Corporation; Biogen; Biotechnica International; California Biotechnology; Centocor; the Cetus Corporation; Collaborative Research; Damon Biotech; the Dow Chemical Company; E. I. Du Pont de Nemours & Company; the Exxon Research & Engineering Company; G. B. Fermentation Industries; Genentech; the Genetics Institute; the Genex Corporation; Hoffmann-La Roche; Johnson & Johnson; Life Technologies; Meloy Laboratories; Molecular Genetics, Inc.; the Monsanto Company; Pharmacia; Phillips Petroleum; Schering-Plough Corporation; G.D. Searle & Company; the Shell Oil Company; the Standard Oil Company (Indiana); and Transgene (France).

of America's most basic industries are currently going through a wrenching period of economic change and adjustment, high technology—from computer chips to bioengineering—is on the rise. In Congress, Republican and Democrat alike are eager to find ways to boost the new industry, and at the state and local level, all kinds of inducements are being offered to private companies in the hopes of spawning new Silicon Valleys.

Biotechnology promises new miracle drugs, genetic cures for certain kinds of disease, and holds out the hope of feeding the world's hungry. For all of these good and high-sounding reasons, the new genetic technologies of medicine and agriculture offer an especially attractive platform for aspiring politicians. All that is needed in the area of agricultural biotechnology, for example, is a nudge from the "right kind" of constituent—a farmer, a seed company, a corporation, or a local university—to convince a governor, a congressman, or a senator "to really get behind this kind of thing," or at least not stand in its way. At the national level, there is also a growing, Olympics-like patriotism associated with encouraging American innovation, maintaining our "competitive edge," and winning the high-tech race in the international marketplace. Yet once the high-technology bandwagon gets rolling, it could be very difficult to stop or even slow down in case of unforeseen consequences. And there will be consequences.

Although the U.S. Congress expressed some nervousness over genetic engineering in the mid-1970s when fears of runaway organisms first emerged, these concerns eventually gave way to worries over the condition of the nation's economy and how new technology might help revitalize America. In Congress in the early 1980s, an informal caucus of Democrats—a group later dubbed the Atari Democrats—advocated a high-technology strategy for keeping America competitive with Japan and other countries. These Democrats urged increased federal spending for biotechnology research and development as well as other strategies. "Companies wanting to pool research and development in joint ventures in which risks or costs make it prohibitive for a single firm to proceed alone," wrote Atari Democrat and Senator Paul Tsongas in a March 1983 *New York Times* Op Ed article, "should be freed of antitrust constraints."

Republican members of Congress—such as Silicon Valley wunderkind Ed Zschau—also advocated "reforming" anti-trust laws to encourage new corporate joint ventures in research. He also supported other measures such as tax and regulatory policies to provide incentives for risk-taking investors and the use of more aggressive trade negotiations to open up overseas markets. Zschau, who came to Congress in 1982 after successfully founding and operating Systems Industries, a company that makes disk memory systems for mini-computers, now heads up the House Republican task force on high-technology initiatives.

Other Congressmen were worrying about America's competitive position internationally. In August 1981, the House Science and Technology Committee and the Senate Commerce Committee asked the Congressional Office of Technology Assessment (OTA) to conduct a study on biotechnology to determine where American companies stood relative to those in other countries. "Given the present and probable course of development of the biotechnology industry," asked House Science and Technology Committee chairman Don Fuqua, "in what areas is the U.S. likely to be in a competitive position relative to other countries in the years ahead (and) in what areas are we likely to lag behind?" Fuqua also wanted to know what the United States could learn from other countries and how biotechnology might be used to spawn other new high-technology industries.

Not to be outdone by Congress or the OTA, the White House Office of Science and Technology Policy (OSTP) commissioned a similar, interagency study of American competitiveness in biotechnology in October 1982. That study, which included the suggestions of ten government agencies and industry representatives, sought ways in which the federal government could help industry move forward with biotechnology.

Both studies came up with similar findings and options. The OSTP study, used internally at the White House, examined about a dozen policy areas including export controls, export licensing procedures, patent issues, and regulatory issues. Among its recommendations were the possible removal of certain limitations on the export of microorganisms, fewer FDA restrictions on exportable substances used in biotechnology research, and an increase in federal funding for plant-genetics research.

The OTA study, entitled *Commercial Biotechnology: An Inter-*

national Analysis, was completed in January 1984, and concluded that the United States was the world leader in most areas of biotechnology, but that Japan was coming up fast. In fact, said the report, "Japan may very well attain a larger market share for biotechnology products than the United States because of its ability to rapidly apply results of basic research available from other countries." The OTA study further explained that "The Japanese consider biotechnology to be the last major technological revolution of this century," and that the commercialization of biotechnology in that country "is accelerating over a broad range of industries."

While the OTA report offered no clear-cut recommendations for Congress, its message was clear: the continued preeminence of American companies in commercial biotechnology was not assured. And the report left the door wide open for legislation in any of ten areas the OTA examined—from patent law and regulatory issues to federal research funding and antitrust law. "To improve the competitive position of the United States," said the report, "legislation could be directed toward any of the 10 factors, . . . although coordinated legislation directed toward all of the factors might be more effective in promoting U.S. biotechnology efforts."

ALTERING THE GRAIN TRADE

Although Congress and the White House have been eager to advance America's international position vis-à-vis biotechnology, there are some interesting and potentially disruptive economic and political developments that could return to American shores with a no-holds-barred advancement of biotechnology. "Think of Russia getting wheat to tolerate freezing," says Ray Valentine of Calgene, "it would change the geo-politics of the world."

If the Soviet Union were to develop or be sold a biotechnology capability that would overcome some of its present environmental limitations and agricultural shortcomings in the grain area, the repercussions on the international market, and grain exporting countries like the United States, Australia, and Canada, would be phenomenal. The Soviets are aggressively pursuing biotechnology with the idea of reducing their dependency on grain imports. According to E. F. Hutton analyst Zsolt Harsanyi, "the USSR is putting a tremendous effort into single-cell protein." By 1990, he claims, the Soviets could be self-sufficient in all animal feed. "If they succeed in producing single-cell protein on a large scale," Harsanyi asks, "what impact will that have on our grain export policy?" But Russia isn't the only place where agricultural biotechnology could affect America's position in the export market.

"Much of the soil in the southern hemisphere," explains Du Pont's Ralph Hardy, "suffers from aluminum toxicity which limits phosphate uptake" necessary for plant growth. In this situation, Hardy speculates, biotechnology might produce plants or microbial inoculants that would enable "more effective scavenging of phosphate (by crops) in high aluminum soils." Such a development, says Hardy, "could make South America an even more significant producer of grains and thereby more formidable competitor with U.S. crop agriculture." A similar scenario, he says, might be proposed for Africa and animal production, "where biotechnology may decrease animal diseases and enable Africa to develop possibly a significant meat packing industry."

ROLLING OUT THE GREEN CARPET

In the United States, meanwhile, politicians and government officials at the state and local level have been actively courting biotechnology businesses. In Maryland, for example, a $3.5 million Center for Advanced Research in Biotechnology (CARB) is being created through a joint venture involving the University of Maryland, the federal government's National Bureau of Standards, and the Montgomery County government. The local government will build the center and lease the site—situated on fifty acres in Maryland's burgeoning "science corrider" along Interstate-270; the university will make available its scientists and expertise; and the nearby National Bureau of Standards will provide scientists with access to its "super computer" for molecule modeling and other biotechnology research. The idea is to lure industry, already present in the Washington, D.C., region because of federal research contracts emanating from agencies such as NIH, FDA, and USDA's Agricultural Research Center in Beltsville. Slated for completion in 1986, the CARB has the enthusiastic support of Maryland Governor Harry Hughes and University of Maryland president John S. Toll. If we can provide the strong science base for the industry to flourish, said

Toll at the project's unveiling in February 1984, "the state of Maryland can become a Biotechnology Valley that is the national focus of an industry analogous to the Silicon Valley of California."

Maryland, of course, isn't the only state using its science resources to build economic investment around biotechnology. New Jersey is making a concentrated effort to build a reputation in biotechnology by working with its existing base of pharmaceutical companies. Wisconsin and Michigan are making a special push for biotechnology in the food processing and agricultural industries, and both states have explored the possibility of encouraging public pension funds to invest in biotechnology ventures.

While some venture capitalists such as James Vaughn, vice president of the Milwaukee-based Lubar & Company, believe that Wisconsin has to go "an extra mile" with inducements to overcome an anti-business legacy of its former populist governor Robert La Follette, a new regime of industry/government cooperation is clearly unfolding in the state. In recent years, for example, Wisconsin has eased the stringency of its security regulations, making it easier for companies to raise money; reduced the tax load on industry by indexing the state income tax to inflation; and adopted the federal approach to capital gains taxation, over-turning the state's former taxation of capital gains as ordinary income. Wisconsin officials have also instructed their universities to be more aggressive in the area of technology transfer to industry, and have encouraged their university deans to provide industry ready acess to their campuses for research assistance and consultation.

Other states have been reluctant to put regulatory restrictions on the biotechnology industry. In the summer of 1982, the California legislature passed a bill entitled the "California Recombinant DNA Safety Act," without a dissenting vote. That measure specified that all state-funded gene-splicing work would have to comply with the NIH guidelines. However, under pressure from some of the state's genetic engineering firms, Governor Jerry Brown, then running for the United States Senate, vetoed the bill. Genentech, the nation's leading biotechnology company headquartered in Berkeley, and several other California companies also involved in genetic engineering, opposed the bill and urged Brown's veto. Robert G. Walters, a California lobbyist for Monsanto who also worked against the bill, called it "ambiguous" and "inflexible." In his veto message, Governor Brown said the bill would impose "certain legal restrictions that could inhibit the growth of an innovative technology, which is fundamental to advances in agriculture, medicine, and pollution control."

Although the biotechnology industry is being wooed increasingly by state and local politicians, a few businessmen in the new industry complain that their industry lacks a strong national advocate. "There is no champion in Congress for biotechnology," charged Cetus Corporation president Ronald Cape at the Biotech '84 Conference in Washington, D.C. The only thing Washington politicians are doing with biotechnology, he said, is considering how to regulate it. "The regulatory issue," Cape said, "attracts politicians like bees to honey."

In Cape's view, this situation is crazy, especially since the United States is now engaged in a "biotech race," which he compared to the space race touched off by the Russian satellite Sputnik in the 1960s. "The U.S. is shooting itself in the foot," he said, "and is not going to win the race the way it is going now." What is needed, in Cape's view, and what he proposed in his Biotech '84 keynote address, is a National Biotechnology Agency—"a supporter," he said, on the order of the National Aeronautics and Space Administration (NASA). Cape's new federal agency would be a "proponent" of the industry, helping to fund basic research, but not regulating research or product development. And in making his appeal, Cape, like OTA, raised the issue of the up-and-coming Japanese industry. "They did it to us in cars," he said, "they did it to us in tape recorders and TVs. Now, they say, they're going to do it to us again in biotechnology."

THE CHANGING POLITICAL LANDSCAPE

As America embarks on a new era of economic development fueled by the powers of biotechnology, a new politics of development will invariably follow at the state, national, and international levels. In the realm of agriculture, some existing industries, such as the American seed industry, have already been dramatically changed as a result of the perceived opportunities in biotechnology, and so has its politics. Old agricultural coalitions are taking new forms, and new trade associations and political action committees are being organized. Politicians of all stripes are pulling the levers of power to move high technology forward, and some see agricultural biotechnology as one way to expand business activity at home and national prestige abroad.

Internationally, the geopolitics of agricultural production, trade, and distribution stands to be completely transformed by the new agrigenetic technologies—a development that with time will inevitably spill over into larger political and humanitarian arenas as well. At home, regional shifts in crop and livestock production will also occur at the hand of biotechnology, affecting the political position of particular farm groups such as the dairymen or the feed-grain producers, as well as the general political position of farmers as a national interest group.

In the new arithmetic of agricultural politics and agricultural biotechnology, however, one trend is clear: major chemical, pharmaceutical, and agribusiness corporations are broadening their political prowess through their new agrigenetic holdings—whether in the form of a seed company, biotechnology firm, or university research contract. They stand to play at least as powerful a role as they have in the past. Moreover, by wielding a technology that sweeps broadly across chemicals, pharmaceuticals, and agriculture—the essential ingredients of health and sustenance—the political position of some of these corporations, and their access to government, may be measurably enhanced.

Most worrisome, perhaps, is that industry and government will be working more in unison than they ever have in the past. And such teamwork and new partnerships—whether at the local, state, or national levels—seems to be increasingly under the patriotic banner of international competitiveness and winning the biotechnology race. Although America may gain a measure of technological supremacy by winning this race, its people may lose a foothold of democratic control. For by allowing an amalgamation of industry and government to preside over the technologies of health and sustenance, we may well have created something more than the ability to be productive and competitive.

CHAPTER

17

IN SEARCH OF NEUTRAL SCIENCE

Biotechnology is biology
for economic development.
—*Cornell University brochure*

Two of the things that strike you about Davis, California, are first that it is very clean—California clean, with that sun-drenched adobe look that permeates many buildings throughout the state—and secondly, that bicycles are on a near-even footing here with automobiles.

Davis is a university town, home of one of the nation's leading agricultural schools, which is part of the University of California and the land-grant system of universities created by Congress in the 1860s. "UC Davis," as it is called, with its various departments of biology-based science, is one reason why this town has become a commercial mecca for agricultural biotechnology. There is brainpower here and cheap student labor—labor that knows something about science. Entrepreneurs and corporate representatives make regular pilgrimages here to visit with faculty to discuss their latest research. Out on Fifth Avenue, in what appears to be the Davis version of an industrial park, are the offices and labs of two new biotechnology companies—Calgene and Plant Genetics.

In the summer of 1981, an imbroglio of science and commerce was set in motion when Ray Valentine, forty-six-year-old professor of molecular biology at the university, decided to establish Calgene with the help of his university position and some money from the Allied Corporation. At issue in this controversy was the mixing of commercial interests with those of scholarship and academic integrity in the pursuit of agricultural technology.

At UC Davis, Ray Valentine worked as a professor in the Department of Agronomy and Range Science and as a researcher in the university's Plant Growth Laboratory, which is also part of the state and federal supported agricultural experiment station. Throughout his research career, Valentine worked with soil microbes that help certain kinds of crops use atmospheric nitrogen. He was keenly interested in the nitrogen-fixation process and soon became known among scientists worldwide for his work on the genetics of nitrogen fixation. Due partly to Valentine's expertise in this area of research, UC Davis received a five-year, $3.9 million grant for nitogen-fixation research in 1977 from the National Science Foundation (NSF). However, when funding cutbacks at NSF made it likely that Valentine's existing grant would not be funded a second time, he began looking around for other possibilities.

In 1979, he submitted a formal proposal to the university for the creation of a nonprofit research foundation to apply genetic engineering techniques to agriculture. In Valentine's view, such a foundation could accept funds from industry and subsequently generate some of its own revenues through patent royalties on the sale of its research discoveries. This plan, however, found no support in the university. Valentine charges that the university "sat" on his proposal for nearly two years.

In the meantime, both Valentine and Dr. Charles Hess, Dean of the College of Agriculture and Environmental Sciences, began looking to industry as a possible source of funding for Valentine's university research. Says Dean Hess: "At this time industry was becoming interested in genetic engineering of the type we were doing, and was interested in establishing a window on this new technology. . . . Both Ray Valentine and I contacted a number of firms including Allied Chemical to see if they would have an interest in supporting the research we were doing. Negotiations were initiated with Allied, both on a scientific and on a contractual basis." While the search for university financing continued, Valentine and an associate, Norman Goldfarb, founded a genetic-engineering company in November 1980. They called their new company Calgene, and set up shop in Davis. The new company—in the mold of Valentine's earlier idea for a university-affiliated research institute—was instead created as a for-profit operation that would focus on the commercial applications of agricultural genetics research.

About six months later, on June 28, 1981, Allied Chemical made a five-year, $2.5 million grant to the university to conduct agricultural research, including genetics work on nitrogen fixation in which Ray Valentine would be a principal investigator. A week after that, Allied bought a 20 percent share of Calgene. Allied spokesman Jim Davis said the deal with Calgene was to better position the corporation in the emerging field of applied genetics research. Part of the work Allied wanted Calgene to conduct—genetic work on nitrogen fixation in crops—was also the primary focus of the research Allied had funded at the university. In both places, the work was to be done under the direction of Ray Valentine.

When the details of the Allied grant to UC Davis and the Allied partnership with Calgene became public, a controversy ensued. In late July 1981, Dean Hess proposed a set of conflict-of-interest guidelines for agricultural-experiment-station researchers that would prohibit them from accepting gifts, grants, or research contracts from companies in which they held an equity interest or served as consultants. The guidelines also proposed that researchers not hold equity in any private companies whose research projects were the same as those on which the researchers worked at the experiment station. By August 1981, Dean Hess presented Ray Valentine with three options: leave Calgene, resign from the experiment station, or resign from Allied's grant to the university. Valentine initially resigned from the experiment station and the Allied grant, but Dean Hess accepted only his resignation from the grant. Hess then notified Allied that the university would not undertake that portion of the research that Valentine would have supervised, thus giving up about $1 million of the Allied grant money.

Although the most glaring conflicts of interest between Valentine, Calgene, Allied, and the university were removed by Valentine's resignation from the Allied grant, other, more subtle areas of conflict, as well as the possibility of Valentine's influence on graduate education and the research program, remained. In fact, by the spring of 1982, the chairman of the university's bacteriology department, JaRue Manning, charged that four of Valentine's five graduate students had transferred out of Valentine's lab because he had told them they would have to clear their research projects with Calgene. Valentine denies that he said this. Other students charged that Valentine ordered them to change their thesis work as a condition of further funding.

Although this particular controversy was never resolved in favor of one side or the other, the possibility that graduate education could be influenced by faculty associations with genetic engineering companies, corporate sponsors, or both is very real. Dean Hess, in fact, says he "readily admit[s] that the potential for this type of conflict of interest certainly does exist, and it is an area that we have a lot of concern about."

In the aftermath of the Calgene/UC Davis controversy, Ray Valentine believes it was a wise decision to move ahead with his own genetic-engineering company, even if that meant establishing new university and private-sector relationships.* "I'm now convinced more than ever," said Valentine in early 1982, "that I was right." Valentine thinks university and private-sector ties should be close, particularly in the face of diminishing federal research funds. He has also argued that America's engineering, electronics, and automobile industries have been hurt by the export of American knowledge in those areas to international competitors. "We are just about to do the same with genetic engineering," he says. "Are we going

*Today, Ray Valentine serves on the Calgene Science Board along with six other scientists from UC Davis. Calgene's Science Board also has two faculty members from UC San Diego, two from the University of Washington, one from UCLA, and one from Stanford University. While the Allied Corporation initially supported Calgene, it terminated its agreement with the biotechnology company in 1984 when it sold its fertilizer business. However, Calgene is also supported by the Continental Grain Company, one of the world's largest grain traders, and is now performing product research under contract to Kemira Oy, Finland's largest chemical company; Rhone-Poulenc Agrochemie, a major French chemical concern; and Nestlé, the Swiss food giant. Meanwhile, in one of Calgene's recent brochures, entitled "The Second Green Revolution is Here Today," the company's proximity to the university is emphasized. Calgene scientists, says the brochure, "live and work in Davis—in the heartland of California agriculture and home of the University of California at Davis, unsurpassed among crop science institutions of the world."

to export it like the auto industry and watch others apply it to make their economy hum?"

However, Valentine's worry about exporting genetic engineering know-how does not concern university officials such as Allen G. Marr, dean of graduate studies at UC Davis, who says "the university is not in the business of conducting foreign policy." Marr and others like him are more concerned with the sanctity of the university, and the effect that close ties between faculty and the private sector may have on the university system, such as the amount of time and energy that professors with "outside involvements" in companies like Calgene can devote to their university duties. They are also concerned with the competitiveness and secrecy that professors working with private firms to produce commercial products might foster within the university. In such an environment, they worry about the "free exchange" of ideas that is at the heart of the university system. They are concerned that the "mission" of the university, and that of certain of its departments and research institutes, could be diverted from the pursuit of general knowledge and scientific advancement. They worry that such involvements might lead to educational and research goals of a more purely commercial orientation.

CONSOLIDATING BIOSCIENCE

Today in the United States, the issue of commercial involvement in the university, and in public-sector research generally, is being sharpened with the arrival of biotechnology. While it is true that American universities and public research institutions have been besieged by business interests in the past, particularly in the fields of engineering, physics, and chemistry, and especially during times of technological innovation, there are several reasons why the new infusion of corporate biotechnology dollars into universities and public research programs is particularly troubling.

First of all, the sciences that now provide the knowledge base for the new businesses and products of biotechnology—unlike the sciences that have supported industrial advancement in physics, engineering, and computer science—are *biologically based* sciences, whose research products are often associated with public health and safety, and in the case of agriculture, human sustenance. Thus, the commercial involvement in, and influence upon, public research agendas and individual scientists in these fields is not something to be taken lightly; such involvements will undoubtedly have important ramifications for the public health and welfare.

Secondly, there are also questions about public subsidy and the role publicly funded scientists and institutions play in national policy-making. Tufts University social scientist Sheldon Krimsky says we should be concerned about the commercial funding of university research because "universities and their faculty are a national resource. Our government depends upon the expertise in academe for public policy formation." Krimsky also notes that universities are "recipients of substantial government support, which implies some responsibility and accountability."

Not long ago, the university and all publicly-supported institutions were believed to carry a broad social responsibility inherent in their charters. Public dollars meant public responsibility, and presumably, public benefit. In one sense, the university was looked upon as society's conscience on a broad range of questions, expected to weigh and carefully consider the best uses and applications of knowledge. Historically, the public looked to the university and public-sector scientists for guidance and direction—and sometimes, when faced with new information, even to articulate new values. Generally, we have tried to create universities and public research institutions as places of neutral inquiry and analysis, where society could turn for unbiased answers.

Prior to World War II, it was considered unethical for a professor even to associate with commercial interests. Today, as the Calgene/UC Davis example makes clear, biology and microbiology professors are becoming private interests themselves, establishing their own firms while maintaining university positions. Major corporations, too, are establishing a foothold in the new biological sciences through research contracts, department endowments, and the "hiring away" of talented public-sector scientists.

In 1980, for example, Hoechst A.G., a large West German chemical concern, made a ten-year, $70 million contract with the Massachusetts General Hospital, a Harvard University affiliate, for the purpose of creating a department of molecular biology to conduct biotechnology research. In 1981, Du Pont made a five-year, $6 million contract with the Harvard Medical School for genetic research. In 1982, Celanese made a three-year, $1.1 million contract with Yale University for basic enzyme research, and Monsanto made a five-year, $23.5 million contract with Washington University of St. Louis to conduct basic and product-oriented research in cell biology.*

Other companies have gone directly to universities and individual scientists for specific kinds of agricultural and biotechnology research. The FMC Corporation, for example, made a $190,000 contract with Harvard biologist Frederick M. Ausubel for work on nitrogen fixation (Ausubel and a colleague at MIT, Dr. William Orme-Johnson, are also working with BioTechnica International, a Cambridge biotechnology firm focusing on nitrogen fixation in plants). These examples, and others like them, illustrate that the once-valued "separation of power" between business and academe; the once-respectable distance between public-sector research and private-sector application—is now disappearing rapidly in the biological sciences. And that has implications for all of us.

The changes that are now occurring on many of the nation's campuses, at its public research institutions, and even at the highest levels of science policy-making—including the National Academy of Sciences, the White House Office of Science and Technology Policy, and the U.S. Department of Agriculture—are more than just temporary incursions into biological science for new commercial applications. It total, these trends suggest both an increasing commercialization of research agendas in the biological sciences as well as an increasing centralization in biogenetic decision-making.

OF BIO-DOLLARS & DIVIDED LOYALTIES

During the mid-1970s, while scientists at a number of universities were conducting basic research in genetics and molecular biology, most of the laboratory-based industries were preoccupied with applied research and product development—spending little of their money on basic research in the biological sciences. But basic break-

*It should be noted that while these specific contracts may be oriented in the near term toward developing pharmaceutical and chemical products, very often basic findings in molecular biology have "carryover" value to other areas, including agriculture. Du Pont, Celanese, Monsanto, and Hoechst are all involved in at least one agricultural product area.

throughs, such as the beginnings of gene splicing and genetic coding, were occurring in places like Harvard, UC Berkeley, and Stanford—breakthroughs that had the potential for transforming the medical, pharmaceutical, and agricultural industries. When the tremendous commercial potential of these scientific developments in molecular biology began to surface, a number of companies were, as former *Science* editor Philip Ableson put it, "caught flat-footed." Few had the in-house expertise to capitalize on the new discoveries. By the time of the U.S. Supreme Court's historic 1980 ruling on the patenting of man-made microorganisms and Genentech's skyrocketing debut on Wall Street, a frenzied search for talented scientists was well under way. But there weren't enough of these scientists to go around.

According to Douglas Rogers, an investment banker with the New York firm of Kidder, Peabody, there are an estimated five hundred scientists in the United States with the ability to lead genetic-research efforts; about three hundred of them are believed to be on university campuses. "It's a very limited pool of very bright people," says Rogers.

In fact, the demand for molecular biologists, plant physiologists, and biochemists is so great that it is outstripping the ability of the universities to produce them. The U.S. Department of Labor has estimated that there is currently a shortage of as many as 10,000 scientists in fields such as genetic engineering and microbiology, and that the demand for such talent will remain high into the future. It is not uncommon for graduate students in molecular biology *without* Ph.D.s to land jobs in industry with starting salaries of $35,000 or more annually.

Agricultural scientists who graduate from places like Cornell and Purdue are hotly pursued by industry; in recent years, as many as 100 companies have converged on Cornell to recruit prospective plant breeders, molecular biologists, geneticists, and other types of researchers. Faculty are recruited as well.

A good professor of plant breeding or bacteriology with research ability can earn as much as $80,000 in industry (compared with $25,000 to $35,000 in academe), and topnotch scientists can command corporate research contracts in the $100,000- to $200,000-a-year range. University of Wisconsin associate dean Robert W. Hougas explained in 1980 that his College of Agricultural and Life

Sciences lost the services of biochemist Julian Davis when Biogen offered to triple Davis' $40,000 university salary. Quite simply, there is money to be made in biotechnology.

But the ground-level business opportunities in biotechnology, and the fatter salaries available in the private sector, have begun to drain away university faculty members and graduate students. The lure of "hitting it big" in biotechnology has been a prime reason in drawing university scientists to the private sector. Examples of university-scientists-turned-millionaires, such as Genentech's Herbert Boyer and others, have certainly had an impact. "The universities cannot compete with industry on financial or even on other grounds," says Nobel Laureate Dr. Paul Berg of Stanford University, a scientist widely respected for his work in genetic engineering. "It raises a concern in my mind that we may lose a better part of a generation of molecular biologists, all biologists, and geneticists. . . . Who will then achieve the breakthroughs of tomorrow?"

However, some universities, in the interest of holding their faculty, have begun to be more generous to faculty in their shared-time arrangements with corporations, and even more flexible in drawing conflict-of-interest guidelines. But the growing popularity of shared-time arrangements—in which professors have one foot in the classroom or university research lab and another in the business world—has raised questions about just where the scientists' loyalties lie.*

Anthony J. Faras, co-founder and co-chairman of Molecular Genetics, and a professor of microbiology at the University of Minnesota's Medical School, explains how carefully his company has separated the two entities: "We were very careful not to move even a pipette between my lab [at the university] and the company," he says. "The only thing we moved was the expertise to apply the technology to a commercial application."

For some university officials faced with a growing number of shared-time arrangements, however, the issue is not merely one of physical separation, but to what end one's intellectual and creative

*Another potential avenue for divided loyalties and conflict of interest lies with university scientists serving on scientific advisory boards of biotechnology companies and major corporations. Literally hundreds of university scientists now serve on such boards, and most are paid for their services.

energies are directed. Says Harvard University president Derek Bok: "Our concern with (shared time arrangements) is not only how many days the professor will be on the premises or in his laboratory if he assumes executive responsibilities, but what he thinks about when he gets up in the morning or eats lunch. It is creative imagination that makes the difference."

Some of the scientists who split their time between commercial and academic duties do not see a conflict. Winston Brill, describing how he came to be both director of research for Cetus Madison Corporation and a half-time professor of bacteriology at the University of Wisconsin, explains: "I'm a professor. I love being a professor. I love interacting with students, and teaching students. . . . Then Cetus came along and made a very interesting offer: to start a subsidiary of Cetus specifically dealing with plants and agriculture in Madison. And there was actually a two-year period of discussions with my department, with the deans (and) with the president of the university." Now, he says, "things have now been worked out where I feel there is a very, very comfortable situation between myself having these two hats."

Yet for some, there remain fundamental differences between the university and the business communities, which tend to create problems when the two are joined together. "Close cooperation between universities and industry could lead to harmful tensions induced by competing value systems," says Philip Abelson. "The value system and the mode of conducting research and development in industry are quite different from those of academia.* To survive," he explains, "a company must make a profit. It must evolve with the changing times."

Accordingly, in industry the pressure of the bottom line "inevitably dictates" research policies, says Abelson. As a result, the goal is not the pursuit of knowledge, but rather attaining a proprietary advantage.

• • •

*For example, Abelson finds the "frenetic tempo" of industry research—which is often turned on and off at a moment's notice—to be "incompatible" with the more in-depth and studied tempo of graduate education, whose students need to pursue a line of inquiry patiently in their thesis work.

ACADEMIC FREEDOM VERSUS CORPORATE SECRECY

One of the thorniest problems now confronting universities involved in biotechnology ventures with corporations is how such ventures might interfere with the exchange of information among scientists, the publication of research findings, and the general intellectual give-and-take that comes with and often stimulates scientific inquiry and new discovery. For example, Jonathan King, a professor of biology at MIT, recalls a meeting at which a major corporate patent lawyer announced to the assembled group of scientists that before attending their next meeting they should have their notebooks certified by a notary public, and should consider restricting their conversations with their scientific peers lest they give away some patentable discovery. In this kind of atmosphere, says King, "you put a barrier in between the free flow of scientific information," which generally cramps the development of knowledge.

Donald Kennedy, president of Stanford University, has also noted a similar muting of the free and open exchange between scientists; an exchange he sees as necessary in the verification of scientific discovery and the forward progress of research. In comparing notes with others during 1981, Kennedy learned of at least three or four incidents in which scientists attending meetings refused upon questioning to divulge the details of one scientific technique or another, claiming that it was a proprietary matter. "If a scientist is not free to communicate his or her research," says Kennedy, "then somebody else can't repeat the experiment," which, he points out, is how science moves forward.

Surprisingly, however, some university administrators appear quite willing to delay their scientists' publication rights in order to get industry funds. For example, in a survey of some 120 research directors, half from universities and half from business, Donald Foweler, general counsel for the California Institute of Technology, found that 82 percent of university administrators thought it appropriate to withhold the reporting of research discoveries during the time needed to obtain patent protection.

For industry, the heart of the problem is the desire to maintain secrecy about research they are funding until they have managed to patent it or move it closer to commercialization. In doing so, industry believes it is only protecting its own best interests. However, some feel there is absolutely no place for secrecy in the university, and that in the long run such practices will thwart scientific progress. Arthur Kornberg, a Nobel Laureate and professor of biochemistry at Stanford University, worries about "the danger of secrecy" inherent in the development of biotechnology. He says that secrecy "helps no one and it hinders progress."

In fact, there are many examples in which the free exchange of information and specimens among scientists has led to the development of useful commercial products that might not have been developed in a rigid environment of secrecy. Dr. Eloise Clark of the National Science Foundation, testifying before a Congressional committee in 1981, pointed to how Professor Hugh Iltis of the University of Wisconsin, in the course of NSF-supported field work on wild potato populations in Peru, also gathered some rare specimens of a wild tomato. Upon his return to this country, Iltis mailed seeds of the tomatoes to Professor Charles Rick, a famed tomato geneticist of the University of California at Davis. Rick later crossed the seeds with commercial tomato varieties to produce a hybrid that yielded large, red tomatoes that contained a significantly elevated sugar content—a development worth literally millions of dollars to the tomato industry.

Since the emergence of commercial biotechnology in the early 1980s, there have been some attempts by university and business leaders to address the various problems associated with industry/university collaboration. In March 1982 for example, the heads of five major universities and eleven corporations held a privately financed and closed meeting at Pajaro Dunes, California. And in December 1982, more than four hundred corporate and academic leaders met in a more open conference at the University of Pennsylvania in Philadelphia. At both of these gatherings, there was a reluctance on the part of the industries and universities represented to draft any broad guidelines or rules for governing industry/university research ventures. Rather, the consensus was to allow individual universities and companies to deal with their own unique situations on a case-by-case basis. Both groups, wrote Barbara J. Culliton of *Science* magazine, concluded that university/industry collaboration is "good for universities, good for business, good in the name of technology transfer, for the United States."

After a series of 1982 Congressional hearings chaired by then Representative Albert Gore and Representative Doug Walgren of Pennsylvania on the subject of industry/university relations, their respective subcommittees recommended that: (1) universities prepare guidelines for industrially sponsored research, requiring the public disclosure of all faculty consulting and contractual agreements; and (2) that full-time faculty be discouraged from holding equity in or directing commercial ventures that coincide with their academic research.

However, some businessmen suggest that if too much of a fuss is made about industry-supported university research, eventually many companies won't be interested in funding such research at all. Agrigenetics' David Padwa, for example, laments the "public versus private" polarization that is now occurring between corporations and universities, and offers his own prediction about what might happen. "We could create such a painful interface between the academic and business communities, that industry's attitude could be 'Oh, the heck with it. . . . They don't understand our concerns. . . . Let's just try to hire a few of those guys and do it in our own lab.' " And he adds that there are "more than a few good scientists who are sick of spending 50 percent of their time writing grant proposals and scratching for funds, who are impatient with deteriorating or inferior equipment in their labs, and (who) . . . will jump at the chance to double their salaries, work with the most modern scientific equipment, with sophisticated computer backups, sizeable long term budgets, and the promise of real interdisciplinary approaches."

Meanwhile, by January 1984, the Congressional Office of Technology Assessment, acknowledging that American and foreign companies "have invested substantial amounts of money in U.S. universities for biotechnology research," concluded that for the most part such agreements were "working well" and that "fears concerning conflicts of interest and (the) comingling of government and industry funds have diminished."

THE RISE AND FALL OF THE LAND GRANTS

The land grant universities were initially authorized by Congress during the Civil War as "colleges for the benefit of agriculture and the mechanic arts." In 1862—the same year Congress passed the Homestead Act giving land grants of 160 acres to family farmers—President Lincoln signed the Morrill Act, which made grants of federal land to the states for the purpose of establishing colleges of agriculture and technical science.*

Today, the three-part complex of the land-grant universities, agricultural experiment stations, and the agricultural extension service forms a huge and sprawling publicly-funded research and educational system comprising thousands of scientists, students, and technical personnel. The land-grant universities alone enroll more than 1.5 million graduate and undergraduate students, and include schools ranging from Purdue, Auburn, and MIT to Montana State University, Lincoln University, and Delaware State College. With its legions of plant pathologists, agricultural economists, animal scientists, and other assorted specialists, the land-grant system, through its application of science, has been responsible for revolutionizing and, some say, industrializing agricultural production. Yet this system has not been without its critics.

The first stinging criticism of the nation's land-grant universities came in 1972–73 with Jim Hightower's book *Hard Tomatoes, Hard Times*. Hightower, who was then director of the Washington, D.C.-based Agribusiness Accountability Project—and who is today Commissioner of Agriculture for the State of Texas—found that the

*The land-grant universities were originally envisioned as "people's universities"—places of higher education for the common man, where "tillers of the soil" would have their own colleges of agriculture. By the mid 1880s, some forty land-grant universities were established. In 1887, the Hatch Act was passed, authorizing direct federal payments to any state that would establish an "agricultural experiment station" in connection with its land-grant college. With this legislation, Congress charged the land-grant system with the responsibility of conducting research and experimentation for the benefit of agriculture, including "researches basic to the problems of agriculture in its broadest aspects," such as "improvement of the rural home and rural life," and "the maximum contribution by agriculture to the welfare of the consumer." (In 1890, a second Morrill Act was passed by Congress, establishing "separate but equal" land-grant colleges for blacks in seventeen southern and border states, which have to this day been very unequal participants in the land-grant system, particularly in terms of federal funding.)

The final component of the "land-grant complex"—the agricultural extension service—was created by the Smith-Lever Act of 1914 to bring the discoveries and science of the land-grant universities and experiment stations directly to the people through the services of "extension agents." By the early 1920s a national agricultural research and extension system was firmly institutionalized.

publicly supported land-grant schools conducted agricultural and nutritional research more for the benefit of private companies than they did for consumers, farmers, or the public generally.*

Following Hightower's exposé, others continued to reveal detailed accounts of conflicts of interest and corporate influence throughout the land-grant system, including, for example, incidents of university pesticide research influenced by chemical-industry funding, and charges that university research on mechanical harvesting equipment would be used to displace farmworkers† and weaken the power of the United Farmworkers Union.

In addition to the complaints of liberal activists, the USDA and the land-grant system have also heard criticism from the scientific establishment. The most famous of these critical reviews was the "Pound Report," a 1972 National Academy of Sciences study that sharply criticized the USDA's research program as sluggish. But the agricultural research establishment wasn't entirely to blame.

Following World War II, agricultural and biological research became "poor country cousins" relative to military, space, national health, and a host of other federally funded research progams. Funding increased dramatically for research in mathematics, physics, chemistry, and engineering. During the 1960s, the Russians' stunning success with Sputnik propelled and expanded public investment in these fields, and new super-agencies such as NASA were created. Then, during the 1960s and 1970s, mounting public concern with cancer and environmental pollution brought increased funding for

*Hightower found, for example, that Iowa State University scientists helped to develop the widespread use in cattle feed of the hormone diethylstilbestrol (DES), a substance later found in clinical testing to be carcinogenic; that Texas A & M University assisted Union Carbide with its experimental work in developing Temick, a highly poisonous insecticide which in the 1980s was found responsible for water contamination in potato-growing regions of New York and Wisconsin; and that the Kansas State University Extension Service conducted research to determine how mirrors and reflected light might be used in supermarkets to make fruit and vegetable produce look better.

†As this book goes to press, a landmark lawsuit is being heard in a California court, charging the University of California system with neglecting some of its federally required rural and social responsibilities in the course of developing certain kinds of agricultural mechanization equipment. One of the basic charges of this lawsuit and other such actions is that the land-grant universities and their researchers have failed to serve a broad public interest as required under their legislative charters.

the medical sciences. With the creation of the National Institutes of Health, the USDA began to lose part of its research turf as work on animal cell biology took on importance for medical and health reasons.

By the mid-1970s, however, there were warnings from institutions such as the National Science Foundation that the nation was neglecting basic research in the biological and agricultural sciences.*

In 1978, as a result of continued criticism, the USDA grudgingly initiated a small $15-million-a-year competitive grants program—a program created apart from Congressionally-mandated Hatch Act funds distributed to the land grant universities on a formula basis. This new competitive grants program was designed to interest molecular biologists in research related to food production from plants and animals.

By 1980, however, the Reagan Administration had come to power, and that's when the land-grant system came in for a new round of sharp review. Two individuals concerned about the quality and direction of agricultural research—Dennis Prager, a physicist in the White House Office of Science and Technology Policy (OSTP), and John Pino with the Rockefeller Institute and the National Academy of Sciences Board on Agriculture—organized a small conference of individuals to discuss "critical issues in American agricultural research." In June 1982, a small group of fifteen individuals† representing a handful of universities, corporations, and the USDA,

*Following the 1974 World Food Conference in Rome, President Ford asked the National Academy of Sciences to prepare a report on how the United States could mobilize its R & D effort to address the problem of world food supply. The NAS study, entitled *The World Food and Nutrition Study,* recommended increased funding for research in plant genetics, biological nitrogen fixation, photosynthesis, and disease- and pest-resistance crop breeding. A second report issued by the OTA at about the same time was more blunt, charging that the United States "has lost its preeminent position in basic research to increase food production." The OTA also recommended high-priority action in photosynthesis, nitrogen-fixation, and cell-culture research.

†Attending this meeting were Dennis Prager, John Pino, Perry Adkisson of Texas A & M, James Bonnen of Michigan State, Winslow Biggs of the Carnegie Institution, Representative George Brown, Jr. (D-CA), Irwin Feller of the Institute for Policy Research and Evaluation, Ralph Hardy of Du Pont, James Kendrick of the University of California at Berkeley, Terry Kinney, Jr., of the Department of Agriculture, Lowell Lewis of the University of California at Berkeley, Judith Lyman of the Rockefeller Foundation, James Martin of the University of Arkansas, John Marvel of Monsanto, and Peter van Schalk of the Department of Agriculture.

met at the Winrock Conference Center in Morrilton, Arkansas. The result of this "workshop," a report entitled *Science for Agriculture,* sharply criticized the USDA-led agricultural research establishment.

The Winrock Report, as it came to be called, charged the USDA system with exercising ineffective scientific leadership; using a piecemeal approach in basic agricultural research; failing to fund the most important areas of science; losing its role on the "cutting edge" of agricultural science; and distributing federal research funds "largely on the basis of geopolitics rather than need or expected return." *Science* magazine reported in 1982 that the OSTP/Rockefeller study invited the USDA "to prune dead wood from the system, where necessary by closing down facilities."

In the interest of the "efficient allocation" of research resources, the Winrock Report suggested that Congress consider selling some of the federally-supported agricultural research experiment stations to the states, to private industry, or both. The report also implied the need for a shift in funding away from the traditional block-grant approach*—in which each state gets a specified amount of federal agricultural research funding—to more federal funding through a "competitive grants" program disbursed on the basis of a research project's scientific worth and a careful peer review of its merits.

One proposal growing out of the OSTP Winrock report was the creation of a National Institute of Agriculture modeled after the National Institutes of Health. Such an institute, it was argued, would coordinate agricultural research, eliminate parochial politics, and establish clear research priorities and new areas of emphasis, such as plant molecular biology. Officials from major corporations, such as Du Pont's Ralph Hardy, and from major universities, such as the University of California's James Kendrick, supported the idea of establishing a National Institute of Agriculture. In an October

*As matters stand today, the USDA spends about $500 million annually on in-house research through its Agricultural Research Service, and another $140 million annually—in block grants to the states—to agricultural experiment stations at the land grant universities. The competitive grants program accounts for about $16.5 million annually, and is available to all research institutions. However, the "agricultural research share" of all federal R & D is a paltry 1.25 percent, and only 2 percent of the USDA's budget. State governments taken together add about $700 million in research funding for use by the state agricultural experiment stations. Nevertheless, this public money for agricultural research is still considerable, especially if it could be spent in a more focused and nationally coordinated manner.

1982 editorial, *The New York Times* added its support for the idea, suggesting that such an institute would "consolidate the chaotic slew of separate labs in a few major centers, organized by scientific discipline, not by crop or Congressional whim."

Not everyone was taken with the idea of a National Institute of Agriculture, however. Some critics charged that under such a "nationalized" program—which would sooner or later abandon or greatly diminish the block-grant system in favor of an open-to-all, competitive grants system—the large land-grant universities and well-known private universities (some already well-endowed by corporations) would be strengthened, leaving smaller, "backwater" land-grant universities and colleges more or less in the scientific dust, scraping to obtain a modicum of support from state governments.

Experience with NIH funding suggests that an institute-type arrangement for USDA and the land-grant system would mean that a few universities would garner most of the funds. In 1980, for example, less than 1 percent of all institutions funded by NIH grants—ten to be exact—received 20 percent of all support. Under a similar plan for agriculture, such land grant universities as UC Davis, the University of Minnesota, the University of Wisconsin, Cornell, and Purdue, as well as private institutions such as Harvard, Washington University, Stanford, MIT, and others would get the lion's share of competitive grant money because of their size, their larger numbers of expert scientists, and possibly their corporate affiliations. In the end, a handful of corporate/university research ventures—those with the talent to compete for the grants—would be strengthened, while the less luminary land-grant institutions would be weakened by a lack of talent and funding.

While a National Institute of Agriculture has not been created, more public money has been moved into the competitive grants program specifically earmarked for biotechnology research, and the USDA has created some new, strategically-placed research institutes, such as the Plant Gene Expression Center at UC Berkeley, promised to be one of the world's largest biological technology centers. Organized in part to draw top scientists to publicly funded agricultural biotechnology research, with a focus on long-term projects that private companies would not attempt, the new Center has been purposely located in the San Francisco Bay Area because of the large number of genetic engineering firms there.

Meanwhile, corporate money has continued to consolidate around some of the most important land-grant schools and private universities in the country.

AG SCIENCE FOR SALE

Cornell University,* one of the nation's better-positioned land-grant schools, was designated "a center for advanced technology" by New York state in 1982. In the following year, a $7.5 million package of gifts and grants was made to Cornell by Union Carbide, Corning, and Eastman Kodak to help establish a new Biotechnology Institute (each of the three companies will contribute up to $2.5 million over the next six years to support the institute's research program. Cornell will contribute up to $4 million of its own money each year to support the institute).

"Cornell and the industrial world share a common interest in the new biology," said Cornell President Frank Rhodes in announcing the grants. "We are persuaded that the next ten years hold the promise of a revolution in biotechnology that can make a profound benevolent contribution to human needs. . . . I am confident that the three major corporations that have joined us . . . can create an institute of international stature and a program of great scientific value."

In 1983, Standard Oil of Ohio made a five-year, $2 million grant to the University of Illinois to establish a Center in Crop Molecular Genetics and Genetic Engineering that will work on long-range research to improve food production, principally with corn and soybeans, through genetic engineering (the SOHIO grant was the largest private grant ever made to the university's College of Agriculture).

Rohm & Haas, a Philadelphia-based pharmaceutical corpora-

*Located in Ithaca, New York, and founded in 1863, Cornell is today one of the world's leading agricultural research centers, with one of the largest collections of plant scientists found anywhere. It is home to some thirty-five different centers, institutes, and laboratories that either teach or study some phase of biology and agriculture. By one count, there are at least twenty buildings at the Ithaca campus that are devoted to the pursuit of biology and agriculture, and another dozen field and research stations thoughout New York. In 1978, the Boyce Thompson Institute for Plant Research decided to move its privately-financed laboratory and its fifty-five Ph.D.s from Yonkers to Ithaca, adding yet another important agricultural research draw to the Cornell campus.

tion, made a $1 million grant to the University of Pennsylvania to establish a Plant Science Institute to conduct basic research in plant development, photosynthesis, plant diseases, and plant reproduction. Rohm & Haas, a company very much involved in agricultural product development, is working on a chemical spray for producing hybrid wheat seed, and has made research arrangements with several major universities in an effort to gain testing and marketing rights to new university-developed wheat lines for use in conjunction with its hybrid wheat spray. Cornell, the Virginia Polytechnic Institute, the University of Nebraska, and the University of Minnesota are among the universities that Rohm & Haas representatives have contacted. According to one former North Carolina State University wheat breeder, "Practically every university with a wheat program has been approached by Rohm & Haas representatives."

The Pennsylvania State University has a new "cooperative" program built around a team of eleven of its molecular biologists who exchange information on the latest developments in molecular biology with corporations paying an annual fee of $15,000 each. Among the first corporations to join the Penn State program, initiated in 1982, were Gulf Oil, Westinghouse, Schering-Plough, Wyeth Laboratories, IBM, CIBCO, and Amax. And there are still others.

Monsanto has made a $4 million contract with the Rockefeller University for research into photosynthesis. Chevron is funding plant-genetics research at eight universities with grants ranging from $20,000 to $100,000, and the Agrigenetics Corporation, now owned by Lubrizol, has more than $20 million tied up in research contracts with scientists at eighteen universities, institutes, and research centers.*

IN SEARCH OF NEUTRAL TURF

When corporate money flows into a university or public research institute, corporate influence inevitably follows. Says Edward E.

*Agrigenetics' university research contracts, which typically run for five years, are currently in effect at Columbia University, the Charles F. Kettering Research Laboratories, the University of Pennsylvania, Cornell, the University of Colorado, the University of Virginia, the University of Tennessee, Purdue, the Boyce Thompson Institute for Plant Research, North Carolina State University, the University of California, Oregon State University, and the University of Illinois.

David, Jr., president of Exxon's research and development division: "to the industrialist, paying for research implies ownership of the results." Speaking of the ten-year, $70 million grant that the West German pharmaceutical giant Hoechst made to the Massachusetts General Hospital to create an entire department of molecular biology, Burke K. Zimmerman of the Cetus Corporation says that "Essentially, everyone in that lab is an indentured servant to Hoechst."*

In the Monsanto/Washington University arrangement, where Monsanto's $23.5 million grant will be used for biotechnology research on basic cell biology, a committee of four university and four company scientists will decide how the money is to be spent. According to Congressman Albert Gore, "you don't have to know algebra to figure out how that committee works. No research can be done unless the company gives permission." Washington University spokesman Dr. David Kipnis disagrees with this assessment, saying that the choice of any project will depend on scientific peer review and the scientific quality of the project.

Whatever the specific arrangement, corporate dollars do have an impact on university research, often influencing the kind of research undertaken. In one survey, which asked the chairmen of twenty-five departments in the College of Agriculture at UC Davis what the most important determinants of research topics were, the most popular response was: "Money can influence, or dictate, what (kind of) research gets done." Other responses in that survey included: "The university gave me an office with no tools. I had to go where the money was, which was the chemical companies"; and, "If industry pays the tab, they've got a right to call the tune."

Industry doesn't pay the entire bill, however. Large and longstanding public subsidies of land, buildings, and laboratory facilities contribute to the support of every researcher in a land-grant university (as well as researchers in private universities who receive other public funds). So even though a particular scientist may be working on a project funded by a major corporation, there are still public subsidies present which contribute to his work and that of

*For a more detailed account of the Hoechst/Massachusetts General arrangement, see Katherine Bouton's "Academic Research and Big Business: A Delicate Balance," in *The New York Times Magazine,* September 11, 1983.

his support staff and graduate assistants. In fact, according to Paul Barnett who has studied chemical industry grants to the University of California for California Rural Legal Assistance, a non-profit legal assistance organization, industry can obtain from five to ten dollars' worth of publicly supported research for each dollar of grant support they give. And according to the *Wall Street Journal,* the corporate sector already provides "more than $1 billion a year to universities and experiment stations for agricultural research."

In the past, corporate money for university-based agricultural research has typically comprised no more than 6 to 7 percent of a university's total agricultural research budget. In 1979, for example, roughly $3.6 million of the University of California's $63.7 million agricultural budget came from private sources, including $689,000 from companies producing pesticides. But critics have argued that it is not so much the proportion of corporate funds to total funds that counts, as it is the way in which such funds influence research priorities and how the corporate money is leveraged in specific research areas to achieve commercial ends.*

Moreover, in cases where the research may impact public health and safety, a few well-placed grants from corporate sponsors might mean the difference between the release of a safe product and one that is dangerous. Consider, for example, the case of Dr. Charles Hine, the Shell Chemical Corporation, and the pesticide, dibromochloropropane, or DBCP.

Dr. Charles Hine is an M.D. who has been employed by the School of Medicine of the University of California at San Francisco since 1947. During Hine's tenure at UC San Francisco, the Shell Chemical Corporation provided more than $400,000 of financial support for his university research. In addition to this money, Shell also paid Hine as a consultant to supply health data from his private laboratory to Shell's Agricultural Chemical Division. One of the chemicals that Hine tested in the university laboratory at the UC San Francisco School of Medicine was Shell's brand of DBCP, called "Nemagon."

DBCP is a nematicide, or soil fumigant, that kills microscopic worms called nematodes that eat crop roots. DBCP is a pesticide

*See, for example, *The Path Not Taken,* A Case Study of Agricultural Research Decision-Making at the Animal Science Department of the University of Nebraska, Center for Rural Affairs, Walthill, Nebraska, April 1982.

that has been used widely in California orchards and vineyards. In the mid-1970s, however, questions arose about the health effects of DBCP on chemical-factory workers. Later, some workers in the Occidental Chemical Company who canned the pesticide in California and other states were found to have either reduced fertility or to be completely sterile. Then, in the late 1970s, high residues of DBCP were found in California food and drinking water, and state health officials ordered forty municipal wells shut down in one area where extremely high levels were found.

At a 1977 California inquiry on DBCP, Dr. Hine failed to disclose the results of his university testing of the chemical on rats, which had shown damage to their testicles. Hine's reason for not disclosing this data was that his research was supported by grants from Shell, and had been solely reported to them on a confidential basis. Hine reported his research on DBCP to Shell as early as 1958. In 1980, the Oil, Chemical, and Atomic Workers Union filed a lawsuit against Hine on behalf of twenty workers allegedly sterilized by DBCP. The suit—still pending in 1984—charges that Hine suppressed evidence on the toxicity and health effects of DBCP.

In addition to Hine, more than a dozen other University of California scientists—chemists, nematologists, toxicologists, and occupational health specialists—had also performed research and testing on Shell's brand of DBCP. All of these scientists found the chemical to be safe, effective, and with little residue on tested crops. However, like Dr. Hine, these scientists had been partly supported by some thirty Shell grants totaling $47,800. In fact, in one 1966 letter to the chairman of the UC Davis department of nematology, which accompanied a Shell grant, Shell executive W. E. McCauley, wrote: "Facts developed [in the research] are to be used in support of label registration and the development of sound recommendations, where justified. More specifically, we are interested in the development of data to support the use of Nemagon Soil Fumigant." McCauley added in his letter to the department chairman that the research "should be discussed in greater detail with our local representative."

The Shell DBCP case is not an isolated one.* Nor is the chemical

*For more examples of this kind of chemical industry influence see Paul Barnett, "The Pesticide Connection," *Science For The People*, July/August 1980; Lewis Regenstein, *America The Poisoned* (Acropolis) 1982; and Keith Schneider, "The Next Step Is Out The Door," *The New Farm*, March/April, 1983.

industry the only one to permeate the land-grant system. Seed companies and seed-certification agencies, for example, have also had very close associations with the land-grant universities and agricultural experiment stations—associations that raise questions about the development of biotechnological products in the future, particularly since genetically engineered crops and microbes will be tested and distributed by many of these same seed-industry channels and university institutions.*

"WITHOUT CONSPIRACY OR MALICE"

Beyond the question of individual universities and professors' specific associations with one particular corporation or industry, is the overall influence of corporate-sponsored research in one general field. Consider, for example, what happened in 1969 when the State of California went looking for independent geologists, geophysicists, and petroleum engineers to testify on behalf of its half-billion dollar oil-spill suit against the Union Oil Company and three other companies. This lawsuit came about as a result of marine oil leaks from

*In California, for example, a new asparagus seed developed at the University of California in 1977 was distributed disproportionately to private interests in one part of the state which gave "gifts" to the university. "UC 157" was a new variety of asparagus developed at the university under the auspices of the Foundation Seed and Plant Materials Service. Research funds for the development of UC 157, which amounted to some $74,000, came from the California General fund and from voluntary "gifts" to the university from the San Joaquin Asparagus Growers Association.

UC 157 was regarded as a superior asparagus variety in its growth habit, yielding ability, uniformity, color, spear formation, and other characteristics. The new variety also exhibited a certain degree of resistance to a bothersome fungus disease. The demand for UC 157 in the vegetable-growing areas of California during 1978 ran extremely high, with requests for 17,000 pounds of the seed against an available base of only 1,300 pounds. Consequently, the university's seed and plant service had to allocate the limited seed to asparagus growers. By 1979, the seed organization's allocation system came under the eye of the California Auditor General, whose report to the California legislature concluded: "The foundation has inequitably allocated UC 157 asparagus to growers. Asparagus association growers in San Joaquin County who provided gifts which helped fund asparagus research received over five times more seed than did non-members in that county." As a result of the Auditor General's inquiry, the university's seed organization changed its county-based allocation system to one more nearly based on individual growers' acreage, and set aside 15 percent of the seed for new growers without previous acreage.

Union's offshore oil well in the Santa Barbara Channel. California's chief deputy attorney general at that time complained publicly that experts at both state and private universities turned down his requests to testify. State officials, later explaining the difficulty they had in obtaining expert testimony, said that "petroleum engineers at the University of California campuses of Santa Barbara and Berkeley and at the privately supported University of Southern California indicated that they did not wish to risk losing industry grants and consulting arrangements." Part of the problem in this case was a scarcity of unaffiliated scientists, as most petroleum engineers in academia did extensive consulting for the oil industry. But something else was at work here too.

"If a sufficiently large and influential number of scientists or engineers become financially involved with industry," says Tufts University's Sheldon Krimsky, "problems related to the commercial applications of the particular area of science/engineering are neglected." When that happens, he says, the scientific community in question, "becomes desensitized to the social impacts of science," and that leads to a conservative shift in attitudes and behavior. "The new values emphasizing science for commerce become internalized and rationalized as a public good," he continues, "the disciplinary conscience becomes transformed. It happens incrementally, without conspiracy or malice."

Krimsky believes that given time, the same sort of social desensitization could come to the sciences behind biotechnology. It is his hypothesis that any sizable academic-industrial association in a particular field "will slowly change the ethos of science away from social protectionism and toward commercial protectionism." And he suggests that this can happen even at the highest levels of science.

In fact, Krimsky found in a study of 345 scientists affiliated with 50 different biotechnology companies, that 62 of them were members of the National Academy of Sciences (NAS), a number which represents 25 percent of that prestigious institution's membership in four key areas: biochemistry, cellular and developmental biology, genetics, and medical genetics.* "This is particularly significant,"

*Krimsky believes that his 25 percent NAS figure is conservative, representing the lower bound of such affiliations in that he did not survey hundreds of other companies engaged in biotechnology research where similar affiliations also exist.

Krimsky said, explaining the results of his survey at a 1984 genetics conference in Boston, "in that the NAS is frequently called upon to render decisions on the social uses of science and technology."

SCIENCE AT THE TOP

The "Academy,"* as it is sometimes called, is the nation's most influential science institution, a premier body of eminent scientists who do the research and the learned crystal-ball gazing that often determine national policies in a range of areas. Operating from a marble headquarters building in Washington, D.C., on Constitution Avenue near the Lincoln Memorial, the Academy studies everything from diet and cancer to the assassination of presidents.

It is often the official harbinger of new developments in science and technology, having an impact on the course of research, what is studied, and how federal funds are distributed to public institutions and researchers far and wide. Whenever there is a natural catastrophe or man-made crisis, it is often the Academy that is first asked to respond, and it is usually the Academy's interpretation that becomes the official explanation or accepted plan of action. Yet the Academy does not always take well to being upstaged, particularly when science is thrust into the public arena in a popular fashion, or in ways that expose the scientific establishment as laggard in citing problems. Such was the case in 1962 when Rachel Carson's explosive book about pesticides, *Silent Spring,* was published.

Shortly after the book appeared, calling attention to the acute toxicity of some pesticides and their indiscriminant use throughout the country, several Academy scientists engaged in pesticide questions or related fields attacked it. Three members of the Academy's Pest Control and Wildlife Relationships Committee—which had

*The National Academy of Engineering and the Institute of Medicine comprise the National Research Council, which is the official science advisory body of the federal government. However, the "Academy," with 1,450 member-scientists, is the largest, oldest, and most influential part of the council. To be elected to the Academy as a member is to have arrived as a scientist. The Academy operates under a Congressional charter that requires it to respond to requests from Congress and federal agencies. Most of its $75 million annual budget is the result of research contracts with the federal government. In fact, a federal judge in Washington once called the NAS an "ally of the government."

already released two reports much less critical of pesticides than *Silent Spring*—were among Rachel Carson's loudest and most prominently positioned critics in the scientific community. I. L. Baldwin, chairman of the pest control committee, wrote a long, critical review of *Silent Spring* in *Science* magazine, charging that the book was not "a judicial review or a balancing of the gains and losses" of pesticide use, and that much sounder information could be found in the "balanced judgements" of Academy reports. Another member of the pest control committee, economic entomologist George C. Decker—who had also been a frequent consultant to the chemical industry—called *Silent Spring* "science fiction," comparable in its message to the TV program "Twilight Zone." And at congressional hearings, Mitchell R. Zavon, a consultant for the Shell Chemical Company and also a member of the Academy's pest control committee, characterized Carson as one of the "peddlers of fear" whose campaign against pesticides would "cut off food for people around the world." Two other Academy scientists engaged in food and nutrition research, William J. Darby and C. Glen King, were also critical of Carson, suggesting that her work suffered from ignorance and bias and that she had ignored the sound appraisals of pesticides conducted by responsible bodies such as the Academy.*

Despite these attacks, the concerns of Rachel Carson and *Silent Spring* were eventually vindicated in at least three non-academy reports, including one from the President's Science Advisory Committee in 1963. As for the academy, Phillip M. Boffey, who studied the internal workings of the NAS and the pesticide/Rachel Carson affair in his 1975 book *The Brain Bank of America,* had this to say:

*Academy scientists, however, were not the only interests attacking Rachel Carson and *Silent Spring*. (There were also a few Academy defenders of Carson and her book). Norman Borloug, the famous Nobel Prize winning plant breeder, once called *Silent Spring* a "half-science, half-fiction novel," and blamed the book for instigating a "vicious, hysterical propaganda campaign against the use of agricultural chemicals." The chemical industry mounted a year-long attack on the book, and the National Agricultural Chemicals Association doubled its public-relations budget. The Nutrition Foundation circulated thousands of critical reviews of the book, one of which was written by Academy pest control committee chairman I. L. Baldwin, and there were other activities and reaction designed to discredit Carson and her book. For more details see Frank Graham, Jr., *Since Silent Spring,* Boston: Houghton-Mifflin Co., 1970.

> the Academy's total contribution to the national debate over pesticides which emerged during the 1960s was obstructive to efforts at reform. The Academy consistently lagged behind other eminent scientific groups in defining pesticide hazards, and its reports were frequently marred by bias or by unwillingness to take controversial stands. To be sure at least two reports, the 1967 fire ant report and the 1971 report on cholorinated hydrocarbons in the marine environment, were sharply critical of pesticide contamination. The committee, whose members attacked Rachel Carson in 1962–63; the committee which backed the Agriculture Department's dubious handling of pesticide regulation in 1965; and the committee which refused to help USDA devise a strategy for phasing out persistent pesticides in 1969 all threw their prestige behind "business as usual" and opposed the reformers.

Since the publication of *Silent Spring* and the initial controversy over pesticides, the NAS has prepared some useful and important studies on pesticides and other agricultural topics, including one 1975 study on pesticides that did vindicate Rachel Carson on some counts, and an important 1972 study entitled *Genetic Vulnerability of Major Crops.* Yet the membership of the academy today, and that of many of its boards and study committees, do not comprise a neutral-based institution of science. Consider, for example, the academy's Board on Agriculture, which is currently involved with three studies dealing with biotechnology: one concerning a national strategy for biotechnology in agriculture, another focusing on public-private sector interaction in biotechnology research, and a third studying the environmental and public health implications of genetic engineering.

The NAS Board on Agriculture is currently headed by William L. Brown, a highly regarded plant breeder who is the recently-retired chairman of Pioneer Hi-Bred International. Besides Brown, thirteen other scientists also serve on the board, nine of whom are from major universities or agricultural experiment stations. Of the four remaining scientists, John A. Pino is from the Inter-American Development Bank, Ralph Hardy is from Du Pont, and Charles C. Muscoplat is from Molecular Genetics, Inc.* In addition to Brown,

*As of November 1984, the membership of the NAS Board on Agriculture included: William L. Brown, former Chairman of the Board of Pioneer Hi-Bred

Hardy, and Muscoplat, each of whom represent, or were recently affiliated with, companies with interests in agricultural biotechnology, at least two other members of this board are also involved with biotechnology companies. Lawrence Bogorad, a sixty-one-year-old professor of biology from Harvard University, also serves on the board of directors of Advanced Genetic Sciences, a Massachusetts biotechnology company in which he also owns stock. Virginia Walbot, also serving on the NAS Board, is an associate professor of biological science at Stanford University who also serves on the Science Board of Calgene. While the five members of the NAS Board on Agriculture with previous or current industry affiliations do not comprise a majority (and there may be others since private consulting arrangements and advisory positions are not always public knowledge), they are not independent scientists, and so the NAS Board on Agriculture is not an entirely neutral scientific body, especially in the area of agricultural genetics and biotechnology.

In the interest of maintaining public confidence in the findings of future Academy studies in the areas of biotechnology, food, and agriculture, it would be important for the Academy to establish clear conflict-of-interest guidelines that would prohibit, for example, any Academy member from serving on a study committee who is directly or indirectly involved with a business related to the subject under study, who holds a financial or proprietary interest in such a business, or has within the last two years worked for such a business. And while it can be reasonably argued that the business perspective is important for such studies, such points of view can be obtained through workshops and correspondence with industry officials and scientists. Industry representatives should not participate in the research or the writing of Academy reports.

• • •

International; John A. Pino, Vice Chairman of the Inter-American Development Bank; Lawrence Bogorad of Harvard University; Eric L. Ellwood of North Carolina State University; Joseph P. Fontenat of VPI; Robert G. Gast of Michigan State University; Edward H. Glass of the New York State Agricultural Experiment Station; Ralph W. F. Hardy of the E.I. Du Pont de Nemours & Company; Charles C. Muscoplat of Molecular Genetics; Eldor A. Paul of UC Berkeley; Roger L. Mitchell of the University of Missouri; Vernon W. Rutan of the University of Minnesota; Champ B. Taner of the University of Wisconsin; and Virginia Walbot of Stanford University.

WHEN IN DOUBT, HIRE CAST

In the past, whenever the official imprimatur of a National Academy of Sciences report or some other government study has not served the interests of agribusiness, it has enlisted the services of private "scientific" organizations such as the Council for Agricultural Science and Technology (CAST), a Des Moines, Iowa–based consortium of twenty-six food and agricultural science societies. Despite its scientific appearance, and members such as the American Society of Agronomy and the Crop Science Society of America, approximately 57 percent of CAST's $300,000-a-year operating budget comes from 200 agribusiness corporations and trade associations such as Dow Chemical, Eli Lilly, Union Carbide, the Fertilizer Institute, and the National Agricultural Chemicals Association.

Not surprisingly, CAST has defended the safety of pesticides such as DDT, aldrin, dieldrin, dioxin, chlordane, and heptachlor while questioning the value of biological pest control. It has also defended the use of food additives such as nitrites in cured meats, and feed additives such as DES. In 1979, however, CAST ran into trouble with one of its studies on the use of antibiotics in animal feed.

Today, about 40 percent of all antibiotics produced by the pharmaceutical industry in the United States is added to the feed of cattle, pigs, and poultry. Since the early 1950s, it has been known that the addition of streptomycin or tetracyclines to livestock feed has both a beneficial antibacterial action and a growth promotion effect on farm animals. However, by 1972, the FDA discovered that such feed additives created an ideal environment for the development of antibiotic-resistant strains of bacteria, some of which could infect humans. In 1977, FDA Commissioner Donald Kennedy sought to restrict the use of certain drugs in feed, only to be restrained by Congress until new studies on the issue were completed by the NAS, CAST, and others.

CAST convened a twenty-five member task force to study the matter, and was ready to take on the FDA with its findings until six of the scientists on the task force resigned in protest over CAST's heavy hand in preparing the final report. One of the protesting scientists, Dr. Richard Novick, a microbiologist at New York City's

Public Health Research Institute, said that the CAST study was "a watered-down version of what we reported."

"Where we noted that a given antibiotic increases the resistance of the remaining organisms," said Novick, "it was reworded by the council to indicate that it 'might' have that effect, when we know very well that it definitely does." Novick also noted that a preliminary version of the CAST report suggested that the wholesale use of antibiotics in feeds was safe and beneficial—"an implication that is contrary to fully accepted standards of medical practice," he said. According to Novick, CAST's study was almost completely funded by industry, especially pharmaceutical companies, and certain members of the task force itself had very close connections with the producers of the controversial feed additives, especially American Cyanamid and Pfizer. Despite these accusations and the clouded credibility of the CAST report, it was used in the 1979 congressional debate on feed additives, a controversy that still exists in the 1980s.

Today, CAST continues to present itself as an organization dedicated to advancing "the understanding and use of agricultural science and technology in the public interest," specializing in providing information on disputed issues in agriculture. It has issued more than 70 task-force reports since its founding in 1972, most de-emphasizing the dangers of certain agricultural ingredients and technologies. On CAST's new list of studies in progress, to be published in the next few years, is one on the question of seed-resource conservation and another on the use of genetic engineering in agriculture.

THE SMALL CIRCLE OF SCIENCE POLICY

In many ways, Washington, D.C., is a very small town when it comes to the formation of science policy. It is not uncommon, for example, to find scientists such as Ralph Hardy, William Brown, or John Pino serving as advisors or expert witnesses at OTA, NAS, the White House, or before various congressional committees. In some cases, such experts may wind up serving on several different government or professional panels studying the same problem during the same time frame. The viewpoints of such individuals, and presumably the industries they represent or are affiliated with, reverberate several times in advisory documents and reports used in policy-making proceedings.*

In Washington, there is also a revolving door in science, with individuals moving back and forth between government service and the industries the federal government seeks to regulate. In 1979, for example, Dr. Zsolt Harsanyi headed one of the first government studies on genetic engineering after public controversy broke out on the topic in the mid-1970s. That study, entitled *Impacts of Applied Genetics,* was completed in April 1981 by the Congressional Office of Technology Assessment (OTA) under Harsanyi's direction. Shortly after the OTA study was released, Harsanyi went to work at E. F. Hutton as a vice president and analyst specializing in biotechnology investment. In 1982, coming full circle, Harsanyi served as a representative of industry on an advisory panel for OTA's next study of biotechnology, this one involving U.S. competitiveness in the international marketplace.

In many policy-making and scientific proceedings in Washington and other places, it is increasingly a handful of industrial and scientific representatives cut from a very similar cloth who set and establish the tone and content for much offialdom that eventually influences legislation and government action. In the course of these various studies and scientific proceedings, consumers, farmers, and environmental interests are not equally represented, nor are their views usually accorded the same weight as scientists or industry representatives. Citizen interests are, for the most part, treated as minority interests, and not surprisingly, the policies and conclusions

*For example, while Bill Brown was Chairman of the Board at Pioneer and Chairman of the Academy's Board on Agriculture, he also initiated and chaired a three-year conclave of university and industry scientists called the Plant Breeding Research Forum. The Forum convened annual meetings of 30-to-40 plant scientists and economists during the summers of 1982, 1983 and 1984, to discuss, respectively, the future of plant breeding research, germplasm conservation, and genetic engineering. Under Pioneer's sponsorship, the Forum printed reports and other literature which was distributed to the public, the media, and national policy makers. (Pioneer hired one of the world's largest public relations firms, Hill & Knowlton, to help with advance work, press releases and other activities.) In Washington, a smaller delegation of six scientists from the Forum, would brief Congressional committees and staff with each year's findings. While much of the Forum's work was constructive, and many of its recommendations publicly beneficial, some of its lobbying and public relations efforts for increased federal funding of basic plant science research would directly benefit private companies.

which result from such proceedings usually reflect that fact, if such interests are reflected at all.

LOSING OUR GRIP

As societies everywhere embark on a new era of food production fueled with the powers of biotechnology, it is fair to ask how bioscience policy decisions will be made, and under what kinds of commercial and political influence.

With the recent surge of commercial interest in biological and genetics research on campus, for example, some observers have begun to wonder whether universities can maintain their traditional role of neutrality in weighing new scientific advances. "If we now turn to the academic community," asks Congressman Albert Gore, "and find professors who have equity positions in corporations seeking to capitalize on each new discovery, then where *can* we turn for neutral advice?"

"Universities will be negotiating less with government bureaucrats, and more with the giants of the business world," says the Washington-based Carnegie Foundation. "Increasingly, academic decisions are being shaped by the decisions of corporate board rooms." Such trends do not bode well for the independence and objectivity of all science, and is especially worrisome for the future integrity of the biological and agricultural sciences essential to national sustenance and public health.

The agricultural and biological sciences have been financially supported by the American people for more than 100 years. Since the nation has invested billions of dollars in our land-grant universities, our state agricultural experiment stations, as well as public research institutes, private universities and USDA, there is, without question, a sizeable public interest in that system. What kind of agricultural and biotechnological research should be conducted in those institutions, and under what kinds of arrangements with the private sector should, therefore, be a matter of some public discussion and policy debate.

With the development and use of pesticides, medicated feeds, and animal health drugs in agriculture following WWII, "farming" took its first steps into the far reaches of laboratory science, thereby shifting some agricultural power to the men and women of science. With these new ingredients of productivity, it became necessary to have the stamp of approval from the scientific establishment for agricultural commerce, and in that process, science was politicized. Now with biotechnology, the dependence on scientific sanction in agriculture has gone one step further, and will likely become even more political. Whether through federal funding cutbacks, streamlining the nation's land-grant university system, or choosing the "right" scientists to conduct a delicate study of some controversial agricultural topic, science policies can and are being influenced, but not, for the most part, by consumer, farm, environmental, or other non-agribusiness interests.

Clearly, a much broader public participation in the development of biological, agricultural, and biotechnological science policies will be needed in the years ahead if we are to continue feeding ourselves and others in a practical, sustainable, and democratic fashion. For unless this happens, we will, by default, leave an ever-narrowing field of university/industry joint ventures and a handful of national science policy members the power to feed a nation, and increasingly, much of the world.

CHAPTER

18

FAUST IN THE WHEAT FIELD

Consider for a moment a fable about a hypothetical meeting of world leaders.

Assume that the time is September 1947 and the place Manitoba, Canada. The wheat fields are turning golden brown, nearing the harvest; the weather is clear and the sky blue, with the temperature a comfortable 72 degrees.

Near the little town of Morden, just across the border from the United States, a small delegation of business executives and government leaders from Europe, the United States, Canada, and Australia meet with an unidentified man dressed in a black pinstripe suit. He is handsome, in his late 40s, about six feet tall with dark hair and a slender, masculine build. He is standing off to one side of a long, walnut table located in a clearing in the middle of a wheat field. The table is attractively set for a luncheon, and the man in the pinstripe suit is discussing some last-minute details with a large maître d' now directing the preparation of the service. As the officials from the various delegations arrive—all men in their fifties and sixties—they are cordially and graciously greeted by the dark-haired host and escorted to their places at the table. Within minutes of the first arrivals, the entire gathering is complete and the host—standing at the head of the table—formally welcomes his guests and bids them to take their seats. Smartly-dressed waiters begin to move swiftly around the table serving the very best wine and a main course of delicately prepared venison.

The reason for this unusual luncheon in the middle of a wheat field is to discuss the changing economic demography following the war—and particularly population growth and the business of producing food. And after the meal is finished, the dark-haired host rises to speak.

"Gentlemen," he says, "I propose a transaction for the future." He goes on to explain just how the world will grow for the next thirty years or so; that there will be a population boom and a great and continuing need to produce food. In this new era of growth, he explains, there will be money to be made in growing crops and raising livestock, particularly in supplying the things needed to sustain farming, ranching, and food processing. Gradual and sustained agricultural growth lies ahead, he tells the group, and so do increasing profits, capital spending, and rising productivity. To participate in this new age of food prosperity, he explains, it will be necessary to make a certain kind of covenant—but one that is guaranteed to succeed.

"Gentlemen," continues the man, "I bring you power." He then paints in eloquent and vivid detail a bright future of new agricultural technologies such as farm machines that will plant and harvest more tons of food than they ever thought possible. He also speaks of new crops, rising fertilizer production, modern irrigation systems, wonder drugs for livestock, special chemicals to control pests, and highly automated factories where tons of vegetables can be handled with the latest freezing and processing techniques.

"This vision of productivity will be yours," he says, "if you hold fast to the technology. There will, however, be some necessary adjustments. Farms will have to be enlarged and farm numbers reduced. More people will move to your cities. Resources will be used heavily, too. But you will be in power, through the legislation you write, the tax incentives you create, the credit you extend. You will dispense the productivity. This technology will give you power over food production, and thereby a means to produce wealth, provide jobs, and lead your people. If you stand firmly behind it, you will profit in social and economic standing, in increased political power, and in widening world influence."

At the conclusion of the man's presentation, there is some discussion. A few of the business executives want to know about the long-term soundness of these investments, about their return on equity, and the security of their portfolios. The politicians present want some assurance about feeding restless populations, and about variables such as elections and even the weather. All questions are answered deftly by the luncheon host, with the utmost assurance and tightest political logic. He waits for their reaction.

For a moment, there is silence around the table. A light breeze waves across a nearby field. Cigar smoke mixes in the air with the smell of ripening grain. Then a contagious confidence moves swiftly around the table, as if the group were simultaneously thinking as one. There is head-nodding and agreement all around. Doubts turn to smiles and then to the laughter of first-discovery.

The discussion soon ends and the meeting adjourns. The men shake hands with one another and engage in the small talk of business and diplomacy as they walk to their limousines at the field's edge. The bargain was made.

When advanced nations bit into the apple of high-tech agriculture at the close of World War II, an enormous financial and industrial apparatus began moving in one direction to produce food from the elements. Politicians, government bureaucrats, agricultural economists, lawyers, research scientists, bankers, investors, and corporate executives all turned toward one kind of agriculture.

Of course more productivity came, but the price was high. Social costs ensued, environmental degradation occurred, and resources were squandered. Farm systems changed too. Costs of production rose, economies of scale expanded, and many farmers and small-town businesses went under. Yet, in the industrialized countries at least, few people worried about food production. There was abundance everywhere; technology was delivering.

But something else was beginning to become visible: high-tech systems were vulnerable to "quirks" such as a ubiquitous disease susceptibility once found in hybrid corn lines, an increasing dependence on energy and capital, and a growing array of sophisticated and interdependent "inputs." Crop yields in some places—even in the United States—were beginning to level off or even decline, but these problems were masked as more technology was applied. The cost of agricultural production for farmers continued to rise, and the number of farms continued to dwindle.

The dilemma and full cost of the post-war technological bargain in agriculture—more yield for more technology—began to become clear when organic and biological practices were offered in the 1970s: the entire yoke of high-tech agriculture had to be thrown off in order to begin anew. Pieces of the technology could not be shed individually because they worked together. The agricultural system of interdependent parts—chemicals, genetics, capital, and machinery—was an integral package; all were necessary in the high-tech, high-yield equation. It would take time for a farmer—as well as the corporate apparatus supplying him—to "switch over" to organic methods, or even to such halfway steps as integrated pest management. Most commercial farmers would have no part of it. They were (and still are) too much on the economic margin for trial and error.

Modern agriculture was on a treadmill; advanced societies had cut a Faustian bargain on the altar of high-technology, and much of the agrarian "soul" and its values had been given away in the process. There was no turning back.

But now comes biotechnology. Societies everywhere are perhaps now in a position to renegotiate the technological covenants of how they produce food, how agricultural productivity is achieved, and how new potential benefits will be allocated. But the bargaining this time will not be easy, since the apparatus of the past is large and ever-present, now in every corner of the world.

A WINDOW OF OPPORTUNITY

Biotechnology and genetic engineering hold enormous beneficial possibilities for agriculture, the environment, and food production worldwide. With speed and accuracy, these technologies promise to remedy all manner of agricultural problems confronting every society attempting to feed itself. In raw food crops, nutritional qualities can be maintained or carefully improved, a wider variety of nutritionally sound fruits and vegetables might be possible, and the threat of "nutritional erosion" through breeding can be overcome or eliminated. New crops may be genetically designed that won't require pesticides, or which will use less water and make their own fertilizer. Livestock can be engineered to produce more and better quality meat, milk, and eggs on less feed. Farm animals can also be helped to genetically fight off disease and assisted in other ways so they won't need antibiotics, hormones, or vaccines. A new and better use of ecological pest controls in agriculture may reign because of better biological understanding made possible through biotechnology.

Nations that have had difficulty feeding themselves in the past, may find it easier to do so in the future because of the economic savings and "good-trait" replications promised by biotechnology. Food prices, too, may be held down by the economics of these technologies. Genetic diversity may improve throughout agriculture as older, "commercially extinct" varieties and breeds are made viable again, and gene mapping and conservation are improved by the ability of biotechnology to screen and store genetic material. Farmers may benefit economically as costs are reduced and choices expand; even farms of ten acres or less may find new economies and new opportunities.

While all of this may sound too good to be true, it is nonetheless possible. New knowledge *can* be applied in a beneficial direction. All that is necessary is the institutional will and political force to move the system in the right direction.

As with all new technology, however, there is some danger that the beneficial side of biotechnology will move through society only once, as a "window of opportunity" to be seized at the right time by the right social and political forces. Inevitably, a very powerful determinant in shaping this technology will be capital and how it is applied. Just as the pesticide establishment has today grown large over all of agriculture, casting shadows of huge and inordinate proportion from Boston to Bhopal, biotechnology—sent off in the wrong direction with large investments of capital and political ideology—could become equally obdurate and oppressive.

CHOICE, RISK, & RESPONSIBILITY

Today, societies everywhere face a turning-point decision about the next generation of agricultural technology; a technology that will take as its starting point the genes in crops, livestock, insects, and microbes. Agricultural genetics has always played a role in food production, but the difference today is that genes can be seen, fashioned, and directed to do what man wants them to. The decisions about how to use these genetically based starting points in agriculture will determine the capitalization, politics, and ecology of food production over the next several decades. To choose unwisely or too precipitously could prove to be both costly and dangerous.

It is one of the inherent risks—and some might say failings—of "progress" that technologies are sometimes embraced without a full understanding of their power. It is on this path of progress—when we choose to move forward without full knowledge—that we place a great deal of trust in business, government, and science to make the right decisions for us and our children. Yet because these decisions are often made by commercial and political interests for commercial and political reasons, it is often the public that is put at risk, rather than the corporation, the venture capitalist, or the career politician. In the nuclear power industry, for example, the public is still bearing the uncertainty of the technology's safety and its mounting financial costs, while those who launched it appear insulated from the consequences. In agriculture and food production, no less than nuclear power, the risks involved are high-stakes risks indeed.

Because agricultural biotechnology is a mass-production technology, there is much talk about reducing the costs and lowering the prices of food; about producing good things cheaply and in abundance. Interestingly, this sounds much like the promise of nuclear power in the 1950s to produce electricity that was "too cheap to meter."

It will therefore be important to all societies that there be some way of determining how much risk we are adding to our agricultural system with each new increment of genetically engineered pedigree; each new level of biotechnology assisted chemical sophistication. Although it may seem to scientists and government officials that we are now light years away in our knowledge and abilities from the 1970 Southern Corn Leaf Blight, we are only in fact one unknown "genetic widow" away from a new agricultural catastrophe. Even as this book goes to press, USDA officials are still pondering the cause of an Avian influenza epidemic that devastated poultry flocks in the heart of the East Coast's "poultry belt."

There is also a certain amount of responsibility that comes with using any new technology. Plant breeders, genetic engineers, and molecular biologists, whether they work in the public or private sector, have a responsibility to society to consider the larger social and economic ramfications of what they do. There is, for example, a responsibility that goes along with introducing new genes into a crop variety, designing "superbreeds" of livestock, or spending years of research on a new pesticide possibility. Scientists, among others, should make it their business to be aware of the general direction their research is taking them, and how the products of their research might be used in society—how, for example, one product may change whole economies, or how it might impact society. In such analysis, "alternatives thinking" should always figure prominently, and should, in fact, be part and parcel of the research process, present at the inception of a project and asked frequently throughout its duration.

Today some chemical company officials argue that they are getting a bum rap on toxic waste problems that were created decades ago when nobody seemed to care where leftover chemicals were dumped. Yet it is precisely the chemical expertise held by these same companies that should have been brought into play *then* to evaluate the safety of dumping wastes in the ground. Certainly, any astute chemist of the 1930s knew that covering up toxic chemicals with a little dirt would not make the chemistry go away.

It is clear, then, that our institutions of science—be they public *or* commercial—share the responsibility of discovering, revealing, avoiding, and ameliorating the potential impacts of the products they develop. And they, or their employees, should speak loudly when all the facts have not been shared.

UNACCOUNTABLE POWER

For at least the last fifty years in America, food production power has been shifting from a land-based, farmer-controlled system to a capital-based, business-dominated one. Yet many Americans continue to abide in the notion that it is important to keep farmers on the land, as Jefferson advised, for reasons of citizenship and political stability. And there is, to this day, a strong popular sentiment favoring farm policies that uphold this idea.* But that farmers

*In November 1979, Louis Harris and Associates conducted a national poll of public opinion for the U.S. Department of Agriculture. In 7,000 interviews with a representative cross-section of American adults, the Harris pollsters made some surprising findings, among them the revelation that 60 percent of those surveyed indicated a preference for "a country which has a large number of small farms,"

will or can remain our political bulwark on the land, ensuring fair food prices and stewarding our national resources, seems an ever-receding idyll.

Today, agriculture, like other sectors of the economy, is experiencing unprecedented economic consolidation. Farmers are going out of business by the thousands. Mergers, takeovers, and conglomeration in the food and farm-supply industries are occurring worldwide. Such changes are simply the facts of life, we are told. But something else is going on here too. Biotechnology is revolutionizing food production. Suddenly economic consolidation in the food industry is magnified a hundredfold. And genes are the magnifiers. They are now at the center of food production; multinational corporations own them, and nation-states want them. Still, we are told there is nothing to worry about. But there is.

The genetic centralization of food production *is* something to worry about. The stuff of national sustenance is involved: the sustaining, essential ingredients of life; the most basic components of food-making. These are the food determinants at the innermost sanctum of biology. Indeed, these are the ingredients of power; these genes that command the faculty of chloroplasts to bottle the sun so we can have food. They are, in their workings, the closest of a kind to anything we dare call sacrosanct. We owe them more respect than to patent their trade.

while 19 percent held a preference for "a country which has a relatively small number of large farms."

"Some Americans see the small family farm as an economically significant reminder of an outdated, romanticized way of life," said Harris, reporting to the USDA on the results of his survey, "but the public's preference is for 'a country which has a relatively large number of small farms.' " Harris explained that "Significantly, there is a broad-based consensus on this issue, with strong support for the small family farm in evidence in every region of the country and in every significant demographic subgroup of the population."

Harris also learned that "a substantial [67%] majority" of the American people were willing to support federal action to assure continued viability of the small family farm. "The broad support we observed for the small farm," said Harris in his report, "does not appear to be entirely the product of the traditional regard with which this institution is held. Rather, it seems that people perceive a larger national interest is to be served by assuring the continued viability of small, family-owned and operated farms." Similarly, a February 1985 *USA Today* poll found a 71% majority who thought the "federal government has an obligation to give financial help to troubled farmers," and a 76% majority who agreed that "farmers represent an important way of life that needs (the) financial support of the federal government."

Yet our institutions of government, science, and commerce have agreed that it is fair and good to proprietize this realm; that it is reasonable for commerce to have a return on investment here. But to whom do we give this power, and what do we get in return? Consider, for example, the recent historical resumé of commerce, and particularly that of the multinational corporations most likely to wield the biogenetic ingredients that will hold sway over famine.

Particularly worrisome here is that among the agrigenetic potentates-to-be are some of the very same corporations that have created health and environmental problems for societies in the past: companies that have prematurely released unsafe drugs for public consumption; companies whose products have caused untolled toxic pollution and cancers now latent in the land; and companies that have bribed public officials or have operated outside recognized channels of business decency and political acceptability.

News accounts and government reports bearing the most unbelievable revelations of corporate behavior raise concerns about how these corporations will use and apply biotechnology in the future. In July 1983, for example, it was learned that officials at Eli Lilly—the nation's seventh largest drug company—knew that its arthritis drug Oraflex had been associated with twenty-nine deaths in Europe before it was approved for sale in this country. Lilly officials didn't bother to tell the FDA of the deaths while their drug was being reviewed. The drug is now banned in the United States.

In April 1983 the public learned that Dow Chemical officials had scientific information on dioxin (a substance found in herbicides such as 2,4,5-T and the Viet Nam defoliant Agent Orange) as early as 1965 that raised questions about its safety to humans but withheld this information from the federal government for more than fifteen years.

Other news accounts focus on incidents of corporate falsification and misrepresentation, some of which have jeopardized public health and safety, or others that have intentionally misled stockholders and the general public. In October 1983, a nation increasingly concerned with toxic chemicals learned a tale of pesticide data manipulation at the Illinois-based Industrial Bio-Test Laboratories, a lab used by the federal government to test hundreds of pesticides and other chemical products now on the market.

In one of the largest financial fraud cases in American history, the U.S. Securities and Exchange Commission (SEC) found that the

Stauffer Chemical Company—a company now heavily involved in the American seed industry, agricultural chemicals, and plant biotechnology—used improper accounting methods to inflate its profit by $31.1 million in 1982, a fraudulent 25 percent increase. Stauffer agreed to issue new financial reports for 1982 and 1983, showing sharply reduced profits, and the company signed a consent degree in federal court agreeing not to violate SEC regulations in the future. SEC spokesman Chiles Larson said of the Stauffer finding, "This is one of the more significant financial fraud accounting cases the commission has brought in recent years. These numbers are pretty gross, both in terms of the size and in terms of the offense. . . . When people start cooking the books, it's pretty serious stuff. Those numbers are something that people rely upon."

In a similar case, the Baltimore spice-manufacturing firm of McCormick & Company—a company involved in biotechnology research of spice plants—admitted to falsifying records and using fictitious accounting practices to inflate sales and profits between 1977 and 1980, amounting to more than $46 million in phony sales and $4 million in phantom profits.

In 1981, Cargill, the multinational grain merchant that is also in the seed business, pleaded guilty to filing false United States corporate income tax returns for 1975 and 1976. Cargill has also settled, out of court, a few not-so-flattering lawsuits—one alleging price fixing for paint resins and another alleging that the company sold contaminated feed to a beef processor.

Still further news accounts implicate the federal government as an accomplice in questionable corporate activities. "U.S. Is Aiding Drug Companies in Bangladesh," reported an August 19, 1982 front-page story in the *Washington Post,* which explained that the U.S. State Department had asked Bangladesh to reconsider a new national policy designed to ban hundreds of ineffective and dangerous drugs, including some that were known to cause serious health problems.

Companies such as Ciba-Geigy, Hoechst, Squibb, Syntex, Dow, and Upjohn have been accused of inadequate labeling and side-effect warnings on drugs sold in the Third World, or of dumping drugs in Third World countries. Cliquinol—a powerful anti-diarrheal drug banned in Japan and withdrawn from American markets in the early 1970s after it was linked to abdominal pain, brain damage, and blindness—can today be bought at roadside stands in Indonesia, or purchased without prescription or a label warning in the Philippines.

A similar pattern can be found with pesticides. Allied, American Cyanamid, BASF, Bayer, Chevron, Ciba-Geigy, Dow, Du Pont, FMC, W.R. Grace, Occidental Petroleum (Hooker), Monsanto, Rohm & Haas, Schering-Plough, Shell, Stauffer, Union Carbide, and Velsicol produce or sell in Third World countries pesticides that are either banned, heavily restricted, or under review in the United States. Moreover, Chevron, Monsanto, ICI, Ciba-Geigy, Castle & Cooke, Velsicol, and Amvac have been identified as "pesticide dumpers" in the Third World.

At home the pesticide record hasn't been much better. An Allied Corporation subsidiary unleashed the pesticide Kepone into Virginia's James River in the mid-1970s, where it is now sinking into the river's silt; Occidental Petroleum's Hooker Chemical division poisoned the Love Canal; Eli Lilly is responsible for the still festering problem of DES. And there are dozens of other examples—both large and small—involving other companies now venturing into, or fully involved in, agricultural biotechnology.

And while it is not necessarily fair to say that a given company's activities with drugs or pesticides will recur with the new products of agricultural biotechnology, past examples of corporate error and malfeasance are nevertheless, real measures of corporate behavior and product performance. The corporate resumé, in other words, has to be seen as some barometer of the corporate character, and should not be dismissed when considering the future, especially when it comes to food and the ability to produce it.

Beyond the gravity of environmental or genetic mistakes by businesses now racing to make their mark in agricultural biotechnology, is the very real possibility that the corporate holders of genetic technology will also be king makers.

With some corporations, food power has already been used to intimidate entire nations. In 1976, for example, the Continental Grain Company forced the African nation of Zaire to "pay up" an outstanding grain debt by slowing new incoming shipments of wheat to a few well-chosen flour mills, throwing that nation into a temporary food crisis and showing Zairean officials just how important Continental wheat was to their urban population.

With the world population slated to reach the 6 billion mark in the year 2000, and the need to double world food production in the sixteen years beyond that, it is clear that food and the agricultural ingredients behind its making will become so important as to be unavoidably politicized beyond what exists today. Governments—interested primarily in political stability, technological supremacy, trade leadership, and international stature—will be inclined to tread softly on the corporate holders of food genes.

COMMON MAN LEVERAGE: FEW FULCRUMS TO APPLY

Once in America, a body of law and regulation was used to rein-in corporate power. Antitrust laws insured competition and protected consumers from monopolization and price-fixing. By comparison, the political landscape emerging today in the United States and other nations is one of a new level of corporate-government cooperation and partnership that will leave little institutional leverage to the common man for rectifying abuse or misdirection. With a new political enthusiasm for high technology, and the new economic opportunities in these areas expanding worldwide, domestic environmental and agricultural issues—whether in the United States or any other country—may take a permanent back seat to the competitive corporatism and nation-state jousting now taking center stage.

Thus, even though biotechnology clearly promises a universe of genetic opportunities in agriculture of potential benefit to the farmer, consumer, and environment, there are increasingly few institutional mechanisms for insuring that such opportunities are pursued. The science establishment and the university are no longer assured bastions of neutrality. Many politicians have either succumbed to the charms of high technology or are strongly persuaded by the chemical, pharmaceutical, or agribusiness establishments as to future courses of action. Congress, though responsive to the people, is also moved by special-interest PACs.

For these very reasons, farmers, consumers, and environmentalists must fight harder than ever before at the federal, state, and local level for new law, regulation, and public accountability across a broad range of policy areas that impinge on agriculture, the patent system, and science funding. Farmers, university faculty, and local university communities concerned with corporate influence on agricultural research priorities must press college administrators and state legislatures for better public-disclosure procedures. Local citizens groups should expect and demand regular briefings and open-house opportunities from companies doing biotechnology research locally.

Confidentiality and trade secrecy provisions in state and federal laws pertaining to pesticides, food additives, and biotechnology processes should be regularly monitored for improper business claims on public information. Environmental organizations and farm groups should lobby Congress for passage of conventional and practical "sustainable agriculture" bills as well as increased federal research funding for environmentally-sound "biotechnological alternatives" to pesticides.

Politically, it is possible that citizens and business leaders could work together in temporary or permanent coalitions to press for constructive uses of biotechnology in agriculture.

Clearly, not all of corporate America is culpable for transgressions against society or the environment. And some companies that have made mistakes or flouted the law in the past have also done good deeds for various groups in society or produced products of genuine value and importance to the public well being. Today, however, corporations in America and abroad—as well as politicians and government officials of all stripes—are facing a crucial test of their skill and foresight; a test which has to do with their discerning real benefits from short-term aggrandizement, putting broader needs ahead of narrower ones, and breaking with some technological covenants of the past.

The results of that test will unfold in this nation and others during the remaining decades of this century. Given the record of the past there is every reason to be skeptical about the outcome. But it is also worth remembering that in Goethe's version of *Faust,* the protagonist extricates himself from his fateful bargain with Mephistopheles through a selfless service to his fellow man. Perhaps the businessmen and government officials who wield the agrigenetic powers behind the world's food supply will, like Faust, see beyond their bargains for power and turn their energies to a better end.

AUTHOR'S NOTE

The Environmental Policy Institute (EPI), where I am Director of Agricultural Projects, is a nonprofit, public interest organization engaged in research, public education, litigation, and lobbying. EPI influences national policy on energy and the environment by anticipating and responding to environmental threats of local, national, and international significance. EPI has the largest team of public interest advocates in Washington working to promote:

- Energy and water conservation
- Clean air, water, and groundwater
- Reclamation of strip-mined lands
- Protection of farmland, food resources, and public health
- Preservation of genetic diversity in agriculture
- Effective management of America's public lands and energy sources
- Increased use of renewable energy sources
- Safe, energy-efficient transportation, industry, appliances, and buildings
- Reduction of U.S. dependence on nuclear power and foreign oil
- Protection of public health and the environment from the risks arising from the federal, civilian, and military nuclear programs
- Protection of rivers, wetlands, estuaries, and oceans

- International water projects to meet development goals without damaging the environment
- Control of the manufacture, use, disposal, recycling, and cleanup of toxic substances
- A community's right to know the nature and extent of risk from exposure to hazardous materials

EPI is a nationally respected source of information used not only by local citizen organizations but also by government, industry, labor, and the media. EPI helps citizens influence policies affecting their daily lives by informing them when and how their views can be most effectively voiced in Washington. EPI forges national consensus on public policies by building coalitions that are economically, politically, and geographically diverse.

—J.D.

NOTES

Chapter 1

Tatum, L. A., "The Southern Corn Leaf Blight Epidemic," *Science,* March 19, 1971, pp. 1113–16.

Foley, D. C., and Knaphus, George, "Helminthosporium Maydis Race T in Iowa in 1968," *Plant Disease Reporter* 55:855–57, 1971.

Ullstrup, A. J., "The Impacts of the Southern Corn Leaf Blight Epidemic of 1970 and 1971," *Annual Review of Phytopathology,* 1972, pp. 37–50.

Sommers, Charles E., "Crop Management: Southern Corn Leaf Blight Disease," *Successful Farming,* March 1971, pp. 15–17.

"Another Year of Heavy Blight Forecast," *Agricultural Chemicals,* February 1971, pp. 16–19.

Moore, W. F., "Origin and Spread of Southern Corn Leaf Blight in 1970," *Plant Disease Reporter* 54:1102–08, 1970.

Hooker, A. L., "Southern Leaf Blight of Corn—Present Status and Future Prospects," *Journal of Environmental Quality* 1(3), 1972.

U.S. Department of Agriculture, "Statement Issued by USDA on Southern Corn Leaf Blight," Washington, D.C., August 18, 1970, 4417, USDA 2531–70.

National Academy of Sciences, *Genetic Vulnerability of Major Crops.* Washington, D.C., August, 1972.

Yarwood, C. E., "Man-Made Plant Diseases," *Science,* April 10, 1970, pp. 218–20.

Russell, Robert, "Corn Leaf Blight Is Critical," *Evansville Press* (Indiana), August 14, 1970, p. 1.

United Press International, "Blight Epidemic Destroying Corn," *Evansville Courier* (Indiana), August 15, 1970, p. 4.

Muhm, Don, "Iowa Corn Market in Turmoil; Reports of Blight Set Off Trading," *Des Moines Register,* August 16, 1970, p. 1.

"Much of Nation's Crop of Corn Called Periled by Southern Leaf Blight," *Wall Street Journal,* August 17, 1970, p. 18.

"Grains, Soybeans Rise Daily Limit on Corn Blight," *Wall Street Journal,* August 18, 1970.

"Corn-Product Prices Are Raised as Blight Imperils Crop and Pushes Grain Quotes Up," *Wall Street Journal,* August 20, 1970, p. 21.

"Seriousness of Corn Leaf Blight Discussed by Trade Experts at Georgia Meeting," *Feedstuffs,* August 22, 1970, p. 7.

"Corn Futures Declines Spark Lower Prices in Other Contracts," *Wall Street Journal,* August 21, 1970, p. 16.

"Concern About Blight Raises Corn Futures As Much as 6½ Cents," *Wall Street Journal,* August 26, 1970, p. 16.

"Grain Futures Hit the Sky," *Business Week,* August 22, 1970, p. 22.

"A Blight That Sows Panic," *Newsweek,* August 31, 1970, p. 51.

"In the Wake of a Corn Blight," *U.S. News & World Report,* August 31, 1970, p. 62.

"Blighted Corn," *Time,* August 31, 1970, p. 63.

"Corn Blight Threatens Crop," *Science,* September 4, 1970, p. 961.

"Whiplash," *Forbes,* September 15, 1970, p. 58.

Cowden, T. K., Assistant Secretary, Department of Agriculture. Memorandum to James V. Smith, Administrator, Farmers Home Administration, August 27, 1970.

Haspray, Joseph, Deputy Administrator, Farmers Home Administration, Department of Agriculture. Memorandum to T. K. Cowden, Assistant Secretary, August 27, 1970.

"Blight Kills 25% of Corn In Illinois, Official Says," *Wall Street Journal,* August 19, 1970, p. 23.

"Corn Blight Reports Cause Uncertainties in Trade; Some Reevaluation of Plans Evident," *Feedstuffs,* August 22, 1970, p. 1.

"USDA Officials Assess Severity of Corn Leaf Blight in Country," *Feedstuffs,* August 22, 1970, p. 1.

"Corn Market Hysteria," *Des Moines Register,* August 24, 1970.

Gray, Kenneth J., telegram to Secretary of Agriculture Clifford M. Hardin, August 23, 1970.

Hardin, Clifford M., Secretary of Agriculture. Letter to President Richard M. Nixon, September 9, 1970.

"Corn Futures Prices Close Off 6⅜ Cents a Bushel in Chicago," *Wall Street Journal,* September 22, 1970, p. 28.

"Meat, Poultry, Egg Production Cuts Seen Due to Corn Blight, Lifting Prices in 1971," *Wall Street Journal,* October 12, 1970, p. 18.

"As a Killer Disease Hits a Major Crop," *U.S. News & World Report,* October 26, 1970, p. 60.

Natz, Daryl, "Corn Blight Consequences of Concern to U.S. Grain Men," *Feedstuffs,* September 5, 1970, p. 1.

"Southern Mills and Feeders Awaiting Results of Blighted Corn Feeding Trials," *Feedstuffs,* October 3, 1970, p. 5.

"Legislator Calls for CCC Corn Release," *Feedstuffs,* April 17, 1971, p. 4.

"Green Revolution: There is considerable speculation as to whether through our exports of diseased corn we are spreading the blight around the world," *Ramparts,* March 1971, p. 6.

Craig, Jewens, "Occurrence of *Helminthosporium maydis* Race T in West Africa," *Plant Disease Reporter* 55:672–73, 1971.

Scott, D. J., "The Importance to New Zealand of Seedborne Infection of *Helminthosporium maydis,*" *Plant Disease Reporter* 55:966–68, 1971.

American Seed Trade Association, press release. "11 Seed Industry Officials Brief Secretary of Agriculture on '71 Situation," August 21, 1970.

Garst, Roswell, Garst & Thomas Hybrid Corn Company. Letter to Don Paarlberg, Director Agricultural Economics, and Francis A. Kutish, Department of Agriculture Staff Economist, August 27, 1970.

Walker, D. D., Funk Brothers Seed Company. Letter to Richard Lyng, Assistant Secretary of Agriculture, August 27, 1970.

"Some 60% of Seed Corn Slated for '71 Crop Feared Vulnerable to Southern Leaf Blight," *Wall Street Journal,* September 28, 1970, pp. 22.

"Seed Corn Supplies for '71 Crop Could Be Tight, Farm Unit Told," *Wall Street Journal,* November 23, 1970, p. 20.

"Blight Will Be Just as Serious in 1971, Seed Company Official Warns," *Feedstuffs,* November 28, 1970, p. 4.

Steinweg, Bernard, Senior Vice President, Continental Grain Company. Telegram to Ned Bayley, Assistant Secretary of Agriculture, August 20, 1970.

Bayley, Ned D., Director of Science and Education, Department of Agriculture. Letter to Bernard Steinweg, Senior Vice President, Continental Grain Company, September 22, 1970.

Hardin, Clifford M., U.S. Secretary of Agriculture. Letter to Gil Preciado, Secretary of Agriculture, Mexico, August 26, 1970.

"Southern Corn Leaf Blight Roundtable," *Crops and Soils Magazine,* February 1971, pp. 12–21.

"Supply of Seed Corn Resisting Blight Holds to '70 Estimate," *Wall Street Journal,* February 8, 1971, p. 20.

Matsunaga, Spark M., "Kansas' Fields, It Appears, Will Be Cornier in 1971," U.S. House of Representatives *Congressional Record,* February 9, 1971.

U.S. Department of Agriculture, "Southern Corn Leaf Blight." Talk at the 1971 Outlook Conference, G. F. Sprague, Leader Corn and Sorghum Investigations, Washington, D.C., February 24, 1971.

"Corn Growers Struggle To Gather Up Supplies of Blight-Proof Seed," *Wall Street Journal,* March 20, 1971, p. 1.

"Corn Blight Rumors in South Denied; Seed Scarce, Costly," *Feedstuffs,* April 17, 1971, p. 2.

"Funk Boosts Seed Price of Blight-Resistant Corn in Revised 1971 Quotes," *Wall Street Journal,* October 12, 1970, p. 8.

"A Threat to U.S. Food Supply," *U.S. News & World Report,* May 17, 1971, p. 1.

"Corn Blight or a Bumper Crop in 1971? Many Farmers Expecting Record Harvest," *Wall Street Journal,* June 21, 1971, p. 14.

"Farming: The Blighted Corn," *Newsweek,* June 28, 1971, p. 75.

Beardsley, George, "Puts Corn Blight, Not Steel Pact, As Inflation Key," *Chicago Tribune,* July 2, 1971, p. 22.

"First Hand Report on a Big Crop Worry," *U.S. News & World Report,* July 19, 1971, p. 36.

Miller, Raymond J., and Koeppe, David E., "Southern Corn Leaf Blight: Susceptible and Resistant Mitochondria," *Science,* July 2, 1971, pp. 67–69.

Telephone conversation with Art Hooker, Bioscience Director, DeKalb-Pfizer Genetics, Inc., September 24, 1984.

Chapter 2

Author's tape recording and transcript of the Department of Agriculture Structure of Agriculture Hearing, Montpelier, Vermont, November 27, 1979.

Tiley, N.A., *Discovering DNA: Meditations on Genetics and a History of the Science.* New York: Van Nostrand-Reinhold, 1983.

United States Congress, Office of Technology Assessment. *Impacts of Applied Genetics: Micro-Organisms, Plants and Animals,* Washington, D.C., April 1981.

"Recombinant DNA Research: A Brief History," *Chemical Business,* April 6, 1981, p. 39.

"Milestones in Recombinant DNA," *Food Engineering,* May 1981.

"Breeding Better Crops," *The Economist,* December 5, 1981, pp. 106–107.

"Shaping Life in the Lab—The Boom in Genetic Engineering," *Time,* March 9, 1981, pp. 50–59.

Krimsky, Sheldon, *Genetic Alchemy, The Social History of the Recombinant DNA Controversy.* Cambridge, Mass.: The MIT Press, 1982.

"Genetic Engineering: Four Years On," *New Scientist,* June 16, 1977, pp. 631–36.

"Genentech Reports Output of Hormone to Spur Cow Growth," *Wall Street Journal,* March 15, 1981.

"Bean Gene Transplanted," *Boston Globe,* June 30, 1981.

"Block Announces Breakthrough for Moving Genes Between Plant Species," news release, Department of Agriculture, Office of Governmental and Public Affairs, June 29, 1981.

"Gene Splicing on the Farm," *Newsweek,* August 10, 1981, pp. 74–75.
Murray, James R., and Teichner, Lester, "Genetic Engineering and the Future of Agriculture," September 1981.
Schmeck, Harold M., Jr. "Gene Splicing Is Said to Be Key to Future Agricultural Advances," *New York Times,* May 20, 1981.
"Mice Double in Size as Scientists Insert a New Growth Gene," *Wall Street Journal,* December 16, 1982, p. 20.
Tamarkin, Bob, "The Growth Industry," *Forbes,* March 2, 1981.
Department of Agriculture, *A Time to Choose,* January 1981.
Corporate Data Exchange, Inc., *CDE Stock Ownership Directory—Agribusiness,* no. 2, 1978.
Leibenluft, Robert F, *Competition in Farm Inputs: An Examination of Four Industries.* Policy Planning Issues Paper, Office of Policy Planning, Federal Trade Commission, February 1981.
Andrew, John, and Shellenbarger, Sue, "Occidental Plans to Buy Iowa Beef for $800 Million," *Wall Street Journal,* June 2, 1981.
Rowen, James, "Oxy Takes a Very Big Bite," *The Nation,* October 31, 1981.
Sorenson, Laurel, "Occidental Petroleum Slates More Talks with Soviets Aimed at Selling Them Beef," *Wall Street Journal,* October 20, 1981.
Rowe, James L., Jr., "Occidental Oil, Iowa Beef Firm Announce Agreement to Merge," *Washington Post,* June 2, 1981.
"Oxy's Bid to Blend Energy With Food," *Business Week,* June 15, 1981.
"Occidental Petroleum Agrees to Acquire Ring Around Products," *Wall Street Journal,* May 23, 1978.
"Unilever: Seed for a Harvest," *Financial Times* (London), November 6, 1981.
Walsh, John, "Biotechnology Boom Reaches Agriculture," *Science,* September 18, 1981.
"Gene Splicers Unite," *Farm Chemicals,* March 1982.
"Big Firms Gain Profitable Foothold in New Gene-Splicing Technologies," *Wall Street Journal,* November 5, 1980.
"Dow Chemical Hires a Genetics Concern to Use Expertise in Growing Technology," *Wall Street Journal,* October 23, 1980.
"The Miracles of Spliced Genes," *Newsweek,* March 17, 1980.
"Allied Chemical Buys 10% in Bio Logicals," *Wall Street Journal,* November 15, 1980.
Morris, Betsy, "Campbell Soup Is Looking for 'Super' Tomato," *Wall Street Journal,* April 2, 1982.
"Campbell Soup, Others Invest in DNA Plant Technology Corp," *Wall Street Journal,* March 25, 1982.
"Campbell Adds DNA Plant Stake," *New York Times,* March 25, 1982.
"The Race to Breed a 'Supertomato,'" *Business Week,* January 10, 1983.

Chapter 3

Crabb, A. Richard, *The Hybrid-Corn Makers: Prophets of Plenty.* New Brunswick: Rutgers University Press, 1947.

Department of Agriculture, "The Farm Seed Industry," unpublished paper, 1980.

Heckendorn, William, and Edwards, Roy A., Jr., "The Four Types of Seed Trade Associations." In *Seeds, The Yearbook of Agriculture.* Washington, D.C.: U.S. Department of Agriculture, 1961, pp. 517–21.

Leibenluft, Robert F., "Competition in Farm Inputs: An Examination of Four Industries," Policy Planning Issues Paper, Office of Policy Planning, Federal Trade Commission, February 1981, pp. 85–95.

Schachowskoj, Sergej, "The Best Seed in the House," *COR,* magazine of the Minnesota Department of Commerce, June 1983.

Northrup King Company, "Growing . . . Around the World," company brochure, p. 6.

Northrup King Company annual report, 1981.

Interview with George Jones, then president of Cargill Seeds, Minneapolis, Minnesota, December 1981.

Wallace, Henry A., and Brown, William L., *Corn and Its Early Fathers.* Lansing: The Michigan State University Press, 1956.

Hayes, Herbert Kendall, *A Professor's Story of Hybrid Corn.* Minneapolis: Burgess Publishing Company, 1963.

Becker, Stanley L., "Donald F. Jones and Hybrid Corn," Lockwood Lecture, the Connecticut Agricultural Experiment Station, April 9, 1976.

"Roswell Garst—The Innovator," excerpted from Hiram M. Drache's *The Furrow,* as published by the Garst and Thomas Seed Co., 1979, p. 11.

Harpsted, D. D., "Man-Molded Cereal—Hybrid Corn's Story," USDA 1975 Yearbook of Agriculture.

Kastner, Joseph, "The Conundrum of Corn," *American Heritage.* August–September 1980.

"Rich Harvest," *Forbes,* March 1, 1972.

Miller Agrivertical Unit, *1980 Top Agricultural Producers Study* (based on a 1979 survey of 2,000 corn farmers).

DeKalb AgResearch, Incorporated, "The DeKalb Story." In *Employee Benefits at DeKalb,* 1981.

DeKalb AgResearch, Incorporated, "This Is DeKalb," corporate brochure, 1979.

DeKalb AgResearch, Incorporated, annual report 1980.

National Academy of Sciences, *Genetic Vulnerability of Major Crops.* Washington, D.C.: National Academy Press, 1972.

Horsfall, James G., "The Fire Brigade Stops a Raging Corn Epidemic." In *1975 Yearbook of Agriculture.* Washington, D.C.: U.S. Department of Agriculture, 1975.

Harvard Business School, "Pioneer Hi-Bred International, Inc.," Case Studies #4-375-109 (1978) and 4-375-109 (1974).
Kahn, E. J., Jr., "The Staffs of Life, 1—The Golden Thread," *The New Yorker,* June 18, 1984, p. 87.
"Rich Harvest," *Forbes,* March 1, 1972.
"A Sustained Harvest," *Forbes,* October 15, 1978.
"Seed Corn's Long, Hot, Bruising Summer," *Business Week,* August 25, 1980.
Gross, Lisa, "Late," *Forbes,* October 25, 1982.
Davenport, Caroline H., "Sowing the Seeds," *Barron's,* March 2, 1981.
Pioneer Hi-Bred International, Incorporated, annual reports 1973, 1980, 1981, 1982.
"Pioneer in Applied Genetics," Pioneer Hi-Bred International, Incorporated, 1976.
DeKalb AgResearch, Incorporated, annual reports 1976, 1980, 1981, 1982.

Chapter 4

Author visit to the offices of Senator Donald Stewart, Russell Senate Office Building, Washington, D.C., December 1980.
U.S. Congress, Senate Subcommittee on Agricultural Research and General Legislation of the Committee on Agriculture and Forestry, *Plant Variety Protection Act,* 91st Cong., 2nd Sess., S. 3070, June 11, 1970.
U.S. Congress, Senate, *Congressional Record,* April 14, 1930, p. 7017.
Thorne, H., "Relation of Patent Law to Natural Products," *Journal of Patent Office Society,* vol. 6, 1923.
White, Richard P., *A Century of Service: A History of Nursery Industry Associations of the United States.* Washington: American Association of Nurserymen, 1975, pp. 128–33 and 253–59.
"Patents for Plant and Tree Developments Proposed by Senator Townsend, Orchardist," *New York Times,* February 23, 1930.
U.S. Congress, House Committee on Patents, *Plant Patents,* H.R. 11372, A Bill to Provide For Plant Patents, 71st Cong., 2nd Sess., April 9, 1930, p. 5.
Allyn, R. S., *The First Plant Patents.* Brooklyn, N.Y.: Educational Foundation, p. 10.
U.S. Congress, House, *Congressional Record,* May 5, 1930, pp. 8391–92.
U.S. Congress, Senate, *Congressional Record,* May 12, 1930, pp. 8750–51, and Wallace, Paul S., Jr., "Legislative History of the Plant Patent Act, P.L. No 71-245 (S. 4015)" Congressional Research Service, U.S. Library of Congress, April 9, 1982.
U.S. Congress, House, *Congressional Record,* May 13, 1930, p. 8866.

Terry, Dickson, *The Stark Story*. St. Louis: Missouri Historical Society, 1966, p. 85.

U.S. Congress, Senate, *Congressional Record,* April 17, 1930, p. 7200.

Beaty, John Y., *Luther Burbank: Plant Magician*. New York: Julian Messener, 1943; Kraft, Ken and Pat, *Luther Burbank, The Wizard and the Man*. New York: Meredith Press, 1967; Dreyer, Peter, *A Gardener Touched With Genius: The Life of Luther Burbank*. New York: Coward, McCann & Geoghegan, 1975.

Burbank, Luther, *How to Judge Novelties,* January 1911, in Dryer, *op. cit.*, p. 237.

"Edison Urges Tariff Protection for Farmers; Commends Plant Patent Bill as Aid to Crops," *New York Times,* May 5, 1930.

Telephone conversation with Rogert J. Craig, Supervisory Patent Classifier, U.S. Patent and Trademark Office, U.S. Department of Commerce, June 18, 1982, and letter to author from Robert J. Craig, June 29, 1982.

Plant Patents With Common Names. Washington, D.C.: American Association of Nurserymen, 1963, 1969, 1974, 1981, and additional materials.

"Patenting of Plants Promises Big Profits—and Big Problems," *Business Week,* August 26, 1931, p. 26.

U.S. Congress, Senate, Subcommittee on Patents, Trademarks, and Copyrights of the Committee on the Judiciary, *Patent Law Revision,* 90th Cong., 2nd Sess., pt 2, February 1, 1968, pp. 638–885.

Airy, John M., et al., "Producing Seeds of Hybrid Corn and Grain Sorghum." In U.S. Department of Agriculture Yearbook, *Seeds, 1961,* p. 145. For the history of hybrid corn development in the U.S. see Crabb, A. Richard, *The Hybrid-Corn Makers: Prophets of Plenty*. New Brunswick, N.J.: Rutgers University Press, 1947; and "Roswell Garst, The Innovator," from Hiram M. Drache, *Beyond the Furrow,* reprinted in pamphlet form by the Garst & Thomas Hybrid Corn Company, Coon Rapids, Iowa, 1979.

To Promote the Progess of Useful Arts. Report of the President's Commission on the Patent System, 1967, p. 21.

Telephone conversation with Stanley F. Rollin, former USDA seed official, October 14, 1982.

"Allenby White Retires from Northrup King," *Seed Trade News,* June 2, 1982.

Studebaker, John A., "Fifty Years With Breeders' Rights," *Seedsmen's Digest,* May 1982, pp. 22–27.

American Bar Association, 1970 Summary of Proceedings, Section on Patent Trademark and Copyright Law, Statement of Dale Porter, p. 71.

U.S. Congress, Senate, *Congressional Record,* October 2, 1970, p. 34675.

U.S. Congress, House, *Congressional Record,* December 8, 1970, p. 40293.

Office of Management and Budget file on P.L. 91-577, in Office of Legislative Reference, Record Section Depository, New Executive Office Building, Washington, D.C. File #T5-22/69.2.

Virginia H. Knauer, special assistant to the President for consumer affairs, memorandum to Wilfred H. Rommel, assistant director for legislative reference, Bureau of the Budget, December 11, 1970.

Richard G. Kleindienst, deputy attorney general, Department of Justice. Letter to George Shultz, director, Bureau of the Budget, December 18, 1970, p. 3.

Telephone conversation with Bill Dickinson, former Bureau of the Budget examiner, October 14, 1982.

W. Dickinson, NRP Division, Bureau of the Budget. Memorandum to LRD-Ficher re: Enrolled Bill S. 3070, December 22, 1970, pp. 1–2.

Telephone conversation with Dr. Arnold Webber (now president of Colorado State University), May 8, 1984.

Telephone conversation with Richard Lyng (now deputy secretary, Department of Agriculture), May 8, 1984.

Wilfred H. Rommel, assistant director for legislative reference. Memorandum for the President, Subject: Enrolled Bill S. 3070—Plant Variety Protection Act, Sponsor: Senator Miller (R-IA) and five others, December 24, 1970, pp. 6–7.

U.S. Congress, House, Subcommittee on Department Investigations, Oversight and Research of the Committee on Agriculture, *Plant Variety Protection Act Amendments,* 96th Cong., 1st and 2nd Sess., July 19, 1979, and April 22, 1980 (see pp. 42, 62 and 68).

Hornblower, Margot, "Controversy Sprouts Over Attempts to Patent Seeds," *Washington Post,* September 25, 1979.

"Seeds of Trouble" (editorial), *Washington Post,* November 11, 1979.

Cary Fowler, National Sharecroppers Fund. Letter to Jack Doyle, Environmental Policy Center, July 9, 1982.

Lappé, Frances Moore, and Collins, Joseph, with Fowler, Cary, *Food First: Beyond the Myth of Scarcity.* Boston: Houghton Mifflin, 1977.

Mooney, P. R., *Seeds of the Earth, a Private or Public Resource?* Published by Inter Pares (Ottawa) for the Canadian Council for International Cooperation and the International Coalition for Development Action (London), 1979.

Cary Fowler, National Sharecroppers Fund. Letter to Secretary of Agriculture Bob Bergland, January 30, 1980.

National Academy of Sciences, *Genetic Vulnerability of Major Crops.* Washington, D.C.: National Academy Press, 1972.

Wilkes, Garrison, *Bulletin of Atomic Scientists,* 1977.

Wellhausen, E. J., The Rockefeller Foundation. Letter to Dan McCurry, August 3, 1979.

Wehr, Elizabeth, "Diverse Critics Attack Amendments to Extend Plant Patent Protection," *Congressional Quarterly,* April 19, 1980, p. 1032.

Crittenden, Ann, "Plan to Widen Plant Patents Stirs Conflict," *New York Times,* June 6, 1980, p. 1.

Stranahan, Susan Q., "A patent issue involves the world's food supply," *Philadelphia Inquirer,* June 29, 1980.

Telephone conversations with Peter Fenn and Fred Wahl, former aides to Senator Frank Church, October 15 and 20, 1982.

"Church Says Plant Variety Protection Amendment Would Finish Job Started Eight Years Ago," news release, offices of Senator Frank Church, October 15, 1979, p. 1.

George E. Brown, Jr., member of Congress. Letters to Kika de la Garza and William Wampler, November 8, 1979.

U.S. Congress, Senate, Subcommittee on Agricultural Research and General Legislation of the Committee on Agriculture, Nutrition and Forestry, *Plant Variety Protection Act,* on S. 23, et al., 96th Cong., 2nd Sess., June 16 and 17, 1980, p. 68.

"Your Support Still Needed for PVP Amendments," *Seed Trade News,* November 5, 1980, p. 29.

U.S. Congress, House, *Congressional Record,* November 17, 1980.

U.S. Congress, Senate, *Congressional Record,* December 8, 1980, pp. S 15823–24.

Senators Herman E. Talmadge, Jesse Helms, Donald W. Stewart, and Patrick Leahy. Letter to Secretary of Agriculture Bob Bergland, December 5, 1980.

U.S. Supreme Court, Syllabus, *Diamond, Commissioner of Patents and Trademarks, v. Chakrabarty,* No. 79-136. Argued March 17, 1980. Decided June 16, 1980. 14 pp.

"Life: Patent Pending," NOVA #907, WGBH-TV Boston, February 28, 1982.

Chapter 5

Interviews and telephone conversations with friends, former associates, and business acquaintances of David Padwa in Santa Fe, Denver, San Francisco, and elsewhere.

Telephone conversation with David Padwa, chief executive officer, Agrigenetics Corporation, April 11, 1984.

Oppenheimer & Company, Incorporated, prospectus, limited partnership offering, Agrigenetics Research Associates Limited, a Colorado limited partnership, November 2, 1981.

"Agrigenetics Corp.: A Bioengineering Company That May Produce Miracle Seeds in 5 Years," *Olsen's Agribusiness Report,* 3(9), March 1982.

Gross, Lisa, "Nine Meals Away from Murder," *Forbes,* March 1, 1982.

Telephone conversation with Robert Appleman, August 19, 1983.

Agrigenetics Corporation, preliminary prospectus for the sale of 2,000,000 shares of common stock, November 21, 1983.

Interview with R. N. Dryden, president, Agrigenetics Corporation, at Agrigenetics corporate offices, June 23, 1983.

Telephone conversation with R. N. Dryden, August 24, 1983.

"Genetic Engineering: The Three Cultures," text of an address by David

Padwa, chairman, Agrigenetics Corporation, at the Invited Colloquia Series, March 9, 1982, Los Alamos National Laboratories, Los Alamos, New Mexico.

Transcribed remarks of David J. Padwa, at the Battelle Institute Conference on Genetic Engineering, April 19, 1981, pp. 4, 7.

"Agrigenetics Granted Basic Patent," *Seed World,* June 1982.

Buxton, Charles R., Jr., "Agrigenetics Explores Food Frontier," *Denver Post,* July 3, 1983, p. 1E.

Fishlock, David, "How to Pick the Genetic Company Winner," *London Times,* November 13, 1981.

"Kellogg Buys Stock in Seed Firm in Denver," *Supermarket News,* Vol. 32, June 7, 1982, p. 8.

"Biotechnology," *Chemical Week,* June 9, 1982, p. 19.

Sun, Marjorie, "Agrigenetics to Go Public," *Science,* November 25, 1983, p. 903.

"Business Opportunities in Biotechnology: Research to Reality." Summary of Remarks by David Padwa, chairman, Agrigenetics Corporation, New York City, October 26–27, 1981.

Telephone conversation with David Padwa, chief executive officer, Agrigenetics Corporation, August 24, 1982.

Conversation with David Padwa, chief executive officer, Agrigenetics Corporation, Washington, D.C., February 12, 1982.

Telephone conversation with David Padwa, chief executive officer, Agrigenetics Corporation, February 22, 1983.

Telephone conversation with David Padwa, chief executive officer, Agrigenetics Corporation, March 11, 1982.

Telephone interview with Jim Carnes, International Seed Company, February 22, 1982.

Padwa, David, "Let's Clean Up Our Language," *Genetic Engineering News,* January/February, 1982, p. 4.

Speech by David Padwa, chairman, Agrigenetics Corporation, at the National Meeting of the Financial Analysts Federation, Boston, Massachusetts, May 10, 1982.

Interview with Zachary S. Wochok, president, Plant Genetics, Incorporated, at corporate offices, Davis, California, July 13, 1984.

"For Lubrizol, a Bigger Bet on Botany," *Chemical Week,* October 10, 1984, p. 11.

"A Chemical Maker Digs Deeper into Plant Genetics," *Business Week,* October 15, 1984, pp. 44–45.

Telephone conversation with David Padwa, chief executive officer, Agrigenetics Corporation, October 11, 1984.

Conversation with Harold Loden, American Seed Trade Association, Washington, D.C., February 1982.

Telephone conversation with Robert Lawrence, director of applied genetics laboratory, Agrigenetics Corporation, May 2, 1983.

Interview with Dr. Ralph Hardy, research director, life sciences, E. I. DuPont de Nemours & Company, Wilmington, Delaware, May 14, 1982.

Kelly, Jacqueline, "The Science Behind Designer Genes," *Northwest Orient,* February 1983.

Ward, Margaret F., "Agrigenetics Capitalizes on New Ideas," *Daily Camera* (Boulder, CO), August 22, 1983.

"Carbide Drops Out of the Seed Business," *Chemical Week,* February 11, 1981, p. 11.

"Harvesting Biotechnology," *The MacNeil/Lehrer Report,* March 19, 1981.

"Agrigenetics Purchases Jacques Seed Division," *Seed World,* December 1980.

"Agrigenetics Research Park," *Seed World,* 1980.

Chapter 6

Bush, Paul, "Seed Firms Bought by Outside Interests," *Packer Weekly,* December 13, 1980.

"How Sandoz Is Building a Beachhead in the U.S.," *Chemical Week,* July 13, 1983.

Schachowskoj, Sergej, "The Best Seed in the House," *COR* (Minnesota Department of Commerce), June 1983, No. 1681.

Youngblood, Dick, "Northrup Chief Admits Problems—Even His Own," *Minneapolis Tribune,* September 5, 1976.

Youngblood, Dick, "Northrup, King Agrees to Sale to Swiss Company," *Minneapolis Tribune,* September 21, 1976.

Youngblood, Dick, "Northrup, King Took an Offer It Couldn't Refuse," *Minneapolis Tribune,* September 26, 1976.

"Swiss Firm Plans to Buy Northrup King," *Minneapolis Star,* September 21, 1976.

"Insiders Gain from Northrup, King Sale," *Minneapolis Star,* December 16, 1976.

Stevens, Charles J., "Confronting the World Food Crisis," Occasional Paper 27, Muscatine, Iowa: The Stanley Foundation, December 1981.

Morgan, Dan, *Merchants of Grain.* New York: Viking, 1979.

Goldman, Sachs & Company, "The Hybrid Seed Corn Industry: Implications of a Changing Environment," 1974.

Mooney, Pat Roy, "The Law of the Seed: Another Development and Plant Genetic Resources," *Development Dialogue,* 1983:1–2, The Dag Hammarskjold Foundation, Uppsala, Sweden.

Mooney, Pat R., *Seeds of the Earth, A Private or Public Resource?* Ottawa, Canada and London: InterPares, 1979.

Collins, Steve, "A 'Seedy' Business Grows Through R & D," *The Commercial and Financial Chronicle,* September 23, 1974.

"A Rich Harvest for Seed Growers," *Business Week,* January 13, 1975.

Larson, Leon M., "Takeovers Trim Ranks of Seed Firms," *Sunday World-Herald* (Omaha, NE), April 11, 1982.

Meyer, Caroline, "Mail-Order Garden Firms Succulent Merger Targets," *Washington Post,* February 20, 1982.

Bennet, Jennifer. "And Now, from Those Wonderful Folks Who Brought You Twinkies . . . , The Patented Seed: Another Step Toward Corporate Agriculture?," *Harrowsmith* 4(2):1979.

"The Biotech Big Shots Snapping Up Small Seed Companies," *Business Week,* June 11, 1984, pp. 69 and 72.

U.S. Congress, House Subcommittee on Department Investigations, Oversight, and Research of the Committee on Agriculture, *Plant Variety Protection Act* Amendments, H.R. 7107, 96th Cong., 2nd Sess., on H.R. 999, April 22, 1980, pp. 101–102.

U.S. Congress, Senate Subcommittee on Agricultural Research and General Legislation of the Committee on Agriculture, Nutrition, and Forestry, *Plant Variety Protection Act,* 96th Cong., 2nd Sess., June 17, 1980.

Telephone conversation with George Pickering, Pickering Seed Company, Lewisville, Indiana, March 8, 1982.

Interview with Robert Hartmeir, branch manager, seed production operations, Asgrow Seed Company, Gonzales, California, August 11, 1981.

Interview with Jeff Shrum, research director, Asgrow Seed Company, San Juan Batista, California, August 11, 1981.

Harvard Business School, "Pioneer Hi-Bred International, Inc.," Case Study No. 4-375-109, 1974, revised October 1978, p. 9.

Seed Trade News, 1980.

Interview with former owner (who prefers to remain anonymous) of midwestern seed company now owned by a major chemical corporation, March 1982.

Telephone conversation with David Padwa, chief executive officer, Agrigenetics Corporation, August 24, 1982.

Leibenluft, Robert F., "Competition in Farm Inputs: An Examination of Four Industries," Policy Planning Issues Paper, Office of Policy Planning, Federal Trade Commission, February 1981, pp. 85–116.

Butler, L. J., and Marion, B. W. "An Economic Evaluation of the Plant Variety Protection Act," prepared for the Agricultural Marketing Service and Economic Research Service of the Department of Agriculture, December 1983, p. 73.

Ruder Finn & Rotman (Chicago, Illinois), corporate profile, "L. William Teweles & Company, Seed and Plant Science Consultants; Merger and Divesture Specialists," December 1983.

Ruder, Finn & Rotman (Chicago, Illinois), biography, "L. William Teweles, President, L. William Teweles & Company," December 1983.

"Bringing the New Plant Biotechnologies to the Marketplace," *Seedsmen's Digest,* June 1982.

"Acquistion Activity on the Upswing," *Seed Trade News,* December 22, 1982.

Telephone conversation with Dennis Stamp, owner, Wilson Hybrids, Incorporated, Harlan, Iowa, October 28, 1982.

Loden, Harold, executive vice president, American Seed Trade Association. Letter to the Editor, *Washington Post,* November 21, 1979.

Lyons, Richard D., "Vast Agricultural Gains Seen in Plant Genetics," *New York Times,* December 8, 1983.

L. William Teweles & Company, news release, "Plant Genetic Engineering Study Sets Dollar Value, Timetable for Improvements," December 1983.

L. William Teweles & Company, news release, "The Greening of The Giants—Major Corportions Buy Seed and Plant Science Firms," December 1983.

L. William Teweles. Letter to Stan Cath, American Seed Trade Association, October 27, 1982.

L. William Teweles & Company prospectus, "The U.S. Seed Industry 1982–1985," October 1982.

Teweles, L. William, "Bringing the New Plant Genetics to the Marketplace," handout and slide presentation at Energy Bureau conference, "Genetic Engineering in Food and Agriculture," Arlington, Virginia, December 6–7, 1982.

"The Livestock Industry's Genetic Revolution," *Business Week,* June 21, 1982, p. 124.

Randal, Judith, "Breeding the Perfect Cow," *Science 81,* November 1981, pp. 86–89.

Bucklin, Randolph E., "Embryo Transfer: Husbandry for '80's," *Washington Post,* July 29, 1982.

U.S. Congress, House Subcommittee on Investigations and Oversight, Committee on Science and Technology, *Genetic Applications to Animal Husbandry,* 1982.

Bratman, Harris, "Engineering the Birth of Cattle," *New York Times Magazine,* May 16, 1983.

"Biotechnology Struts into the Hen House," *Business Week,* April 11, 1983.

Trevis, Jim, and Bertelsen, Annette, "Genetic Engineering: Promise for Agricultural Industries," *Feedstuffs,* February 1, 1982.

Wennblom, Ralph, "First Sure-Fire Vaccine for Livestock," *Farm Journal,* August 1981.

"Vaccines That Lengthen the Lives of Livestock," *Chemical Week,* November 24, 1982.

"IMC: Slimmer Profits for a Fertilizer Giant," *Chemical Week,* March 17, 1982.

Tables compiled by the Environmental Policy Institute, Washington, D.C., "Biotechnology Companies and Established Corporations Involved in Agricultural Biotechnology" and "Corporate Investments, Research Contracts and

Joint Ventures With Biotechnology Companies for Agricultural and Related Products," September 1984.

"Big Firms Gain Profitable Foothold in New Gene-Splicing Technologies," *Wall Street Journal,* November 5, 1980.

Tamarkin, Bob, "The Growth Industry," *Forbes,* March 2, 1981.

"Dow Chemical Hires a Genetics Concern to Use Expertise in Growing Technology," *Wall Street Journal,* October 23, 1980.

"The Miracle of Spliced Genes," *Newsweek,* March 17, 1980.

"Monsanto's Profits Formula: Basics Plus High Technology," *Chemical Business,* July 27, 1981.

"Genentech Reports Output of Hormone to Spur Cow Growth," *Wall Street Journal,* March 16, 1981.

Singer, Dale, "Monsanto Co. Is Betting Millions It Can Make Genetic Research Pay," *St. Louis Post-Dispatch,* February 16, 1981, p. 1, Sec. B.

"Hartz Sold to Monsanto Subsidiary," *The Daily Leader* (Stuttgart, Arkansas), April 25, 1983.

Schneiderman, Howard, director of research and development, the Monsanto Company, "Some Thought on Biotechnology," *Chemical & Engineering News,* February 1, 1982.

Statement of John T. Marvel, Ph.D., general manager, research division, Monsanto Agricultural Products Company, before the Subcommittee on Investigations and Oversight of the House Committee on Science and Technology, June 9, 1982.

"And Grace Bets $50 Million on Biotechnology," *Chemical Week,* August 11, 1982.

"Grace, Cetus Form Agricultural Venture," *Journal of Commerce,* August 21, 1984.

" 'Possibly the Largest' Bioventure," *Chemical Week,* June 20, 1984, pp. 15–16.

"Cetus, Grace Announce Plans for Joint Venture." *Journal of Commerce,* June 12, 1984.

Pramik, Mary Jean, "Atlantic Richfield Builds Research Team for Agricultural Applications, in Biotechnology," *Genetic Engineering News,* May/June 1982.

Burnham, T. J., "Vast Plant Genetics Breakthrough in View for Crops." *The Fresno Bee,* January 2, 1983.

Interview with Dr. Gene Fox, Director, ARCO Plant Cell Research Institute, at corporate offices, Dublin, California, July 17, 1984.

Morris, Charles E., "Genetic Engineering: Its Impact on the Food Industry," *Food Engineering,* May 1981.

"The Bust in Biotechnology," *Newsweek,* July 26, 1982.

"Gene Splicers Unite," *Farm Chemicals,* March 1982.

"DNA's Role in Agriculture May Top Medical Potential," *Journal of Commerce,* November 18, 1981.

"Biotechnology—Seeking the Right Corporate Combinations," *Chemical Week,* September 30, 1981.

Meltzer, Yale L., "Genetic Engineering: Building New Profits," *Chemical Business,* April 6, 1981, p. 39.

Fox, Jeffrey L., "Biotechnology: A High-Stakes Industry in Flux," *Chemical & Engineering News,* March 29, 1982.

Chapter 7

Splinter, William E., "Center-Pivot Irrigation," *Scientific American,* June 1976.

Wheels of Fortune, A Report on the Impact of Center Pivot Irrigation on the Ownership of Land in Nebraska. Walthill, Nebraska: Center for Rural Affairs, 1975.

DeKalb AgResearch, Incorporated, brochure. "Irrigated Corn Management," 1982.

Fact Book of U.S. Agriculture, Publication Number 1063. Washington, D.C.: U.S. Department of Agriculture, November 1981.

"Top 150 Print Advertisers Over Past 5 Years," 1982 Marketing Services Guide, *Agrimarketing.*

U.S. Congress, Office of Technology Assessment, "Technology, Public Policy, and the Changing Structure of American Agriculture," OTA project proposal, July 1983.

A Time to Choose: Summary Report on the Structure of Agriculture. Washington, D.C.: U.S. Department of Agriculture, 1981, pp. 46–47.

"Fewer, Larger U.S. Farms by the Year 2000—and Some Consequences." Washington, D.C.: U.S. Department of Agriculture, 1970.

Palmer, Lane, "How It Came About," Du Pont *Context,* 11(2).

Crittenden, Ann, "Gene Splicing and Agriculture," *New York Times,* May 5, 1981.

Reimund, Donn A., Martin, J. Rod, and Moore, Charles V., *Structural Change in Agriculture, The Experience for Broilers, Fed Cattle, and Processing Vegetables.* Washington, D.C.: U.S. Department of Agriculture, April 1981.

Quinby, J. Roy, *Sorghum Improvement and the Genetics of Growth.* College Station, Texas: Texas A & M University Press, 1974.

National Academy of Sciences, *Genetic Vulnerability of Major Crops.* Washington, D.C.: National Academy Press, 1972.

May, Earl Chapin, *The Canning Clan.* New York: MacMillan, 1937.

Beaty, John Y., *Luther Burbank: Plant Magician.* New York: Julian Messner, 1943.

Paul, Bill, "It Isn't Chicken Feed to Put Your Brand on 78 Million Birds," *Wall Street Journal,* May 13, 1976, p. 1.

Strange, Marty, and Hassebrook, Chick, *Take Hogs, for Example. The*

Transformation of Hog Farming in America. Walthill, Nebraska: Center for Rural Affairs, June 1981.

Chuck Hassebrook, "Antibiotic Curb Would Help Family Farm," *New Land Review,* Walthill, Nebraska: Center for Rural Affairs, Spring 1979.

Pond, Wilson G., "Modern Pork Production," *Scientific American,* May 1983.

Klober, Kelly, "Why Are Those Boars Bored?," *Prairie Sentinel.* Walthill, Nebraska: Center for Rural Affairs, December 1981.

Telephone conversation with Dr. Royce Bringhurst, strawberry breeder, University of California at Davis, May 27, 1983.

Telephone conversation with Steve Huffstutlar, agricultural marketing specialist at Agricultural Cooperative Program, USDA, Salinas, California, July 21, 1982.

Interview with Michael D. Goldberg, director of agricultural business development and vice president, Cetus Madison Corporation, at Cetus corporate headquarters, Emeryville, California, July 17, 1984.

Remarks by N. L. Reding, executive vice president, Monsanto Company, "Technology and Agriculture: A Partnership for the Future," at the Kansas State Board of Agriculture, Topeka, Kansas, January 11, 1984, p. 6.

Dryden, R. N., Jr., "Farmers' Help," *Nature,* August 18, 1983.

"Production Rules Down on the Farm," *Nature,* June 23, 1983.

Interview with Norman Goldfarb, chief executive officer and chairman of the board, Calgene, Incorporated, at corporate offices, Davis, California, July 13, 1984.

Zwerdling, Daniel, "Down on the Farm," *The Progressive,* September 1983.

Remarks of Tom Urban, president, Pioneer Hi-Bred International, Incorporated, "Shaping the Image of American Agriculture," at meeting of the National Governors' Association, Washington, D.C., February 27, 1984, pp. 4–5.

"The Livestock Industry's Genetic Revolution," *Business Week,* June 21, 1982, p. 124.

U.S. Congress, House Subcommittee on Investigations and Oversight, Committee on Science and Technology Hearings, "Genetic Applications to Animal Husbandry," July 28, 1982.

Butler, L. J., and Schmid, A. Allen, "Genetic Engineering and the Future of the Farm and Food System in the U.S.," Cooperative Extension Service, Michigan State University, 1983.

Kalter, Robert H., "Biotechnology: Economic Implications for the Dairy Industry," *Dairy Marketing Notes,* Cornell University, Department of Agricultural Economics, vol. 3, 1983.

Agriculture 2000, A Look at the Future. Columbus, Ohio: Battelle Memorial Institutes, 1983.

"Open Market in Jeopardy," *Seed Trade News,* May 4, 1983.

Shand, Hope, "Billions of Chickens: The Business of the South," *Southern Exposure,* November/December 1983, pp. 76–82.

DeKalb AgResearch, Incorporated, annual report 1981.

"Genetic-Packaged Cattle May be Wave of Feedlot Future," *Feedlot Manager,* May 15, 1982.

Wisconsin Department of Agriculture, Trade and Consumer Protection, "Processed Vegetable Contract Market News: Peas—Sweet Corn—Snap Beans," February 1, 1980.

Telephone conversation with Barry Treat, senior agronomist, Adolph Coors, Incorporated, December 15, 1982.

Weaver, Samuel H., supervisor of grain research development, Quaker Oats Company. Letter and attachments to Jack Doyle, Environmental Policy Institute, March 12, 1982.

Jesse, Edward V., and Johnson, Aaron C., Jr., *Effectiveness of Federal Marketing Orders for Fruits and Vegetables.* Washington, D.C.: U.S. Department of Agriculture, June 1981.

Chapter 8

Northrup King Company, "Rio Verde, One for the Road," advertisement in *Seedsmen's Digest,* August 1981.

Stevens, Allen, "Varietal Influence on Nutritional Value," in White, P. L., and Selvey, N. (editors), *Nutritional Quality of Fresh Fruits and Vegetables.* Mt. Kisco, N.Y.: Futura Publishing, pp. 87–110.

National Academy of Sciences/National Research Council, Report of the Task Force on Genetic Alterations in Food and Feed Crops, Warren H. Gabelman, chairman, August 20, 1973.

U.S. Department of Agriculture, National Plant Genetics Resources Board, *Plant Genetic Resources, Conservation and Use.* See memorandum, pp. vii–ix, dated January 11, 1979.

Rick, Charles M., "The Tomato," *Scientific American,* August 1978.

L. William Teweles & Company, "Improved Tomatoes: Plant Genetic Engineering to Bear Fruit in the 1980s," news release, December 1983.

Stevens, Allen, "Breeding for Safety in Vegetables and Fruits, professional paper, 1973.

Axtell, J. D., "Breeding for Improved Nutritional Quality," in Frey, Kenneth J. (editor), *Plant Breeding II.* Ames, Iowa: The Iowa University Press, 1981.

Interview with Cathy Cryder-Sower, tomato breeder, Asgrow Seed Company, San Juan Batista, California, August 12, 1981.

Godard, Mary S., and Mathews, Ruth H., "Contribution of Fruits and Vegetables to Human Nutrition," *Horticulture Science,* June 1979.

Interview with Glenn Kardel, seed stock manager, Asgrow Seed Company, San Juan Batista, California, August 12, 1981.

National Research Council, Committee on Diet, Nutrition and Cancer, *Diet, Nutrition and Cancer*. Washington, D.C.: National Academy Press, June 1982.

Russell, Cristine, "Cancer Society Starts Crusade on U.S. Diet," *Washington Post*, February 11, 1984, p. 1.

Brody, Jane E. "Diet to Prevent Heart Attacks Aims to Cut Blood Fat Levels," *New York Times*, May 16, 1984, p. A16.

Kelly, John F., and Rhodes, Billy B., "The Potential for Improving the Nutrient Composition of Horticultural Crops," *Food Technology*, May 1975.

Correspondence from Joan D. Gussow, Chair, Department of Nutrition Education, Teachers College, Columbia Unviersity, September 17, 1984.

Telephone conversation with Dr. Robert Plaisted, potato breeder, Cornell University, January 19, 1984.

Telephone conversation with Dr. S. J. Piloquin, potato breeder, University of Wisconsin, Madison, January 23, 1984.

Telephone conversation with Gary Johnston, potato breeder (retired), University of Guelph, Ontario, Canada, January 23, 1984.

Akeley, R. V., Cunningham, C. E., Mills, W. R., and Watts, James. "Lenape, a New Potato Unusually High in Solids and Chipping Qualities," USDA release announcement, 1967.

"Minutes of Lenape Conference," University of Guelph, Ontario, Canada, December 11, 1969.

U.S. Department of Agriculture, "New Potato Variety Withdrawn," February 1970.

"Health Threat Found in Variety of Potato," *Toronto Globe and Mail*, Feburary 12, 1970.

"Canada, U.S. Ban New Potato Variety," *Toronto Daily Star*, February 13, 1970.

"Scientists' Efforts Halt Distribution of New Potato for Chip Industry," *London Free Press*, March 19, 1970.

"Canadian Research Brings Ban on New Chip Potato," *Toronto Telegram*, March 19, 1970.

"Occurrence of Bitter Lenape Potatoes in Ontario," University of Guelph, Ontario, Department of Agriculture and Food, Canada Department of Agriculture, Information for Extension Personnel, February 1970.

Zitnak, Ambrose, and Johston, Garnet R., "Glycoalkaloid Content of B5141–6 Potatoes," *American Potato Journal* 47(70).

Spiher, Alan T., Jr., "The Growing of GRAS," *Horticultural Science* 10(3).

Reitz, L. P., and Caldwell, B. E., "Breeding and Safety in Field Crops," paper presented at the Crop Science Society Meeting, Las Vegas, Nevada, November 15, 1973.

Gabelman, W. H., "GRAS Legislation: Viewpoints of a Professional Horticulturist and Plant Breeder," *Horticultural Science* 10(3).

Crosby, Edwin, "GRAS Regulations: Concerns of the Food Processing Industry," *Horticultural Science* 10(3).

Miller, Judith, "Agriculture: FDA Seeks to Regulate Genetic Manipulation of Food Crops," *Science,* July 19, 1974.

Senti, F. R., and Rizek, R. L., "Nutrient Levels in Horticultural Crops," *Horticultural Science* 10(3).

Munger, Henry M., "The Potential of Breeding Fruits and Vegetables for Human Nutrition," *Horticultural Science* 14(3).

Miller, Sanford A., director, Center for Food Safety and Applied Nutrition, U.S. Food and Drug Administration. Letter to Jack Doyle, director, Agricultural Resources Project, Environmental Policy Institute, Washington, D.C., July 17, 1984.

Remarks by Mark Novitch, M.D., acting commissioner, U.S. Food and Drug Administration, before the National Food Policy Conference, Mayflower Hotel, Washington, D.C., March 27, 1984.

Monsanto Company brochure, "Genetic Engineering: A Natural Science," 1984, p. 11.

Calgene, Incorporated, "A Conversation with Normal Goldfarb, President, Calgene, Inc.," issued by Calgene in 1984, p. 7.

Calgene, Incorporated, brochure, "The Second Green Revolution Is Here Today," 1982.

Sharp, William R., Evans, David A., and Ammirato, Philip V., "Plant Genetic Engineering: Designing Crops to Meet Food Industry Specifications," *Food Technology,* February 1984, pp. 112–19.

Morris, Betsy, "Campbell Soup Is Looking for 'Super' Tomato," *Wall Street Journal,* April 2, 1982.

U.S. Congress, House Subcommittee on Departmental Operations of the Committee on Agriculture, *Plant Variety Protection, H.R. 13424,* et al., 91st Cong., 2nd Sess., June 10, 1970.

Hollie, Pamela G., "Straining to Be More Than Just Soup," *New York Times,* March 20, 1983.

"The Race to Breed a 'Supertomato'," *Business Week,* January 10, 1983.

Campbell Soup Company, annual report, 1981.

Chapter 9

1980–81 National Gardening Survey, conducted by the Gallup Organization, Incorporated for Gardens for All, The National Association for Gardening.

"The Joy of Gardening," *Newsweek,* July 16, 1982.

Bubel, Nancy, *Vegetables Money Can't Buy but You Can Grow.* Boston: David A. Godine, 1977.

Mayer, Caroline E., "Mail-Order Garden Firms Succulent Merger Targets," *Washington Post,* February 2, 1982.

Proulx, E. A., "Selling the Seeds of Success"; and Cook, Jack, "Northern Seeds for Northern Gardens," *Horticulture,* January 1982, pp. 44–74.

Mooney, Pat Roy, "The Law of the Seed: Another Development and Plant Genetic Resources," *Development Dialogue.* Uppsala, Sweden: Dag Hammerskjold Foundation, 1983, 1–2, p. 102.

Environmental Policy Institute, Washington, D.C. "Seed Company Acquisitions, Mergers, and Related Business Ventures, 1968–1984," September 1984.

Whealy, Kent, Seed Savers Exchange, *The 1982 Winter Yearbook,* p. 1.

Whealy, Kent, Letter to Jack Doyle, May 15, 1983.

Jabs, Carolyn, "Meet the Seed Savers," *Organic Gardening,* June 1981, p. 54.

Interview with Kent Whealy of the Seed Savers Exchange. *Mother Earth News,* January/February 1982, p. 18.

Jabs, Carolyn, "Seed Savers," *Country Journal,* June 1981, p. 36.

Brown, George. Letter to Kika de la Garza, chairman, House Agriculture Subcommittee on Department Investigations, Oversight and Research, November 8, 1979.

Santelli, Carolyn Y. Letter to members of the House Agriculture Subcommittee on Research, July 23, 1979.

"Britain Losing Seed Species," *New York Times,* August 12, 1980.

U.S. Congress, Senate Subcommittee on Agricultural Research and General Legislation of the Committee on Agriculture, Nutrition, and Forestry, *Plant Variety Protection Act,* S. 2, et al., 96th Cong., 2nd Sess., June 17 and 18, 1980, p. 132.

Randolph, Eleanor, "Seed Patents: Fears Sprout at Grass Roots," *Los Angeles Times,* June 2, 1980, p. 10.

Katharine Anderson, director, Botanical Gardens Project, Gardens for All, The National Association for Gardening, Letter to Senator Patrick J. Leahy, June 20, 1980.

Rodale, Robert, "Germ Plasm—Two Words You Need to Know," *Organic Gardening,* February 1980, p. 36.

De Crosta, Anthony, "The Real Scoop on the Plant Patent Controversy," *Organic Gardening,* May 1980, p. 109.

Environmental Policy Institute, Washington, D.C., "Major Patent Holders—Field Crops and Vegetables," compiled from documents and publications of the U.S. Office of Plant Variety Protection, 1982–1984.

Louis, Arthur M., "The Bottom Line on Ten Big Mergers," *Fortune,* May 3, 1982.

"Seed. Feed. Succeed," advertisement, Burpee Corporation, *Newsweek,* April 11, 1983.

"Home Pesticide Sales Offer a Ray of Sunshine," *Chemical Week,* May 18, 1983.

"Chevron Building Research Facility," *Seed Trade News,* April 27, 1983.

Standard Oil Company of California, annual report 1983, p. 23.

Plant Genetics, Incorporated, corporate brochure, 1983–84.

Rogers, Michael, "Synthetic-Seed Technology," *Newsweek,* November 28, 1983, p. 11.

Gibhart, Fred, "Plant Genetics Inc. Reports Successful Vegetable Harvest from Synthetic Seeds," *Genetic Engineering News,* vol. 4, no. 1, January/February 1984.

Linden, Tim, "Planting Seeds for the Future . . . Encapsulation," *Western Grower and Shipper,* January 1984, pp. 12–14.

Chapter 10

Rhoades, Robert E., "The Incredible Potato," *National Geographic,* May 1982.

Davidson, Joanne, "Saga of a French-Fry King," in "Modern Tycoons—How They Made It, How They Live," *U.S. News & World Report,* May 31, 1982.

Henry, Charles, "Simplot's Feedlot of the Future," *Successful Farming,* September 1982.

Krebs, A. V., "Roll Along, Columbia: Corporate Agribusiness in the Mid-Columbia Basin," San Francisco: Urban Center for the Study of Land and Food, 1979, pp. 5–7.

Idaho Citizens Coalition, "Water, Energy and Land: Public Resources and Irrigation Development in the Pacific Northwest—Who Benefits and Who Pays." Boise, Idaho: Idaho Citizens Coalition, 1981, pp. 44–58.

National Potato Council, "1982 United States Seed Acreage Entered for Certification," *Spudletter,* August 1982.

National Academy of Sciences, *Genetic Vulnerability of Major Crops.* Washington, D.C.: National Academy Press, 1972.

Browning, J. Artie. Statement before the U.S. Senate Committee on Agriculture, Nutrition and Forestry Hearings, "Integrated Pest Management," October 31 and November 1, 1977.

Duvick, Donald N., "Genetic Diversity in Major Farm Crops on the Farm and in Reserve." Thirteenth International Botanical Congress, Sydney, Australia, August 28, 1981.

Telephone conversation with Sam Levings, plant breeder, North Carolina State University, June 28, 1982.

Sommers, Charles E., "How Vulnerable Are Corn Hybrids?," *Successful Farming,* November 1979.

"Do American Corn Hybrids Have a Dangerously Narrow Genetic Base?," *Successful Farming,* November 1979.

Corey, Fred P., director, market development, International Apple Institute, McLean Virginia. Letter to Jack Doyle, Environmental Policy Institute, July 23, 1982.

Reichert, Walt, "Agriculture's Diminishing Diversity: Increasing Yields and Vulnerability," *Environment,* vol. 24, no. 9, November 1982, pp. 7–11, 40–44.

Cole, H. H., and Garrett, W. N. (editors), *Animal Agriculture.* San Francisco: W. H. Freeman & Company, 1980.

Pond, Wilson C., "Modern Pork Production," *Scientific American,* May 1983.

Large, E. C., *The Advance of the Fungi.* London: Jonathan Cape, 1940.

Harlan, Jack R., "Crop Monoculture and the Future of American Agriculture," in *The Future of American Agriculture as a Strategic Resource: A Conservation Foundation Conference,* July 14, 1980, Washington, D.C.

Bottrell, Dale R., for the President's Council on Environmental Quality, *Integrated Pest Management,* December 1979.

Congress of the United States, Office of Technology Assessment. *Pest Management Strategies in Crop Protection,* Volumes I and II, October 1979.

Manning, Joe, "Genetically Altered Corn Plants to Be Tested in Top Secret Field," *Milwaukee Sentinel,* October 1983.

Feldman, Moshe, and Sears, Ernest R., "The Wild Gene Resources of Wheat," *Scientific American,* January 1981, pp. 102–12.

Carson, Rachel, *Silent Spring.* Boston: Houghton Mifflin, 1962.

Boraido, Allen A., "The Pesticide Dilemma," *National Geographic,* February 1980.

Pathak, Mano D., "Utilization of Insect-Plant Interactions in Pest Control," in David Pimentel, *Insects, Science & Society.* New York: Academic Press, 1975.

National Academy of Sciences, *Contemporary Pest Control Practices and Prospects;* vol. I: *The Report of the Executive Committee;* and vol. II: *Corn/Soybeans Pest Control.* Washington, D.C.: National Academy Press, 1975.

Telephone conversation with Ray Wilcoxon, USDA Wheat Rust Laboratory, University of Minnesota, St. Paul, November 28, 1982.

Telephone conversation with John Campbell, entomologist, Pioneer Hi-Bred International, Incorporated, October 23, 1984.

Weaver, D. B., Rodriguez-Kabana, R., and Robertson, D. C., "Nematode Resistant Varieties Sought for Infested Alabama Fields," *Highlights,* Alabama Agricultural Research Station and Auburn University, vol. 31, no. 3, fall 1984, p. 7.

Conversation with Cathy Cryder-Sower, Asgrow Seed Company Research Station, San Juan Batista, California, August 11, 1981.

Roberts, Daniel Altman, et al., *Fundamentals of Plant-Pest Control.* San Francisco: W. H. Freeman & Company, 1978.

Van der Plank, J. E., *Plant Diseases, Epidemics and Control.* New York: Academic Press, 1963.

Interview with William L. Brown, chairman, Board on Agriculture, National Academy of Sciences at Joseph Henry Building, Washington, D.C., November 18, 1983.

Walsh, John, "Genetic Vulnerability Down on the Farm," *Science,* October 9, 1981.

"Seed-Borne Viruses Threaten Germ Plasm Collections," *MFS*, March 3, '83.
Wolff, Anthony, "Plants in the Bank" (publication unknown), 1982–83, ›. 8–10.
National Academy of Sciences, *Conservation of Germplasm Resources: An nperative.* Washington, D.C.: National Academy Press, 1978.
Cox, Meg, "A French-Fry Diary: From Idaho Furrow to Golden Arches," *'all Street Journal,* February 8, 1982.

:hapter 11

Interviews with Dr. Ralph Hardy, research director, life sciences, E. I. Du 'ont de Nemours & Company, Wilmington, Delaware, May 14, 1982, and May 18, 1984.
"Du Pont: Seeking a Future in Biosciences," *Business Week,* November 24, 1980.
Tamarkin, Bob, "The Growth Industry," *Forbes,* March 2, 1981.
"The Du Pont Agrichemicals Story," *Farm Chemicals,* March 1981, pp. 13–35.
Carisio, Justin, "Crop Protection Chemicals: Cornerstone of Agricultural Productivity," *Du Pont Context,* vol. II, no. 2, 1982.
Du Pont Company, "A Brief History of the Du Pont Company," Public Affairs Department, Wilmington, Delaware, March 1979.
"Getting Out of Plastics," *Chemical Week,* February 16, 1983.
"Some Players Call It Quits," *Chemical Week,* May 4, 1983.
Hardy, Ralph W. F., "Biotechnology: What, How, Why, Where and When," presentation at Japan Techno-Economics Society (JATES) Forum, Tokyo, Japan, October 30, 1981.
———, "Biotechnology: Approach of a Multinational High Technology Company," presented at the International Conference on Investing in Biotechnology, Royal Lancaster Hotel, London, February 24–25, 1982.
———, "The Outlook for Agricultural Research and Technology," presented at Agriculture in the Twenty-First Century, Richmond, Virginia, April 11–13, 1983.
———, "Biotechnology in Agriculture: Status, Potential, Concerns," seminar series on Environmental Aspects of Biotechnology, U.S. EPA and AAAS, Washington, D.C., April 19, 1983.
———, Remarks at the National Academy of Sciences Convocation on Genetic Engineering of Plants, Washington, D.C., May 23–24, 1983.
———, statement before the U.S. House of Representatives Science and Technology Subcommittee on Investigations and Oversight and the Subcommittee on Science, Research and Technology, June 22, 1983.
"The Second Green Revolution," *Business Week,* August 25, 1980.

Calgene, Incorporated, "The Second Green Revolution Is
corporate brochure, 1982.

Steinhart, Peter, "The Second Green Revolution." *New Yor*
azine, October 25, 1981.

U.S. Department of State and The President's Council on E
Quality, *The Gobal 2000 Report to the President,* The Technica
II, 1980.

Barton, K. A., and Brill, W. J., "Prospects in Plant Genetic E
Science, February 11, 1983.

Wilkes, Garrison, "The World's Crop Plant Germplasm—An
Species," *Bulletin of the Atomic Scientists,* February 1977.

———, "Current Status of Crop Plant Germplasm," *CRC Criti*
in Plant Sciences, Volume I, Issue 2.

Harlan, Jack R., "Genetics of Disaster," *Journal of Environment*
vol. 1, 1972.

———, "Our Vanishing Genetic Resources." *Science,* May 9, 1

Mooney, P. R., *Seeds of the Earth.* Ottawa: Inter Pares, 1979.

Myers, Norman, *The Sinking Ark.* New York: Pergamon Press,

Nabhan, Gary. Correspondence to Jack Doyle, February 28, 198

Telephone conversation with Gary Nabhan, March 28, 1984.

Norman, Colin, "The Threat to One Million Species," *Science,* D
4, 1981.

"Foreign Fields Save Western Crops—Free of Charge," *New Scien*
tober 28, 1982.

Murray, James R., and Hiam, Alexander, "Biological Diversity and
Engineering," presented at the U.S. Department of State Strategy Co
on Biological Diversity, Washington, D.C., November 1981.

Walsh, John, "Germplasm Resources Are Losing Ground," *Science,* (
23, 1981.

Crittenden, Ann, "U.S. Seeks Seed Diversity as Crop Assurance," *Ne*
Times, September 21, 1981.

The NPGS: An Overview, 1982. Published by *Diversity,* special rep
1, April 1982.

Vicker, Ray, "U.S. Scientists Are Collecting Plant Germ-Plasm as
guard Against an Agricultural Disaster," *Wall Street Journal,* Ju
1982.

United States General Accounting Office, "The Department of Agric
Can Minimize the Risk of Potential Crop Failures," April 10, 1981.

The National Germplasm System. Washington, D.C.: Science and Edu
Division, U.S. Department of Agriculture, September 1981.

Plucknett, D. L., et al., "Crop Germplasm Conservation and Devel
Countries." *Science,* April 8, 1983.

"Viruses in Strategic Germplasm 'Ticking Biological Time Bomb,' "
fornia Arizona Farm Press, February 26, 1983.

"Big Firms Gain Profitable Foothold in New Gene-Splicing Technologies," *Wall Street Journal,* November 5, 1980.

"Dow Chemical Hires a Genetics Concern to Use Expertise in Growing Technology," *Wall Street Journal,* October 23, 1980.

"The Miracles of Spliced Genes," *Newsweek,* March 17, 1980.

"Allied Chemical Buys 10% in Bio Logicals," *Wall Street Journal,* November 15, 1980.

"Biotechnology—Seeking the Right Corporate Combination," *Chemical Week,* September 30, 1981.

Fox, Jeffrey L., "Biotechnology: A High-Stakes Industry in Flux," *Chemical & Engineering News,* March 29, 1982.

Crittenden, Ann, "The Gene Machine Hits the Farm," *New York Times,* June 28, 1981.

Geissbuhler, Hans, et al., "Frontiers in Crop Production: Chemical Research Objectives," *Science,* August 6, 1982.

Carson, Rachel, *Silent Spring,* New York: Fawcett, 1962.

Brown, Alan S., "Battling for the Winning Edge in Herbicides," *Chemical Business,* May 3, 1982.

"The Herbicide Families," *Successful Farming,* 1982.

"Behind Eli Lilly's Price War in Herbicides," *Chemical Week,* February 2, 1983.

Arntzen, Charles J. director, plant research laboratory, Michigan State University, "Introducing Herbicide Resistance into Crops via Novel Genetics." Remarks at the National Academy of Sciences Convocation on Genetic Engineering in Plants, Washington, D.C., May 23–24, 1983.

Board on Agriculture, National Research Council, *Genetic Engineering of Plants: Agricultural Research Opportunities and Policy Concerns.* Washington, D.C.: National Academy Press, 1984, pp. 40–44.

Marcus, Steven J., "Low-Dosage Herbicides," *New York Times,* May 19, 1983.

Alsop, Ronald, "Gains in Weed Control Yield Patent, Low-Dose Herbicides," *Wall Street Journal,* March 30, 1984, p. B1.

Marx, Jean L., "Plants' Resistance to Herbicide Pinpointed," *Science,* April 1, 1983.

Adler, E. F., et al. Lilly Research Laboratories, Greenfield, Indiana, "Future Trends in Controlling Weeds and Plant Pathogens," Beltsville Symposium VIII, Agricultural Chemicals of the Future," May 16–19, 1983.

Statement of Robert J. Kaufman, Ph.D., research director, plant sciences, Monsanto Agricultural Products Company, before the Subcommittee on Investigations and Oversight of the House Committee on Science and Technology, June 9, 1982.

"The Hot Market Herbicides," *Chemical Week,* July 7, 1982.

"Calgene, Inc. Announces First Genetically Engineered Gene for Herbicide Resistance," *Seed World,* November 1982.

Valiulis, David, "Plant Biotechnology at Calgene: After the Hype, There's Hope," *California Farmer,* June 16, 1984, pp. 6–7, 36–37.

Interview with Robert Goodman, vice president for science and development, Calgene, Incorporated, Davis California, July 13, 1984.

"Herbicides Follow the Current Trend to Low-Till Farming," *Chemical Week,* May 9, 1984, pp. 12–13.

"Herbicide Markets for Resistant Crop Plants," *Genetic Technology News,* April 1984, pp. 6–7.

Environmental Policy Institute, "U.S. Biotechnology Companies and Corporations Working on Herbicide-Resistance in Agricultural Crops," Washington, D.C., September 1984.

Glaser, Vicki P., "Researchers Tackle Herbicide Resistance," *Biotechnology,* December, 1983.

"Ciba-Geigy Introduces Unique 'Package' for Sorghum," *Farm Chemicals,* July 1979.

Rogers, Michael, "Synthetic-Seed Technology," *Newsweek,* November 28, 1983.

Interview with Zachary S. Wochok, president, Plant Genetics, Incorporated, Davis, California, July 13, 1984.

Plant Genetics, Incorporated, company brochure, 1984.

Gebhart, Fred, "Plant Genetics Inc. Reports Successful Vegetable Harvest from 'Synthetic Seeds,' " *Genetic Engineering News,* vol. 4, no. 1, January/February 1984.

Hughey, Ann, "More Firms Pursue Genetic Engineering in Quest for Plants with Desirable Traits," *Wall Street Journal,* May 10, 1983.

Hilts, Philip J., "The Test Tube Babies of Agriculture," *Washington Post,* April 30, 1983.

Kaufman, D. D., and Kearney, P. C., "Microbial Transformations in the Soil," in L. J. Audus (editor), *Herbicides.* New York: Academic Press, 1976, pp. 29–64.

Cohen, S. Z., et al., "Potential for Pesticide Contamination of Ground Water Resulting from Agricultural Uses," in *Treatment and Disposal of Pesticide Wastes,* American Chemical Society Symposium Series #259, Washington, D.C., 1984.

Spalding, R. F., et al., *Pesticide Monitoring Journal,* vol. 14, no. 2, 1980, pp. 70–73.

State of Maryland, Department of Health and Mental Hygiene, Office of Environmental Programs, "Results of Maryland Groundwater Survey," 1983.

Iowa Geological Survey, reports for Floyd Mitchell County and Big Spring Basin, 1984.

Greenhouse, Steven. "Trading in Monsanto Interrupted," *New York Times,* June 8, 1984, p. D4.

"EPA Support Expected for Monsanto Herbicide," *Journal of Commerce,* June 13, 1984.

Graff, Gordon M., "Keeping the World's Breadbasket Full," *Chemical Week,* February 11, 1976.

Nickell, Louise G., "The Growth Regulators: Controlling Biological Behavior with Chemicals," *Chemical & Engineering News,* October 9, 1978.

Gwyune, Peter, with Carey, John, "Hormones for Profit," *Newsweek,* October 27, 1980, p. 114.

Cooke, Anson R. Union Carbide Agricultural Products Company, Incorporated, "The Role of Plant Growth Regulators in Agriculture," Beltsville Symposium VIII, Agricultural Chemicals of the Future, May 16–19, 1983.

"The Coming Revolution in Agricultural Chemicals," *Chemical Week,* June 15, 1983.

Armstrong, Scott, "The New Agricultural Chemistry: Will It Boost Food Production?," *Christian Science Monitor,* October 18, 1983.

"Plant Growth Regulators: Low Profile, High Hopes," *Chemical Week,* September 12, 1984, pp. 22–23.

Tucker, William, "Of Mites and Men," *Harper's,* August 1978.

"Allied Pushes a Bacterial-Nitrogen Plan," *Chemical Week,* August 11, 1982.

Daum, Rudy M., "Zoecon's Quest for Success Tied to R & D," *Chemical & Engineering News,* August 10, 1981.

"How Sandoz Is Building a Beachhead in the U.S.," *Chemical Week,* July 13, 1983.

Silverstein, Robert M., "Pheromones: Background and Potential for Use in Insect Pest Control," *Science,* September 18, 1981.

"Biological Pesticides May Be Replacing Chemicals in Agriculture in 20 Years," *Seedsmen's Digest,* August 1981.

"Viral Insecticides for Chemical-Resistant Bugs," *Chemical Week,* August 29, 1984, pp. 32–33.

"Time Release Herbicides," *Chemical Week,* May 23, 1984, pp. 32–34.

Chapter 12

Schmeck, Harold M., Jr., "Gene-Splicers Plan Release of Bacteria to Aid Crops," *New York Times,* August 30, 1983.

Rogers, Michael, "Gene Splicing Leaves the Lab," *Newsweek,* August 15, 1983.

Hilts, Philip J., "U.S. Approves Dissemination of Gene-Engineered Microbe," *Washington Post,* September 14, 1983.

May, Lee, "3 Groups Sue to Block Gene Experiment," *Los Angeles Times,* September 16, 1983.

Hilts, Philip J., "NIH Weighing Plans to Release Altered Bacteria," *Washington Post,* September 20, 1983.

"Blocking Gene-Spliced Cells," *Chemical Week,* September 21, 1983.

Hilts, Philip J., "Field Testing a Gene-Engineered Microbe Endorsed," *Washington Post,* September 22, 1983.

Boffey, Philip M., "Plan Gains for First Outdoor Test of Genetically Engineered Plant," *New York Times,* September 23, 1983.

Sharples, Frances E., *Spread of Organisms with Novel Genotypes: Thoughts from an Ecological Perspective.* Oak Ridge, Tenn.: Oak Ridge National Laboratory, 1981.

Barta, Suzanne W. T., "Biological Control in Agroecosystems," *Science,* January 8, 1982.

Brill, Winston J., "Agricultural Microbiology," *Scientific American,* September 1981.

Hardy, Ralph, Statement before House of Representatives Science and Technology Subcommittee on Investigations and Oversight and Subcommittee on Science, Research and Technology, June 22, 1983.

Wallis, Claudia, "Honoring a Modern Mendel," *Time,* October 24, 1983.

Wilford, John Nobel, "A Brilliant Loner in Love with Genetics," *New York Times,* October 11, 1983.

Altman, Lawrence K., "Long Island Woman Wins Nobel in Medicine," *New York Times,* October 11, 1983.

Cohn, Victor, "Long-Neglected Woman Scientist Awarded Nobel," *Washington Post,* October 11, 1983.

Lewin, Roger, "No Genome Barriers to Promiscuous DNA," *Science,* June 1, 1984, pp. 970–71.

Cavalieri, Liebe F., statement of September 26, 1983, submitted with affidavits in *Foundation on Economic Trends v. Margaret M. Heckler,* Civil Action No. 83–2714, United States District Court, filed September 14, 1983.

Olson, Steve, "Fighting Frost Bugs," *Science 81,* November 1981.

Schneider, Keith, "First-Ever Release of Lab-Created Life Form." *California Journal,* February 1984.

Foundation on Economic Trends, et al. v. Margaret M. Heckler, et al., Civil Action No. 83–2714, U.S. District Court for the District of Columbia, with affidavits, filed September 14, 1983.

David, Peter, "Suit Filed Against NIH," *Nature,* September 22, 1983.

———,"Rifkin's Regulatory Revivalism Runs Riot," *Nature,* September 29, 1983.

Fox, Jeffrey L., and Norman, Colin, "Agricultural Genetics Goes to Court," *Science,* September 30, 1983.

Budiansky, Stephen, "Frost Damage Trial Halted," *Nature,* October 13, 1983.

"A Furor Over Gene-Splicing Outside The Lab," *Business Week,* October 3, 1983.

Norman, Colin, "Legal Threat Could Delay UC Experiment," *Science,* October 21, 1983.

Anthan, George, "Can 'Designer Genes' Fight Frost Damage?," *Des Moines Register,* October 2, 1983, p. 2A.

Schmeck, Harold M., Jr., "Gene-Spliced Organisms Are About to Come Out of the Lab," *New York Times*, September 25, 1983.

"Court Threat Delays Experiments in Genetically Modified Bacteria," *New York Times*, October 5, 1983.

"Air Rights for Genetic Creations," *New York Times*, October 10, 1983.

"Another Challenge to Frost-Control Genetics," *Chemical Week*, October 12, 1983.

Raeburn, Paul, "Researchers Opposed in Use of Genetic Bacteria for Crops," *The Salt Lake Tribune*, February 12, 1984, p. 4A.

Powledge, Tabitha M., " 'Ice-Minus' Talks Cancelled After Court Action," *Biotechnology*, March 1984.

Cowen, Robert C., "Congress Must Find Better Ways to Regulate Genetic Engineering," *Christian Science Monitor*, October 25, 1983.

Lyman, Francesca. Interview with Jeremy Rifkin, "Are We Redesigning Nature in Our Own Image?," *Environmental Action*, April 1983.

U.S. Congress, House, Science and Technology Subcommittee on Investigations and Oversight and Subcommittee on Science, Research and Technology, *Environmental Implications of Genetic Engineering*, 98th Cong., 2nd Sess., June 22, 1983.

Gore, Albert, Jr. Letter to James B. Wyngaarden, director, National Institutes of Health, October 21, 1983.

Hilts, Philip J., "Federal Agency on Gene Splicing Proposed," *Washington Post*, November 1, 1983.

Alexander, Martin, "Spread of Organisms with Novel Genotypes," U.S. Environmental Protection Agency and the American Association for the Advancement of Science, seminar on environmental aspects of biotechnology, Washington, D.C., May 17, 1983.

Correspondence from Martin Alexander to Jack Doyle, Environmental Policy Institute, Washington, D.C., September 23, 1984.

Robbins, Anthony, professional staff member, U.S. House of Representatives, Committee on Energy and Commerce. "Release of Genetically Engineered Organisms," address at the annual meeting of the American Association for the Advancement of Science, New York City, May 28, 1984.

U.S. Congress, House, Science and Technology Subcommittee on Investigations and Oversight, staff report, "The Environmental Implications of Genetic Engineering," February 1984.

Pramer, David, and Halvorson, Harlyn O., American Society For Microbiology. Letter to Albert Gore, Jr., March 30, 1984.

Wines, Michael, "Genetic Engineering—Who'll Regulate the Rapidly Growing Private Sector?," *National Journal*, October 15, 1983.

"Gene Splicing Sheds Its Mad-Scientist Image," *Business Week*, May 16, 1983.

"Now Who Will Regulate Man-Made Organisms," *Chemical Week*, July 6, 1983.

Hilts, Philip J., "EPA Will Take Over Regulation of Gene-Engineering Industry," *Washington Post,* August 9, 1983.

Hilts, Philip J., "EPA to Regulate Gene Engineering Firms," *Los Angeles Times,* August 10, 1983.

"More Rules for Gene-Splicers," *Chemical Week,* November 2, 1983.

Budiansky, Stephen, "Regulation issue is resurrected by EPA," *Nature,* 18 August 1983.

Johnson, Irving S., vice president, Lilly Research Laboratories. Letter to Dr. William J. Gartland, Jr., National Institutes of Health, January 31, 1984.

Hopkins, L., manager, biological research, Chevron Chemical Company. Letter to director, Office of Recombinant DNA Activities, NIH, January 23, 1984.

Nicholas, Robert B., chief counsel/staff director, Subcommittee on Investigations and Oversight, Committee on Science and Technology, U.S. House of Representatives, "The Double Helix Comes of Age: Congressional Oversight." Address before the Animal Health Institute, Tampa, Florida, October 19, 1984, pp. 6–7.

Hilts, Philip J., "U.S. Judge Halts Experiment in Gene Altering," *Washington Post,* May 17, 1984, p. 1.

Boffey, Philip M., "Judge Stalls Experiment in Genetic Engineering," *New York Times,* May 17, 1984.

Wines, Michael. "Court Bans 1st Release of Man-Made Microbe," *Los Angeles Times,* May 17, 1984.

Schorr, Burt, "Approval of Testing of Modified Genes is Blocked by U.S.," *Wall Street Journal,* May 17, 1984.

Powledge, Tabitha M., "Industry, EPA Politely Negotiate New Regs," *Biotechnology,* October 1983.

David, Peter, "Living with Regulations," *Nature,* October 27, 1983.

Fox, Jeffrey L., "Despite Doubts RAC Moving to Widen Role," *Science,* February 24, 1984.

"RAC Meeting Settles Little; Future Role Still Unclear," *Biotechnology News,* February 15, 1984.

"Let's Keep RAC in Charge," editorial, *Chemical Week,* February 29, 1984, p. 3.

"Regulatory Chaos Looms For Biotechnology," *Chemical Week,* February 29, 1984, pp. 35–36.

Henderson, Nell, "DuPont Chief Calls for Biotech Policy," *Washington Post,* September 13, 1984.

Monsanto Company, "Genetic Engineering: A Natural Science," company brochure, October 1984, p. 19.

Statement of Dr. Martin Alexander, Cornell University, before the Senate Environment and Public Works Subcommittee on Toxic Substances and Environmental Oversight, "The Environmental Consequences of Genetic Engineering," September 25, 1984.

Chapter 13

Auerbach, Stuart, "Dwindling Reserves Force India to Buy Wheat From U.S.," *Washington Post,* July 22, 1981.

Bickel, Lennard, *Facing Starvation: Norman Borlaug and the Fight Against Hunger*. New York: Readers Digest Press/E.P. Dutton & Company, 1974.

Wellhausen, Edwin J., "The Agriculture of Mexico," *Scientific American,* September 1976.

Plucknett, Donald L., and Smith, Nigel J. H., "Agricultural Research and Third World Production," *Science,* July 16, 1982.

Mellor, John W., "The Agriculture of India," *Scientific American,* September, 1976.

Wallace, James N., "Green Revolution Hits Double Trouble," *U.S. News & World Report,* July 28, 1980.

Perelman, Michael, "The Green Revolution: American Agriculture in the Third World," in Richard Merrill (editor), *Radical Agriculture*. New York: Harper & Row, 1976.

Interview with Dr. M. S. Swaminathan, *ICDA News,* newsletter of the International Coalition for Development Action, November 1981, p. 7.

Morgan, Dan, *Merchants of Grain*. New York: Viking Press, 1979.

Franke, Richard W., "Miracle Seeds and Shattered Dreams in Java," *Natural History,* January 1974.

Brush, Stephen B., "Farming the Edge of the Andes," *Natural History,* May 1977.

Lappé, Frances Moore, and Collins, Joseph, with Fowler, Cary, *Food First*. Boston: Houghton Mifflin, 1977.

Gough, Kathleen, "The Green Revolution in South India and North Vietnam," *Monthly Review,* January 1978.

Mooney, P. R., *Seeds of the Earth. A Private or Public Resource?* Ottowa: Mutual Press, Limited, 1979.

Wharton, Clifton R., Jr., "The Green Revolution: Cornucopia or Pandora's Box?," *Foreign Affairs,* April 1969.

Wilkes, H. Garrison, and Wilkes, Susan, "The Green Revolution," *Environment,* October 1972.

Paddock, William C., "How Green is the Green Revolution," *BioScience,* August 15, 1970.

Jennings, Peter R., "The Amplification of Agricultural Production," *Scientific American,* September, 1976.

Lewin, Roger, "Never-Ending Race for Genetic Variants," *Science,* November 26, 1982.

Rodale, Robert, "The Greening of the Green Revolution," *Organic Gardening and Farming,* July 1976.

"Biotechnology for Developing Nations Examined," *Chemical & Engineering News,* January 31, 1983.

Bodde, Tineke, "Biotechnology in the Third World," *BioScience,* October 1982.

Brady, Nyle C., "Chemistry and World Food Supplies," *Science,* November 26, 1982.

Swaminathan, M. S., "Biotechnology Research and Third World Agriculture," *Science,* December 3, 1982.

Gandhi, Indira, "Scientific Endeavor in India," *Science,* September 10, 1982.

Agricultural Genetics Report, May/June, July/August, and September/October issues. New York: Mary Ann Liebert, Inc., Publishers, 1982.

Biotechnology Newswatch. New York: McGraw Hill, March 7, 1982.

Biotechnology Newswatch. New York: McGraw Hill, November 15, 1982.

Biotechnology Bulletin. London: Oyez Scientific and Technical Services, Limited, November 1982.

"UNIDO Hopes for Biotechnology Center," *Science,* September 30, 1983.

"A U.N. Bid for Genetic Research," *Chemical Week,* September 21, 1983.

Bull, A. T., Holt, G., and Lilly, M. D., "International Trends and Perspectives in Biotechnology," Committee for Scientific and Technological Policy, Organization for Economic Co-operation and Development (OECD), Paris, March/April 1982.

Teso, Bruna, "The Promise of Biotechnology . . . and Some Constraints," *OECD Observer,* no. 118, September 1982.

Abrams, Robert E., executive vice president, International Plant Research Institute, "Bringing New Genetic Technology To Agro-Industrial Projects in the ASEAN Region," an address delivered at the Seminar for Overseas Investment in Food Production and Processing, sponsored by OPIC, Los Angeles, California, April 13–14, 1982.

"Blighted," *The Economist,* March 20, 1982.

Interview with Norman Goldfarb, chief executive officer and chairman, Calgene, Incorporated, Davis, California, July 13, 1984.

"Unilever: Seed for a Harvest," *Financial Times* (London), November 6, 1981.

Blair, John G., "Test Tube Gardens," *Science 82,* January/February 1982.

"Targeting Southeast Asia," *Chemical Week,* November 10, 1982.

Rosson, C. Parr, III, "U.S. Seed Exports," *Seedsmen's Digest,* May 1983.

Guske, Susanne. Unpublished paper, Western Michigan University, Department of Political Science, "An Expanding Private International Legal Regime for Patenting New Plant Varieties: A Preliminary Analysis of UPOV," December 1980.

Kenneth A. Dahlberg, memorandum to individuals concerned about plant patenting and U.S. membership in UPOV, December 30, 1980.

Declan J. Walton, director, IAA, FAO office memorandum to L. E. Haiguet, director, FOR, "Cooperation with the International Union for the Protection of New Varieties of Plants [UPOV]."

"Filipinos Turn Back From Green Revolution," *New York Times,* May 16, 1982.

"Tanzania Self-Reliance: A New National Policy," *The New Farm,* September/October 1983.

Rodale, Robert, "The Road to Morogoro: The Story of How Rodale Press Came to Export Organic Farming," *The New Farm,* September/October 1983.

Sinclair, Ward, "Back to Basics: Tanzanians Study U.S. Method of Farm Composting," *Washington Post,* July 4, 1983.

Plucknett, D. L., et al., "Crop Germplasm Conservation and Developing Countries," *Science,* April 8, 1983.

Lewin, Roger, "Funds Squeezed for International Agriculture." *Science,* November 26, 1982.

Crittenden, Ann, "Gains for Latin American Crops Called No Help to Small Farmers," *New York Times,* May 4, 1981.

Rowen, Hobart, "Population Explosion Forecast," *Washington Post,* July 11, 1984, p. 1.

Feeding the 500 Million. Amsterdam: International Association of Plant Breeders for the Protection of Plant Varieties (ASSINSEL), 1981.

"Fertilizer Subsidy Plan Stirs Controversy in India," *Journal of Commerce,* October 3, 1983.

Chapter 14

Interview with Robert J. Chapman, executive vice president, secretary and director, Genetic Engineering, Incorporated, Northglenn, Colorado, June 23, 1983.

Genetic Engineering, Incorporated, "Genetic Engineering Inc. Announces Breakthrough in Sexing of Bovine Embryos," press release, October 21, 1982.

"Genetic Engineering Identifies Sex of Calf Days After Conception," *Wall Street Journal,* December 29, 1982.

U.S. Congress, House, Subcommittee on Investigations and Oversight, Committee on Science and Technology, Genetic Applications to Animal Husbandry, Washington, D.C., July 28, 1982.

"The Livestock Industry's Genetic Revolution," *Business Week,* June 21, 1982.

King, Wayne, "Microscopic Techniques have a Gigantic Effect on Cattle Breeding Industry," *New York Times,* December 6, 1982.

"Genetic Engineering, Inc., Announces Private Placement," press release, April 18, 1983.

"Genetic Engineering Inc.: The Impossible Is Only the Beginning," *Limousin Journal,* May 1983.

"Vive la difference," *The Economist,* February 19–25, 1983.

"Embryo Transfer Made Easier," *Successful Farmer,* April 1982.

"The Growing Challenge of Genetic Technology," *Advanced Animal Breeder,* May 15, 1982.

Jones, Stacy V., "Patents—Labeling Chromosomes in the Sperm of Bulls," *New York Times,* December 11, 1982.

Genetic Engineering, Incorporated, annual report 1982.

Genetic Engineering, Incorporated, public offering prospectus, D. H. Blair & Company, January 14, 1981.

Carnation Company, annual report 1980, pp. 15, 19.

Lappé, Frances Moore, *Diet for a Small Planet.* New York: Ballantine, 1975.

Axtell, J. D., "Breeding for Improved Nutritional Quality," in Frey, Kenneth J. (editor), *Plant Breeding II.* Ames: Iowa University Press, 1981.

Bhatia, C. R., and Rabson, R., "Relationship of Grain Yield and Nutritional Quality," in *Nutritional Quality of Cereal Grains: Genetic and Agronomic Improvements.* Agronomy Society of America.

Interview with Robert Romig, director of research, Northrup King Company, Minneapolis, Minn., December 17, 1981.

Interview with Charles Krull, director of temperate corn research, DeKalb AgResearch, Incorporated, Dekalb, Illinois, March 4, 1982.

Telephone conversation with Dr. Edwin Mertz, Purdue University, January 10, 1984.

Fox, Jeffrey L., "More nutritious corn aim of gene engineering," *Chemical & Engineering News,* December 7, 1981.

SerVaas, Cory, "Purdue High-Lysine Corn Recipes," *Saturday Evening Post,* December 1983.

Telephone conversation with Carlos Reichert, Crow's Hybrid Corn Company, February 24, 1984.

Telephone conversation with Dr. John D. Axtell, Purdue University, February 23, 1984.

National Academy of Sciences, *Genetic Engineering of Plants: Agricultural Research Opportunities and Policy Concerns.* Washington: National Academy Press, 1984.

Molecular Genetics, Incorporated, corporate brochure, 1981.

Molecular Genetics, Incorporated, common stock prospectus, Kidder, Peabody & Co. et al., March 11, 1983.

Whitehouse, R. N. H., "The Prospects of Breeding Barley, Wheat and Oats to Meet Special Requirements in Human and Animal Nutrition," *Nutrition Society Proceedings,* vol. 29, 1970.

U.S. Department of Agriculture, "Solving Genetic Puzzle May Make Wheat Even More Nutritious," *Major News Releases and Speeches,* December 13, 1982.

"Exploring Genetic Key to Wheat Protein," *Milling & Baking News,* August 16, 1983.

Fishlock, David, "Exxon Looks for Enzyme Factory," *Financial Times* (London), October 30, 1981.

Mateles, Richard I., and Tannenbaum, Steven R. (editors), *Single-Cell Protein.* Cambridge, Massachusetts, and London: M.I.T. Press, 1968.

U.S. Congress, Office of Technology Assessment, *Commercial Biotechnology: An International Analysis,* Washington, D.C., January 1984.

Tannenbaum, Steven R., and Wang, Daniel I. C. (editors), *Single-Cell Protein II.* Cambridge, Massachusetts, and London: M.I.T. press, 1975.

Imperial Chemical Industries, PLC, " 'Pruteen': ICI's Unique Single-Cell Protein Process," corporate brochure, 1983.

Provesta Corporation, *Provesteen,* corporate brochure, 1984.

Inglish, Howard, "Wide Use Seen for Protein Powder," *Washington Post,* May 6, 1984, p. D7.

Chapter 15

Patterson, William Pat, "The Rush To Put Biotechnology to Work," *Industry Week,* September 7, 1981.

Sullivan, Lawrence Anthony, "Antitrust Law and Patents," Chapter 6 in *Handbook of the Law of Antitrust,* St. Paul, Minnesota: West Publishing Company, 1977.

Money, Pat Roy, "The Law of the Seed: Another Development and Plant Genetic Resources," *Development Dialogue.* Uppsala, Sweden: Dag Hammarskjold Foundation, 1983:1–2.

"Pfizer to Get Rachelle Units," *New York Times,* July 6, 1983.

Goldstein, Jorge A., "From Pseudomonas to the Birds: Are Animals Patentable?" *Recombinant DNA Technical Bulletin,* NIH, vol. 6, no. 2, June 1983.

Jackson, David A., "Patenting of Genes: What Will the Ground Rules Be?," in Robert F. Acker and Mosello Schaechter (editors), *Patentability of Microorganisms: Issues and Questions,* American Society for Microbiology, July 1980.

National Academy of Sciences, *Genetic Vulnerability of Major Crops.* Washington, D.C.: National Academy Press, 1972.

Harpstead, D. D., "Man-Molded Cereal—Hybrid Corn's Story," USDA Yearbook 1975, pp. 213–224.

Horsfall, James G., "The Fire Brigade Stops a Raging Corn Epidemic," 1975 Yearbook of Agriculture, U.S. Department of Agriculture.

Kahn, E. J., Jr., "The Staffs of Life: 1. The Golden Thread," *The New Yorker,* June 18, 1984.

Telephone conversation with Dr. James G. Horsfall, Connecticut Agricultural Experiment Station, July 7, 1983.

Telephone conversation with Clyde Willian, Hume, Clement, Hume & Lee, Chicago, Illinois, August 16, 1984.

"Seed Piracy in the Northwest," *Farm Journal,* April 1981.

Interview with Dick Haynes at Farmterials, Incorporated, Baker, Oregon, August 1981.

North American Plant Breeders, press release, "NAPB Wins Landmark Case on Plant Breeder's Rights," January 20, 1981.

Dick Haynes, Farmterials, Incorporated, press release, March 12, 1980.

Dick Haynes, Farmterials radio advertisement, March 30, 1981.

Congressional Research Service, Library of Congress, "Melcher Amendment Concerning Sale of Certified Seed," memos to Charles Benbrook, House Agriculture Committee, September 30, 1981, and October 2, 1981.

Odle, Jack, "Beware of the Bite of Plant Variety Protection Laws," *Progressive Farmer,* January 1982.

Levin, Tamar, "The Patent Race in Gene-Splicing," *New York Times,* August 29, 1982.

Sanger, David E., "Biotechnology's Patent War." *New York Times,* March 19, 1984.

"Patent on Gene Splicing, Cloning Granted To Stanford and University of California," *Wall Street Journal,* December 4, 1980.

Day, Kathleen, "Biotech Firms Testing Limits of Patent Law," *Los Angeles Times,* June 24, 1984.

"Agrigenetics Granted Biotechnology Patent," *Seed Trade News,* July 7, 1982.

Yoxen, Edward, *The Gene Business.* New York: Harper & Row, 1983.

U.S. Office of Plant Variety Protection, "Applicants and Total Number of Certificates Issued as of June 30, 1984."

Tegtmeyer, Rene D., Assistant Commission for Patents, "Patenting of Genetic Engineering Technology," delivered at National Academy of Sciences' Convocation on Genetic Engineering of Plants, Washington, D.C., May 24, 1983.

Reding, Nicholas L., executive vice president, Monsanto Company, "Biotechnology: Huge Potential in Agriculture," *Journal of Commerce,* August 19, 1983.

"Property Right Protection Needed," *Seed Trade News,* July 6, 1983.

"ABC Petitions Patent Office to Reinstate Fast-Track Approvals for Biotech Applications," *Genetic Engineering News,* March 1984.

Monsanto Company, annual report 1980.

U.S. Congress, Senate, "Additional Views of Senators Edward M. Kennedy and Howard M. Metzenbaum on S.255," The Patent Term Restoration Act of 1981, Report No. 97–138, June 16, 1981.

"Patents and Medicine," *Washington Post,* August 5, 1982.

"An Unwarranted Patent Stretch," *New York Times,* August 7, 1982.

Sun, Marjorie, "The Push to Protect Patents on Drugs," *Science,* November 11, 1983.

U.S. Congress, Senate. "The Patent Term Restoration Act of 1981, S.255." Hearings Before the Committee on the Judiciary, 30 April 1981, Statement of Thomas D. Kiley, Vice President and General Counsel, Genentech, Incorporated.

U.S. Congress, Senate. "The Patent Term Restoration Act of 1983." Hearings Before the Committee on the Judiciary, 22 June 1981, Statement of Brian Cunningham, General Counsel, Genentech, Incorporated.

"A Near-Pact on Drug Patents," *Chemical Week,* May 30, 1984.

Hinkle, Maureen, senior policy analyst, National Audubon Society, Letter to George Brown, Sr., Re: H.R.5203, April 19, 1982.

Mott, Laurie, project scientist, National Resources Defense Council, Incorporated. Letter to Senator Masyuki Matsunaga, July 24, 1982.

Carey, William D., executive officer, American Association for the Advancement of Science. Letter to George E. Brown, Jr., subject: H.R.5203, July 30, 1982.

Hinkle, Maureen, National Audubon Society. Memorandum to Scott Flemming, Re: H.R.5203 "Innovating Methodology," August 3, 1982.

Supreme Court of the United States, Syllabus, Ruckelshaus, Administrator, U.S. Environmental Protection Agency *v.* Monsanto Company, Decided, June 16, 1984.

Sinclair, Ward, "Pesticide Bill Change Mystifies Lawmaker," *Washington Post,* August 1, 1984.

"The Hot Market in Herbicides," *Chemical Week,* July 7, 1982.

"Behind Eli Lilly's Price War In Herbicides," *Chemical Week,* February 2, 1983.

Brown, Alan S., "Battling for the Winning Edge in Herbicides," *Chemical Business,* May 3, 1982.

"Herbicides Follow the Current Trend to Low-Till Farming," *Chemical Week,* May 9, 1984.

Telephone conversation with Stanley F. Rollin, October 14, 1982.

Chapter 16

Siegal, Ralph, "Hart Plants Campaign Seeds in Cinnaminson Research Lab." Cinnaminson, N.J. *The Times,* May 30, 1984.

Diamond, Randy, "Hart Labors to Show He's Right." *Daily News* (Secaucus, N.J.), May 30, 1984.

Boyd, Gerald M., "Hart Attacks Rivals; Offers 'Third Way' For Economic Gains." *The New York Times,* May 30, 1984, p. 1.

Glab, Theresa A., "Hart Makes Unplanned School Visit." Cinnaminson, N.J. *Courier-Post,* May 30, 1984, p. 3–A.

Thomas, Evan, "Last Call, and Out Reeling." *Time,* June 11, 1984, p. 26.

Peterson, Bill, "In New Jersey, Sen. Hart Tries to Pull Campaign Back on Track for Tuesday." *Washington Post,* May 30, 1984, p. A–3.

Price, Keith, Vice President, Gurney Seed & Nursery Company, Letter to Hon. Tom Daschle, October 2, 1981.

Waltrip, Jim, Merchandising Manager, Gurney Seed & Nursery Company, Letter to Hon. Tom Daschle, October 1, 1981.

Dworkin, Peter. "Amfac at a Tropical Crossroad." *New York Times,* April 10, 1983.

"Amfac Picks 'Someone to Tighten the Screws'." *Business Week,* December 16, 1982.

Amfac, Incorporated, Annual Report, 1980 and Conversation with Analysts, 1980.

"Henry Field's sold to Amfac Inc." Shenandoah, Iowa *Evening Sentinel,* October 22, 1980.

"Amfac, Buyer of Henry Field's, Welcomed to Shanandoah Thursday." Shenandoah, Iowa *Evening Sentinel,* January 9, 1981.

Conversations with House Agriculture Committee staff members February and July 1982.

Price, Keith, Vice President, Gurney Seed & Nursery Co., Western Union Mailgram to Hon. Tom Daschle, December 11, 1981.

Daschle, Hon. Tom, Member of U.S. Congress, Letter to Keith Price, Vice President, Gurney Seed & Nursery Company, December 17, 1981.

American Seed Trade Association, Washington, D.C., 1980 Roll of Active Members.

National Council of Commercial Plant Breeders, Membership List, 1981.

Conversation with Harold Loden, Washington Representative, American Seed Trade Association, Washington, D.C., August 15, 1982.

"Selling in a More Competitive World." Remarks of Seeley Lodwick, Under Secretary of Agriculture, at the ASTA Corn and Sorghum Industry Research Conference, Chicago, Illinois, December 10, 1981.

"Ease on Captan Ban Big News at ASTA." *Seedsmen's Digest,* August 1981, p. 18.

Conversation with former Senate Agriculture Committee staff member, Washington, D.C., October 8, 1982.

Information compiled by the Environmental Policy Institute, Washington, D.C. on the location, by U.S. Congressional District, of corporate property in the United States seed and agricultural biotechnology industries.

"PACs Gear Up For Their Biggest Year." *Chemical Week,* April 4, 1982, p. 29.

Letter sent to various seed industry members at their home addresses from SEEDPAC and ASTA representatives, August 1982.

"One Trade Group in the Fast Lane." *Chemical Week,* February 2, 1983.

"Gene Splicers Unite." *Farm Chemicals,* March 1982.

Industrial Biotechnology Association. "IBA In The News." 1982 and 1983.

Industrial Biotechnology Association, Member Companies, February 1984.

Norman, Colin. "Democrats Boost R & D." *Science,* April 3, 1983.

Tsongas, Paul E. "Atarizing Reagan." *New York Times,* March 1, 1983.

Young, Leah R. "Republican Pushes Hi-Tech Plan." *Journal of Commerce,* December 16, 1983, p. 1–A.

Fuqua, Hon. Don. Letter to Hon. Ted Stevens, Chairman, Technology Assessment Board, August 19, 1981.

"White House gets study on biocompetitiveness: Will it surface or sink?" *Biotechnology Newswatch,* June 20, 1983.

U.S. Congress, Office of Technology Assessment. *Commercial Biotechnology: An International Analysis,* Summary and Complete Study, January 1984.

Schmeck, Harold M., Jr. "Report Says Japan Could Lead in Commercial Biotechnology." *New York Times,* January 27, 1984.

Salisbury, David F. "How U.S. Can Stay Out Front in Biotech." *Christian Science Monitor,* February 1, 1984.

Pyatt, Rudolph A., Jr. "Biotech Center Planned for Shady Grove." *Washington Post,* February 11, 1984, p. C–1.

Pyatt, Rudolph A., Jr. "Maryland Leads Way in Biotechnology Race." *Washington Post,* February 14, 1984.

"Unusual Partners Launch a Biotechnology Venture." *Science,* Vol. 223, p. 800.

"The States Make a Bid for Biotech." *D & B Reports* (The Dun & Bradstreet Magazine), July/August 1984, p. 46.

"California Lawmakers to Revive Vetoed Measure Making NIH Recombinant-DNA Guidelines Binding in State." *Biotechnology Newswatch,* November 15, 1982.

Chapter 17

Cornell University, Cornell Biotechnology Program, brochure, "Biotechnology: Biology for Economic Development in New York State," March 1983.

Castro, Mike. "Prof Tries On A Business Suit." *Sacramento Bee,* January 4, 1982.

Hess, Charles E. "The Evolution of Calgene, A Potential Conflict of Interests and Its Resolution—From a Dean's Perspective." Statement before the House Science and Technology Subcommittee on Investigations and Oversight and Subcommittee on Science, Research and Technology, Hearings, "University/Industry Cooperation in Biotechnology," June 1982.

"Allied Corporation in Davis Firm." *Sacramento Bee,* July 9, 1981.

Foust, Debra. "Hess Alleges Calgene Conflict." *Sacramento Bee,* August 12, 1981.

See William Boly, "Strained Relations: A UC Davis biologist battles his students," as attachment in Congressional Hearings, p. 172, see Hess, *op. cit.*

"Allied, Calgene Cancel Research Contract Pact." *Journal of Commerce,* February 15, 1984.

Krimsky, Sheldon. "The Corporate Capture of Academic Science and its Social Costs." Paper delivered at the Genetics and the Law Conference, Sponsored by the American Society of Law and Medicine, April 2, 1984, 19 pp.

Bouton, Katherine. "Academic Research and Big Business: A Delicate Balance." *New York Times Magazine,* September 11, 1983.

"Business and Universities: A New Partnership." *Business Week,* December 20, 1982.

"Corporations Bet on Campus R & D." *Business Week,* December 20, 1982.

Sanger, David E. "Corporate Links Worry Scholars." *New York Times,* October 17, 1982.

"A Campus Brain Drain." *New York Times,* June 29; 1981.

Abelson, Philip H. "Differing Values in Academia and Industry." *Science,* September 17, 1982.

Lyons, Richard D. "DNA Research Raises Stature of Big Agriculture Schools." *New York Times,* November 24, 1981.

Ruttan, Vernon W. *Agricultural Research Policy.* Minneapolis: University of Minnesota Press, 1982.

Blaska, David. "Genetics Boom Is Changing the Face of Our Agriculture." *Capital Times* (Madison, Wisconsin), July 1981, p. 7.

Patterson, William Pat. "The Rush to Put Biotechnology to Work." *Industry Week,* September 7, 1981.

"Biotechnology—Seeking the Right Corporate Combinations." *Chemical Week,* September 30, 1981.

"Life: Patent Pending," *NOVA,* WGBH-TV, Boston, Massachusetts. Transcript of broadcast first aired February 28, 1982.

U.S. House of Representatives, Science and Technology Subcommittee on Investigations and Oversight, Hearings, "University/Industry Cooperation in Biotechnology," June 16 and 17, 1982.

In statement of Dr. Eloise E. Clark, Assistant Director for Biological, Behavioral, and Social Sciences, National Science Foundation, Before the House Agriculture Subcommittee on Department Operations, Research, and Foreign Agriculture, March 20, 1981.

Telephone conversation with a university/USDA-employed barley breeder who wishes to remain anonymous, May 25, 1982.

Culliton, Barbara J. "Academe and Industry Debate Partnership." *Science,* January 14, 1983.

"Academe and Industry: Closer." *Chemical Week,* November 24, 1982.

Openheimer & Company Limited Partnership Offering, Agrigenetics Research Associates Limited, November 1981, pp. 19–20; and telephone conversation with David Padwa, CEO, Agrigenetics, Incorporated, August 24, 1982.

Padwa, David. Speech delivered at the American Seed Trade Association's Farm Seed Conference, Kansas City, Missouri, November 8, 1982.

"Biotechnology's Ties to Academia Assessed." *Chemical & Engineering News,* April 26, 1982.

U.S. Congress, Office of Technology Assessment, *Commercial Biotechnology: An International Analysis,* January 1984.

Hightower, Jim. *Hard Tomatoes, Hard Times.* Cambridge, Massachusetts: Schenbman Publishing, 1973.

Marshall, Eliot. "USDA Research Under Fire." *Science,* July 1982, p. 33.

Norman, Colin. "White House Plows into Ag Research." *Science,* September 24, 1982, pp. 1227–28.

"Betting on the Farm." *Scientific American,* May 1983, pp. 88–89.

National Research Council. *World Food and Nutrition Study: The Potential Contributions of Research.* Washington, D.C.: National Academy of Sciences, 1977.

The Rockefeller Foundation and the Office of Science and Technology Policy, Executive Office of the President, United States of America. *Science for Agriculture, Report of a Workshop on Critical Issues in American Agricultural Research.* October 1982, 33 pp.

Lepkowski, Wil. "Shakeup Ahead for Agricultural Research." *Chemical & Engineering News,* November 22, 1982, pp. 8–16.

"The Worm in the Bud." *New York Times,* October 21, 1982.

Extramural Trends. National Institutes of Health, Statistics and Analysis Branch, Division of Research Grants, 1981.

"Cornell Will Start a Biotechnology Institute." *Chemical Week,* April 6, 1983, p. 37.

Telephone conversation with wheat breeder, North Carolina State University, May 19, 1982. Information also based in part on interview with Dr. Thomas Starling, Professor of Plant Breeding and Crop Sciences, and Dr. Robert L. Harrison, Virginia Crop Improvement Association, at VPI, Blacksburg, Virginia, February 8, 1982.

"Wrap-up." *Chemical Week,* February 2, 1983.

Agricultural Genetics Report, July/August 1982.

Edward E. Davis, Jr., Address to the New York City Bar Association, April 1982, as appeared in *Environment,* July/August 1982.

Crittenden, Ann. "Farm Research Aims Disputed." *New York Times,* December 2, 1980.

Barnett, Paul. "The Pesticide Connection." *Science for the People,* July/August 1980.

Fujimoto, Isao, and Fishe, Emmett. "What Research Gets Done at a Land Grant College: Internal Factors at Work." Paper presented at the 1975 Rural Sociological Society Meeting, San Francisco.

Bellew, Patricia A. "Agricultural Research, Once Little Noticed, Grows Controversial." *Wall Street Journal,* November 21, 1984, p. 1.

Anderson, H. H., Hine, C. H., Kodama, J. K., and Willingston, J. S. "An Evaluation of the Degree of Toxicity of 1.2-Dibromo-3-Chloropropane. 1. Chronic Feeding Experiments in Rodents," U.S. Report No. 228. Confidential report to Shell Development Company, Department of Pharmacology and Experimental Medicine, U.C. School of Medicine, 1958, San Francisco.

McCauley, W. E. Letter to M. W. Allen, Chairman of the Department of

Nematology, University of California Davis, March 15, 1966. From Pesticide Development Department, Shell Chemical Company, New York.

"Improvements Needed in the University of California's Foundation Seed and Plant Materials Service." Report of the Office of the Auditor General, State of California, to the California Joint Legislative Committee, November 1979.

Boffey, Phillip M. "Science Academy Looking for Funds." *New York Times*, May 4, 1983.

Boffey, Phillip M. *The Brain Bank of America, An Inquiry into the Politics of Science*, McGraw-Hill: New York, 1975.

Graham, Frank, Jr. *Since Silent Spring*, Boston: Houghton-Mifflin Company, 1970.

National Academy of Sciences, Membership, Board of Agriculture, May 1983.

Prospectus, Advanced Genetic Sciences, Incorporated, Smith Barney, Harris Upham & Co., July 1983.

Calgene, Incorporated. "The Second Green Revolution is Here Today," The Science Board, 1982.

Council for Agricultural Science and Technology. "This is CAST: What It Is, What It Does, and How It Operates," February 1981.

Herrig, Robin Marantz. "CAST-Industry Tie Raises Credibility Concerns." *BioScience*, Vol. 29, No. 1, January 1979.

Webster, Bayard. "6 Scientists Quit Panel in Dispute Over Livestock Drugs," *New York Times*, January 23, 1979.

Marshall, Eliot, "Scientists quit Antibiotics Panel at CAST," *Science*, February 23, 1979.

Letter & enclosure from Charles A. Black, Executive Vice President, Council for Agricultural Science and Technology, to Jack Doyle, Environmental Policy Institute, May 21, 1982.

Chapter 18

Mintz, Morton, "Lilly Official Knew of Deaths Before U.S. Approved Drug," *Washington Post*, July 22, 1983, p. A80.

———,"Lilly Gives U.S. Documents in Oraflex Probe," *Washington Post*, October 1, 1983, p. D11.

Burnham, David, "1965 Memos Show Dow's Anxiety on Dioxin," *New York Times*, April 19, 1983, p. 1.

Randolph, Eleanor, and Brown, Cabot, "Dioxin Perils Long Known, Papers Show," *Los Angeles Times*, July 6, 1983.

Blumenthal, Ralph, "Files Show Dioxin Makers Knew of Hazards," *New York Times*, July 6, 1983, p. 1.

Shabecoff, Philip. "E.P.A. Faults Tests on 200 Pesticides," *New York Times*, May 12, 1983, p. 1.

Klose, Kevin, "Ex-Officials of Chemical-Testing Lab Found Guilty of Falsifying Results," *Washington Post*, October 22, 1983, p. A7.

Isihoff, Michael, "Firm Said to Inflate Earnings," *Washington Post*, August 14, 1984, p. D1.

"Stockholders Sue Stauffer for Overstating Earnings," *Chemical Week*, August 29, 1984, p. 9.

Knight, Jerry, "McCormick & Co. Admits Falsifying Records," *Washington Post*, May 29, 1982, p. D1.

Shellenbarger, Sue, "Bigness Counts in Agribusiness, and Cargill, Inc. Is Fast Becoming a Commodities Conglomerate," *Wall Street Journal*, May 7, 1982.

Charlton, Penny, "U.S. Aiding Drug Companies in Bangladesh," *Washington Post*, August 19, 1982, p. 1.

Wallis, Claudia, et al., "A Double Standard on Drugs?," *Time*, June 28, 1982.

Weir, David, and Schapiro, Mark, *Circle of Poison: Pesticides and People in a Hungry World*. San Francisco: Institute For Food and Development Policy, 1981, pp. 79–80.

Zim, Martin H., "Allied Chemical's $20-Million Ordeal With Kepone," *Fortune*, September 11, 1978, pp. 82–90.

Mesdag, Lisa Miller, "Remember Kepone," *Fortune*, August 22, 1983, p. 193.

Meyers, Robert, *D.E.S.: The Bitter Pill*. New York: Seview Books/Putnam, 1983.

AGRIBUSINESS AND THE FOOD CHAIN

Business Investments in Agriculture, Genetics, and Biotechnology Research

As of January 1985

Corporation with Fortune 500 Rank (1984)	Seed/Livestock, Food/Agriculture Subsidiaries, &/or Foreign Interests	Biotechnology Ventures; Investments; Facilities	Agricultural Genetics Research & Related Activities	General Information; Agricultural/Food Products
Abbott Laboratories Chicago, IL Fortune Rank: 131	in 1977 made a joint venture with Takeda Chemical Industries of Japan called TAP Pharmaceuticals, Inc., for marketing products in U.S. & Latin America	invested $5 million in Applied Molecular Genetics (AmGen) in 1980	interested in crop & livestock applications of biotechnology; has financially supported university research on plant growth regulators, seed treatments, & biological pesticides	10th largest U.S. drug company; produces & sells animal health products, herbicides, plant growth regulators, & some biological pesticides; also sells Similac
Allied Corporation Morristown, NJ Fortune Rank: 26	acquired Nitragin, Inc., in 1982, a $6 million-a-year company specializing in the production & marketing of nitrogen-fixing bacteria	invested $2.5 million in Calgene, Inc., in 1981 (20% share; Allied later terminated agreement in 1984); also holds equity in Bio-Logicals, Inc., of Toronto, Canada; $16.5 million genetics research contract with Genex, Inc., for "broad industrial applications"; & a $10 million investment in the Genetics Institute, Inc.	Allied has 40 scientists involved in agricultural bioengineering research & spends about $30 million annually on ag research; has biotech research lab in Syracuse, NY; expects that its first genetically engineered product will be a bacteria that "fixes" nitrogen more efficiently in soybeans	6th largest U.S. chemical company, is a major producer of fertilizers; acquired Bendix in 1983

Corporation with Fortune 500 Rank (1984)	Seed/Livestock, Food/Agriculture Subsidiaries, &/or Foreign Interests	Biotechnology Ventures; Investments; Facilities	Agricultural Genetics Research & Related Activities	General Information; Agricultural/Food Products
American Cyanamid Co. Wayne, NJ Fortune Rank: 100	owns Lederle Laboratories, which makes drugs & animal health products; also sells Old Spice, Pine-Sol, & Formica consumer products as well as Centrum multivitamins	invested $5.5 million in Molecular Genetics, Inc. (MGI), in 1981 (20% share), & holds contracts with MGI for research on herbicide-resistant corn & some livestock applications. In 1983, Cyanamid made a $1.25 million contract with Biotechnology General Corp. of New York & Rehovot, Israel, to conduct research on bovine, porcine, & chicken hormones	involved in both crop & livestock applications of biotechnology; spends an estimated $70 million annually on agricultural R & D; has financially supported university research on herbicides, nematode control, & livestock drugs	7th largest U.S. chemical company; 3rd largest U.S. producer of animal health products; also produces & sells herbicides, insecticides, plant growth regulators, & fertilizers

Corporation with Fortune 500 Rank (1984)	Seed/Livestock, Food/Agriculture Subsidiaries, &/or Foreign Interests	Biotechnology Ventures; Investments; Facilities	Agricultural Genetics Research & Related Activities	General Information; Agricultural/Food Products
Amfac, Inc. Honolulu, HI	owns 7 U.S. companies in the seed, nursery, &/or garden products business; owns 60,000 acres in sugar plantations & resorts in Hawaii; also involved in mushroom production & potato processing; owns Lamb-Weston, Pacific Pearl Seafood, & Fisher Cheese Co.	made a $1.9 million, 3-year research contract with Hybritech, Inc., in October 1983 for research on pharmaceutical products	involved in horticultural plant breeding; supports genetic research for the development of new varieties of sugarcane by providing $2.4 million annually to the Hawaiian Sugar Planters' Association; Amfac Amycel produces mushroom spawn	largest sugar producer in Hawaii; in 1980, Amfac claimed a 15% share of the mail-order horticultural business & said it was "the largest wholesale ornamental horticulture company in the U.S."
AMOCO (Standard Oil of Indiana) Chicago, IL Fortune Rank:10		in 1978, acquired a 12% interest in Cetus Corp. & is working with Cetus on fertilizers		produces & sells fertilizer outside U.S.
Anheuser-Busch St. Louis, MO Fortune Rank: 53	in 1982 acquired the barley breeding program of North American Plant Breeders, Berthoud, CO; also owns Campbell Taggart Bakeries	in May 1983 made a 3-year research agreement with Interferon Sciences, Inc., to genetically engineer yeast strains for use in fermentation; $600,000 to Washington Univ. Center for Biotechnology for basic research in biochemical mixing & separation techniques	biotech interests appear related to development of yeast strains for both beer- & bread-making & genetics of barley for beer-making & for use in livestock feed	#1 brewer in U.S.; Campbell Taggart is 3rd largest baked foods producer

Corporation with Fortune 500 Rank (1984)	Seed/Livestock, Food/Agriculture Subsidiaries, &/or Foreign Interests	Biotechnology Ventures; Investments; Facilities	Agricultural Genetics Research & Related Activities	General Information; Agricultural/Food Products
Archer-Daniels Midland Company Decatur, IL Fortune Rank: 74	in 1982 acquired the Columbian Peanut Co. (peanut processing); also owns Supreme Sugar Corp.; Gooch Foods (La Rosa & others); Fleischmann Malting Co.; Tabor Grain Co.; & the Farmer City Grain seed company	made collaborative agreement with DNA Plant Technology Corp. in 1984 "to test feasibility of growing selected fresh, quality herbs hydroponically," possibly for mass production in a joint venture	crop genetic interests in corn, sorghum seed, & other crops; company has 4.5-acre vegetable hydroponics research complex in Decatur, IL; also has ongoing research in fermentation microbiology	3rd largest U.S. flour miller; major producer of sweeteners from corn & sugarcane; also sells fertilizers, feeds, feed concentrates, & mineral supplements to livestock & poultry producers
ARCO Atlantic Richfield Los Angeles, CA Fortune Rank: 12	in 1980 acquired Desert Seed Co., El Centro, CA (vegetable seed); Canyon Seed Co. of Canyon, TX (triticale). Also owns Valley Dehydrating Co., El Centro, CA (alfalfa processing & xanthophyll production)	in 1981 established the Plant Cell Research Institute in Dublin, Ca. In 1983, this ARCO subsidiary contracted with Heinz Co. to develop high-solids tomato. ARCO also supplied early funding to International Plant Research Institute (IPRI) & is a major stockholder in International Genetic Engineering, Inc.	ARCO is making a major effort in plant biotechnology, eyeing possibilities in both food & energy crops, including triticale, onions, & tomatoes; e.g., its researchers isolated a protein gene in the Brazil nut that codes for high methionine, an important amino acid	one of world's largest oil corporations; attempted to acquire Gulf Oil Corp. in 1984; is now world leader in onion seed production; also working on microbes to destroy sulfur in coal

Corporation with Fortune 500 Rank (1984)	Seed/Livestock, Food/Agriculture Subsidiaries, &/or Foreign Interests	Biotechnology Ventures; Investments; Facilities	Agricultural Genetics Research & Related Activities	General Information; Agricultural/Food Products
Bayer, AG Leverkusen, West Germany	owns the Helena Chemical Co. (seeds), Mobay Chemical Corp. (pesticides), Cutter Laboratories (makes Ag-Master corn & alfalfa silage inoculants)	in 1984, made drug-related biotech contracts with Genentech, Inc., & Genetic Systems Corp.; also has agreement with Max Planck Institute in West Germany for training Bayer scientists & basic research projects in molecular plant genetics	interests in livestock & plant applications of biotechnology; Miles Labs has also been involved in dairy food technology & vegetable protein research; Mobay Chemical Co. has financially supported university research on fungicides & herbicides	West Germany's largest chemical company; sells pesticides & fertilizer; Miles Labs also sells cattle insecticides & biologicals
Beatrice Foods Chicago, IL Fortune Rank:36	owns Dahlgren & Co. (seeds); acquired Tropicana in 1978 & Esmark for $2.7 billion in 1984; also owns Tindle Mills, Inc., which produces feed supplements for poultry, cattle, & swine	invested $5 million in International Genetic Engineering, Inc., in 1982 for undisclosed product development	company's Dahlgren & Co. subsidiary engaged in plant breeding	world's 2nd largest food company—Fischer nuts, Hunt's sauces & ketchup, Wesson & Sunlight oils, Peter Pan peanut butter, Eckrich & Swift meats, Butterball turkeys—also sells VigorTone feeds & supplements

Corporation with Fortune 500 Rank (1984)	Seed/Livestock, Food/Agriculture Subsidiaries, &/or Foreign Interests	Biotechnology Ventures; Investments; Facilities	Agricultural Genetics Research & Related Activities	General Information; Agricultural/Food Products
Brown & Williamson Tobacco Corporation		made an October 1983 research contract with DNA Plant Technology Corp. for biotechnological plant research on tobacco		
Campbell Soup Camden, NJ Fortune Rank: 107	operates retail garden outlets in Morristown, NJ, Farmington & Newton, CT, & Lexington, MA; holds one of the world's largest private collections of vegetable seed; has developed 30 new tomato varieties since turn of the century	In December 1981 made a $9 million investment & donated a $5 million lab & greenhouse to DNA Plant Technology Corp., which is doing contract work for Campbells on new tomato varieties; in September 1984 Campbell made two new research contracts with DNAP for general tomato research, & one with Calgene, Inc., for "developing high-solids processing tomato varieties with optimum taste, color, & texture"	interested primarily in the genetics of the tomato, cucumber, & other vegetables	largest soup manufacturer in the U.S.; also owns Pepperidge Farms & Vlasic Foods

Corporation with Fortune 500 Rank (1984)	Seed/Livestock, Food/Agriculture Subsidiaries, &/or Foreign Interests	Biotechnology Ventures; Investments; Facilities	Agricultural Genetics Research & Related Activities	General Information; Agricultural/Food Products
Cargill Wayzata, MN	is 5th largest seed producer in U.S.; owns 6 American seed companies; in 1978, paid nearly $69 million to acquire MBPXL Corp., 2nd largest U.S. beef packer		is conducting plant breeding research on hybrid wheat, hybrid corn, hybrid sunflowers, soybeans, cotton, safflower, & barley in U.S. as well as in Argentina, Brazil, & France; also involved in poultry & hog breeding	world's largest grain company; also sells livestock feeds, fertilizers, & is involved in soybean crushing, oilseed processing, meat packing, & commodity trading
Carnation Company Los Angeles, CA (acquired by Nestlé)	in 1983 acquired 5 farm supply distribution centers in the U.S.; also involved in tomato, potato, & refined bean processing	Carnation animal genetics research conducted at Hughson, CA, Carnation, WA, & Watertown, WI; has also financially supported livestock research at Univ. of California, Davis	involved in both crop & livestock genetics	Carnation feeds sold in U.S. & 20 foreign countries; also sells top-rated bull semen & Holstein heifers for upgrading cattle & dairy herds worldwide

Corporation with Fortune 500 Rank (1984)	Seed/Livestock, Food/Agriculture Subsidiaries, &/or Foreign Interests	Biotechnology Ventures; Investments; Facilities	Agricultural Genetics Research & Related Activities	General Information; Agricultural/Food Products
Celanese New York, NY Fortune Rank: 120	acquired 4 U.S. seed-related businesses between 1975 & 1980; then sold these businesses to LaFarge Coppee in 1984; in 1981, acquired Virginia Chemicals for $67 million, one of largest producers of alkyl amines, a herbicide intermediate	in February 1982 made a 3-year biotechnology research grant to Yale University; company is also interested in using biotech to produce chemicals & engineering microbes to clean up chemical wastes	gum from the guar seed is key raw material used by Celanese in its water-soluble polymers; in 1982 Moran Seed Co. was working to improve guar yield through classical breeding techniques; the company processes guar in 6 countries & operates world's largest guar factory in Vernon, TX	8th largest U.S. chemical company; makes herbicide intermediates
Chevron Standard Oil of California Fortune Rank: 11	owns more than 300,000 acres of agricultural land in CA; company's Ortho Consumer Products Division is a leading supplier of garden products & books; 15 million Ortho books published 1973–83; has agrichemical import agreement with Toyo Menka Kaisha (Japan) that may extend to biotech products	in 1978, invested $1 million in Cetus Corp. (17% share) with research project that includes a joint venture on production of high-quality fructose through enzymes; in 1985 will open $38 million Ortho Research Center in Richmond, CA, a significant portion of which will be devoted to plant- & pest-related biotech research	interested in biotechnology & genetic engineering for development of crops, new pesticides, & fertilizers; has financially supported university research on its fungicides, herbicides, & insecticides, cotton defoliation, integrated pest management, & nitrogen fertilizer efficiency	2nd largest U.S. oil company; 7th largest producer of ammonia; also sells pesticides; acquired Gulf Oil Corp. in 1984 for $13 billion

Corporation with Fortune 500 Rank (1984)	Seed/Livestock, Food/Agriculture Subsidiaries, &/or Foreign Interests	Biotechnology Ventures; Investments; Facilities	Agricultural Genetics Research & Related Activities	General Information; Agricultural/Food Products
Ciba-Geigy Ardsley, NY	owns Funk's & Louisiana seed companies in U.S. & 3 in Canada; owns Carters, Dobies, & Cuthberts seed companies in U.K.; also sells seed treatments & "safeners," which protect seeds from herbicide soil residue	owns 80% of ALZA Corp., a genetic engineering firm; built new $7.5 million biotech lab at Research Triangle Park, NC, in 1984; made $42 million dollar deal with Genentech in 1985 to produce and market animal health products	pursuing gene technology for medium & long-term objectives in agriculture; working to develop soybeans resistant to its atrazine herbicides; has financially supported university research on herbicides & fungicides	is 4th largest seller of hybrid corn seed in U.S.; & is leading seller of corn, soybean, & cotton herbicides; also produces insecticides, fungicides, plant growth regulators, & livestock drugs & insecticides
Conagra, Inc. Omaha, NE Fortune Rank: 121	acquired Singleton Seafood (#1 shrimp processor) in 1981; Peavy Co. in 1982 & the Armour Food Co. in 1983; also owns United AgriProducts (pesticide distributor), Greely, CO; & AgriBasics Co. (feeds & fertilizer), Knoxville, TN; acquired 5 smaller feed & pesticide companies in 1984			largest U.S. flour miller; largest producer of poultry products; also sells frozen foods, livestock feeds, & antibiotics, pesticides, & fertilizers, & is engaged in commodities trading

Corporation with Fortune 500 Rank (1984)	Seed/Livestock, Food/Agriculture Subsidiaries, &/or Foreign Interests	Biotechnology Ventures; Investments; Facilities	Agricultural Genetics Research & Related Activities	General Information; Agricultural/Food Products
Continental Grain Company New York, NY	owns 1 Canadian seed company & holds contracts with 2 Australian seed companies	invested $1 million in Calgene, Inc., in 1981		one of world's largest grain companies
Crown Zellerbach Corporation San Francisco, CA Fortune Rank: 129		has a 1981 research contract with Native Plants, Inc., for tissue culture research on Alder (*Alnus Rubra*) tree from Pacific Northwest for clonal propagation		one of the nation's leading paper & forest products companies

Corporation with Fortune 500 Rank (1984)	Seed/Livestock, Food/Agriculture Subsidiaries, &/or Foreign Interests	Biotechnology Ventures; Investments; Facilities	Agricultural Genetics Research & Related Activities	General Information; Agricultural/Food Products
DeKalb AgResearch, Inc. DeKalb, IL	between 1971 & 1980 acquired: J. J. Warren Companies (poultry breeding); Lubbock Swine Breeders (hog breeding); Lindsay Mfg. (irrigation equipment); Ramsey Seed Co. (seed pelleting), & Heinhold Companies (commodity trading)	in 1982 invested $600,000 in Bethesda Research Labs & formed major joint venture in crop genetics with Pfizer called DeKalb-Pfizer Genetics; in 1985 DeKalb-Pfizer signed research agreement with Calgene to develop & market herbicide-tolerant hybrid corn varieties	involved in both crop & livestock genetics; sells corn, sorghum, soybean, alfalfa, & sunflower seed in U.S. & 40 countries abroad; also sells poultry & swine breeding stock; DeKalb egg-laying breeding stock sold in more than 55 countries	DeKalb-Pfizer is #1 U.S. producer of hybrid sorghum seed & #2 producer of hybrid corn seed; also engaged in oil & gas business

Corporation with Fortune 500 Rank (1984)	Seed/Livestock, Food/Agriculture Subsidiaries, &/or Foreign Interests	Biotechnology Ventures; Investments; Facilities	Agricultural Genetics Research & Related Activities	General Information; Agricultural/Food Products
Diamond Shamrock Dallas, TX Fortune Rank: 83	acquired Shell Chemical's animal health business in 1979	in 1983 invested undisclosed sum in the Animal Vaccine Research Corp., a U.S. biotech company. In July 1983, joined Shamrock's agrichemical & animal health businesses with Japan's Showa Denko to form SDS Biotech Corp. to sell agrichemical products worldwide	has financially supported university research on Ectirin cattle eartags (insecticide) & Bravo 500 fungicide, as well as studies on the toxicity of Dacamine-40 (2,4-D herbicide) on specific soybean varieties	sells chlorothalonil fungicides for cotton, phenoxy herbicides for wheat & barley, & some insecticides; also sells feeds, larvacides, insecticides, & antibiotics for livestock
Dow Chemical Midland, MI Fortune Rank: 25	in 1981 acquired the drug business of Richardson-Merrell, Inc., for $260 million; Dow's ag chem sales recently strong in Australia, New Zealand, & Europe (1983)	established in-house biotechnology program in 1981; also involved with Collaborative Genetics, Genex, & Genentech	involved in both crop & livestock applications of biotechnology; spending at least $40 million annually on agricultural research; has researched photosynthesis & nitrogen fixation	2nd largest U.S. chemical company; produces & sells herbicides, insecticides, a nitrogen stabilizer & is researching plant growth regulators

Corporation with Fortune 500 Rank (1984)	Seed/Livestock, Food/Agriculture Subsidiaries, &/or Foreign Interests	Biotechnology Ventures; Investments; Facilities	Agricultural Genetics Research & Related Activities	General Information; Agricultural/Food Products
Du Pont Wilmington, DE Fortune Rank: 7	in 1981 acquired the agrichemicals division of the Charqeurs Reunis Group of France & its national distribution network; in 1985, acquired the crop protection chemical business of Ruhr-Stickstoff of West Germany	In 1981 committed $120 million to life sciences R & D; in September 1984, opened a new $85 million Life Sciences research complex. Owns New England Nuclear, Inc., a biotechnology supply company; in 1984 made a $4.5 million equity investment in Biotech Research Laboratories "to gain options on current & future products"	has made a major commitment to ag biotech research; working on herbicide-resistant crops, nitrogen fixation, & plant growth regulators; has financially supported university research on Marlate 50 cattle insecticide, sugarcane growth regulators, & herbicides	largest U.S. chemical company; acquired Conoco Oil Co. in 1981; produces & sells insecticides, herbicides, fungicides, & nematicides in more than 100 countries

Corporation with Fortune 500 Rank (1984)	Seed/Livestock, Food/Agriculture Subsidiaries, &/or Foreign Interests	Biotechnology Ventures; Investments; Facilities	Agricultural Genetics Research & Related Activities	General Information; Agricultural/Food Products
Eastman Kodak Rochester, NY Fortune Rank: 28	agricultural research conducted under its Eastman Chemicals Division	in March 1983 made a 6-year, $2.5 million research grant to Cornell University's Biotechnology Institute; in June 1984 invested $8.4 million in ICN Pharmaceuticals (Covina, CA) to facilitate research in medical & innovative agricultural technologies; in July 1984 announced plans to build a multimillion animal nutrition research center in Tennessee		world's largest producer of photographic film & accessories

Corporation with Fortune 500 Rank (1984)	Seed/Livestock, Food/Agriculture Subsidiaries, &/or Foreign Interests	Biotechnology Ventures; Investments; Facilities	Agricultural Genetics Research & Related Activities	General Information; Agricultural/Food Products
Elf Aquitaine Paris, France & Stamford, CT	in 1981 acquired Texasgulf, Inc., a major producer of fertilizer; holds 67% interest in Rousselot, world's leading maker of gelatines & Europe's #1 producer of livestock protein feeds & additives (over 200 products); sells seeds through Elf Bio-Industries	has in-house biotech research underway at its new Labege Research center in Toulouse; in 1981, signed research contract with Native Plants, Inc. (UT) for tissue culture research on Endod plants for insecticidal compounds & other purposes	biotechnology research is focused on obtaining energy from cellulose & improving nitrogen-fixation in plants	largest oil/chemical company in France; Rousselot's livestock services to farmers include "veterinary advice on stock breeding techniques" & computer software (Bestmix) "which formulates feed in response to a host of variables, including technical specifications relating to the species . . ."

Corporation with Fortune 500 Rank (1984)	Seed/Livestock, Food/Agriculture Subsidiaries, &/or Foreign Interests	Biotechnology Ventures; Investments; Facilities	Agricultural Genetics Research & Related Activities	General Information; Agricultural/Food Products
Eli-Lilly Indianapolis, IN Fortune Rank: 130	owns Elanco Products Co., an $800-million-a-year agricultural products subsidiary; in 1984 signed exclusive agreement with Phytogen (CA) to manufacture & market that company's Rhizobium seed treatments worldwide	in 1982, invested $5 million in International Plant Research Institute (IPRI) & contracted for plant genetics research; also holds equity in the Genetics Institute, is working with Genentech, and is funding research on photosynthesis in several major crops at Phytogen	involved in both crop & livestock genetics; working in-house to develop & commercialize genetically made bovine growth hormone; is also working on photosynthesis, herbicide-resistant crops, & plant growth regulators; spends more than $60 million annually on ag research	2nd largest U.S. producer of animal health products such as growth hormones, antibiotics, & animal insecticides; also a major producer of fungicides & herbicides (used on more than 50 crops)
Ethyl Corporation Richmond, VA Fortune Rank: 223	in 1981 acquired Hardwick Chemical Co., a producer of insecticide intermediates	in 1981 invested $1 million in Biotech Research Labs, of Rockville, MD		produces insecticides & intermediates for fungicides & herbicides

Corporation with Fortune 500 Rank (1984)	Seed/Livestock, Food/ Agriculture Subsidiaries, &/or Foreign Interests	Biotechnology Ventures; Investments; Facilities	Agricultural Genetics Research & Related Activities	General Information; Agricultural/Food Products
FMC Corporation Chicago, IL Fortune Rank: 119	"FMC agricultural chemical sales to the People's Republic of China . . . have grown 17 percent annually since 1978 . . ." (1983 annual report)	in 1981, built a new $30 million agrichemical research facility in Princeton, NJ; in 1982 surveyed over 20 biotech companies for possible joint ventures & other research collaboration	doing in-house genetics & tissue culture research in agriculture; has financially supported university research on corn, sorghum, sunflowers, & cotton as well as shrimp preservation & the use of Pounce insecticide on a broad range of agricultural & horticultural plants; has also supported nitrogen fixation research work of F. Ausubel at Harvard Univ. at $190,000/yr	sells insecticides, nematicides, & fungicides as well as specialty harvesting equipment & agricultural sprayers

Corporation with Fortune 500 Rank (1984)	Seed/Livestock, Food/Agriculture Subsidiaries, &/or Foreign Interests	Biotechnology Ventures; Investments; Facilities	Agricultural Genetics Research & Related Activities	General Information; Agricultural/Food Products
General Foods White Plains, NY Fortune Rank: 39	acquired Oscar Mayer & Co. for $470 million in 1981	has been contracting research with the Cetus Corporation since 1978; in 1980, invested $500,000 in Engenics, a center for biotechnology research at Stanford Univ. (participating with 5 other companies); made a 2-year, $594,000 research contract with DNA Plant Technology, Inc., in December 1982 for "biotechnological plant research"; & in March 1984, contracted with International Plant Research Institute to develop "improved quality food processing materials"	interests appear related to plant genetics & food processing	leading U.S. producer of roasted coffee (40% of company sales); 3rd leading U.S. producer of breakfast cereals

Corporation with Fortune 500 Rank (1984)	Seed/Livestock, Food/Agriculture Subsidiaries, &/or Foreign Interests	Biotechnology Ventures; Investments; Facilities	Agricultural Genetics Research & Related Activities	General Information; Agricultural/Food Products
General Mills Minneapolis, MN Fortune Rank: 64	acquired J. Lynch & Co., Inc., in 1981, a grain merchandising firm		conducting in-house research in plant genetics at the company's James Ford Bell Technical Center in Golden Valley, MN	2nd leading U.S. producer of breakfast cereals; experimenting with hydroponically grown lettuce in climate-controlled facility
W. R. Grace New York, NY Fortune Rank: 50	owns American Breeders Services, largest dairy breeder in U.S.	in 1981, acquired a 16% share of AGRI, Inc., a biotechnology company working on genetically engineered livestock vaccines, & in June 1984, established the Agracetus Corp., a $60 million venture with Cetus Corporation's Madison operation to develop, manufacture, & market biotechnology-based products for agriculture	involved in both crop & livestock genetics. Agracetus plans to genetically improve corn, cotton, soybeans, wheat, & rice; to develop new agricultural microbes; & to produce new livestock vaccines & growth hormones	5th largest U.S. chemical company; leading producer of ammonia & phosphate fertilizers; also sells Far Better Feeds for cattle & Walnut Grove feeds for hogs

Corporation with Fortune 500 Rank (1984)	Seed/Livestock, Food/Agriculture Subsidiaries, &/or Foreign Interests	Biotechnology Ventures; Investments; Facilities	Agricultural Genetics Research & Related Activities	General Information; Agricultural/Food Products
H. J. Heinz Pittsburgh, PA Fortune Rank: 98	acquired Star-Kist Foods in 1963 (tuna & pet foods); Ore-Ida Foods in 1965 (potatoes & frozen foods); W. Darlington & Sons, U.K., in 1969 (farms); the Hubinger Co. in 1975 (corn derivatives); Weight Watchers International in 1978; Gagliardi Brothers in 1980 (meats); Weldon Farm Products in 1981	in December 1982 entered a 5-year research agreement with ARCO's Plant Cell Research Institute to develop high-solids tomato varieties; in May 1984, entered a 3-year research agreement with Biotechnica International, Inc., "to develop new processes leading to products of interest to the food & animal feeding industry"	has in-house tomato & cucumber breeding programs; company's Stanley Wine subsidiary (Australia) using grafting/tissue culture techniques to produce new grape lines; also working to improve potato lines for Nadler subsidiary in West Germany; Italian subsidiary, Plada, began calf-rearing program in 1983 for its veal operation	largest U.S. ketchup manufacturer; also a leading producer of tuna, potatoes, soup, & baby food; in 1982, "Heinz research teams worked with suppliers to develop new farming & fishing techniques, superior crop strains, & better ways to harvest, ship, & store raw materials"
Hercules, Inc. Wilmington, DE Fortune Rank: 144	acquired Pure Culture Products Inc., a biotechnology-based food ingredients company, from Amoco in November 1984	formed a joint venture with IGI Biotechnology, Inc., in November 1984 to manufacture & market whey-based products & sugar-based polysaccharide gums for use in food industry		

Corporation with Fortune 500 Rank (1984)	Seed/Livestock, Food/Agriculture Subsidiaries, &/or Foreign Interests	Biotechnology Ventures; Investments; Facilities	Agricultural Genetics Research & Related Activities	General Information; Agricultural/Food Products
Hilleshog Kabdsjribaum, Sweden	owns International Forest Seeds, Birmingham, AL (pine seed)	in March 1980 acquired a 14% share of Advanced Genetic Sciences, Inc. (AGS), & in 1982-83 spent $575,000 in R & D contracts with AGS to develop new varieties of rapeseed, wheat, & barley		
Hoechst, AG West Germany		has been working on genetically engineered vaccine for hoof-and-mouth disease since late 1970s; in 1980 made a 10-year, $70 million contract with the Harvard-affiliated Massachusetts General Hospital to create Dept. of Molecular Biology for biotech research; also expanding biotech research in Germany & Japan	company's American division has financially supported university research on herbicides & insecticides for cotton & other field crops	world's largest drug company; its Animal Health Division is a world leader in livestock vaccines

Corporation with Fortune 500 Rank (1984)	Seed/Livestock, Food/Agriculture Subsidiaries, &/or Foreign Interests	Biotechnology Ventures; Investments; Facilities	Agricultural Genetics Research & Related Activities	General Information; Agricultural/Food Products
Hoffman-La Roche Basel, Switzerland		became a 15% owner of the Agrigenetics Corp. in 1982–83; research collaboration with Agrigenetics includes: biochemistry of plant lectins; the role of plant antiviral agents; & the effects of cloned interferons on combating viral diseases in plants	involved in both crop & livestock applications of biotechnology; working to develop genetically engineered bovine growth hormone at its Nutley, NJ, research station; has also financially supported university research on the feeding of sorghum to cattle	one of the world's top 20 drug companies
Imperial Chemical Industries (ICI) Billingham, Cleveland, UK	in 1983, acquired 2 U.K. fertilizer companies; in 1984, merged its animal health operation with that of the Wellcome Foundation to form Coopers Animal Health, potentially the world's top producer of animal health products	in 1984, established a joint venture to apply genetic engineering to agricultural crops with the Swedish agribusiness group Cardo; working on pyrethroid insecticides & a pheromone for pink bullworm	has been working on single-cell protein livestock feed called "Pruteen" since 1970; now employing biotechnology in that process & for other ag applications; spent $60 million on agrichemicals research in 1983	is the largest public company in Great Britain; 4th largest chemical company in Europe & is also the world's 5th leading producer of agrichemicals—sells herbicides, insecticides, fungicides, & plant growth regulators

Corporation with Fortune 500 Rank (1984)	Seed/Livestock, Food/Agriculture Subsidiaries, &/or Foreign Interests	Biotechnology Ventures; Investments; Facilities	Agricultural Genetics Research & Related Activities	General Information; Agricultural/Food Products
INCO Toronto, Ontario Canada		in 1981, invested $12 million in Biogen (20% share); also owns equity in Plant Genetics, Inc.		one of world's leading mining companies; produces nickel, copper, precious metals, & cobalt
International Minerals & Chemicals Corporation (IMC) Northbrook, IL Fortune Rank: 231	in 1982, acquired the fertilizer & crop chemical outlets of the First Mississippi Corp. in IA & IL; in 1984, acquired Sterwin Laboratories of Millsboro, DE, a manufacturer of poultry biologics, & also acquired a 50% share of the Chinhae Chemical Co., a Korean fertilizer producer; also has 13-year agreement to mine & supply phosphate fertilizer to Zen-Noh, Japan's largest farm co-op	working with Genentech, Inc., to develop livestock hoof-and-mouth vaccine & is also involved in a partnership with Biogen to develop & commercialize genetically engineered bovine growth hormone; IMC has the exclusive marketing rights to species-specific growth hormones developed by Biogen, & will share profits with worldwide sale of Genentech's hoof-and-mouth vaccine; company has recently built new animal science lab in Terre Haute, IN	in-house research on plant nutrition & fertilizer technology, as well as animal health & nutrition products; has also financially supported university research on its Ralgro livestock implants (growth promotants) & on feeding sorghum to cattle	world's largest private producer of phosphate rock & potash; also produces anhydrous ammonia; North America's largest producer of phosphate & potassium feed nutrients, & growth-promotants for cattle & feed-lot lambs; also produces antibiotics & vaccines for poultry

Corporation with Fortune 500 Rank (1984)	Seed/Livestock, Food/Agriculture Subsidiaries, &/or Foreign Interests	Biotechnology Ventures; Investments; Facilities	Agricultural Genetics Research & Related Activities	General Information; Agricultural/Food Products
International Multifoods Minneapolis, MN Fortune Rank: 293	owns Guildersleeve (IL) & Lynk Brothers (IA) seed companies; also holds 45% interest in Mexican company that produces & markets livestock feeds as well as sorghum seed		seed companies involved with hybrid corn & sorghum	5th largest U.S. flour miller (wheat, durum, & rye); also sells Supersweet poultry & livestock feeds in U.S. & Canada as well as livestock drugs under the Asborn & Tevco labels
ITT New York, NY Fortune Rank: 21	owns the W. Atlee Burpee Seed Co. & O. M. Scott & Sons (plans to divest these units in 1985)		company's O. M. Scott & Sons has financially supported university research on crop response to fertilizer & irrigation, fertilizer product evaluation, & soil chemistry & fertility in south Texas	one of the world's largest manufacturers of telecommunications equipment
Kellogg Company Battle Creek, MI Fortune Rank: 143	in 1982 acquired Pure Packed Foods, Inc., a company capable of producing nondairy milk substitutes (possibly to be used with breakfast cereals)	in 1982 became a 5% owner of the Agrigenetics Corp. after $10 million purchase of company's stock	joint research projects with Agrigenetics are focused on increasing cereal grain protein levels, mold resistance in corn, & higher-yielding hard starch corn lines	largest breakfast cereal producer in the U.S. (40% share of the market)

Corporation with Fortune 500 Rank (1984)	Seed/Livestock, Food/Agriculture Subsidiaries, &/or Foreign Interests	Biotechnology Ventures; Investments; Facilities	Agricultural Genetics Research & Related Activities	General Information; Agricultural/Food Products
Kemira OY Helsinki, Finland & Framingham, MA	acquired L&K Fertilizers Ltd. (U.K.) in 1982; owns soil testing co. in Finland & fertilizer supply terminals in Malaysia, Thailand, & Guatemala	made a November 1983 research contract with Calgene, Inc., to develop herbicide-resistant varieties of turnip rape; also conducting biotech research at its Espoo Research Centre in Finland	biotech interests appear related to pesticides & fertilizers	is Finland's largest chemical company; is a major producer of fertilizers & ag chemicals
Kirin Brewery Japan		invested $1 million in Plant Genetics, Inc. (CA), & has put up $2 million for licensing & research agreement for marketing Plant Genetics' "synthetic seed" products & techniques in Asia & the Pacific Rim		

Corporation with Fortune 500 Rank (1984)	Seed/Livestock, Food/Agriculture Subsidiaries, &/or Foreign Interests	Biotechnology Ventures; Investments; Facilities	Agricultural Genetics Research & Related Activities	General Information; Agricultural/Food Products
Koppers Company Pittsburgh, PA Fortune Rank: 202	is involved in the forest products business (lumber & paper)	invested $10 million in Genentech in 1979 & $25 million in Genex Corp. (25% share); also owns 5.8% of Engenics (Stanford Univ research center); in 1982, invested $1.7 million in DNA Plant Technology Corp., & in February 1984 formed joint venture with DNAP to develop & commercialize diagnostic kits for plant diseases in turfgrass, citrus fruits, & other ag products	biotechnology interests include plant genetics & the production of Phenylalanine, a specialty chemical used in a sugar substitute	involved in mining, construction, & engineering
John Labatt, Inc. London, Ontario, Canada	in 1984, acquired the Silverwood & Royal Oak dairy operations; is also a 60% owner of McGavin Foods, the largest bakery in Western Canada; & owns other food companies in U.S. & Canada	working in-house to genetically engineer yeast strains; in 1981 acquired a 30% share of Allelix, Inc., a Canadian biotech company	"We believe we can genetically manipulate industrial yeast strains as well as, if not better than, anyone else."—company brochure	largest Canadian brewer; also involved in wine, packaged foods, dairy products, milled grain products, livestock feeds, & mushroom production

Corporation with Fortune 500 Rank (1984)	Seed/Livestock, Food/Agriculture Subsidiaries, &/or Foreign Interests	Biotechnology Ventures; Investments; Facilities	Agricultural Genetics Research & Related Activities	General Information; Agricultural/Food Products
LaFarge Coppee France	attempted to acquire Agrigenetics in 1984; will own 4 U.S. seed companies with those acquired from Celanese	operates an in-house biotech company, Orsan, which is also working on soil microbes	interests include corn, soybean, sunflower, & sorghum seed	largest cement company in France
Land O'Lakes, Inc. Arden Hills, MN Fortune Rank: 161	owns 1 U.S. seed company; in 1980 acquired Dawson Mills, a large soybean processing & marketing co-op, & Lake-to-Lake Dairy cooperative; in 1981, merged with Midland Cooperatives, Inc., 9th largest co-op		is engaged in livestock breeding & embryo transplant research with hogs; does plant breeding with soybeans, alfalfa, & grass species	3rd largest U.S. farmer cooperative; sells wide range of farm & food products

Corporation with Fortune 500 Rank (1984)	Seed/Livestock, Food/Agriculture Subsidiaries, &/or Foreign Interests	Biotechnology Ventures; Investments; Facilities	Agricultural Genetics Research & Related Activities	General Information; Agricultural/Food Products
Lubrizol Corporation Wickliffe, OH Fortune Rank: 344	with the September 1984 acquisition of the Agrigenetics Corp., now owns 13 companies in the U.S. seed industry	in 1979 purchased a 20% share of Genentech & in January 1984 acquired a 28% interest in Sungene Technologies Corp. for $14 million	working with biotechnology to develop improved seeds for major crops, specialty chemicals derived from crops, & vegetable oils derived from hybrid sunflower	world's largest manufacturer of lubricating-oil additives; wants to be #1 in oil-seed production
Martin-Marietta Corporation Bethesda, MD		engaged in algal physiology & biochemistry research with biotechnology at company labs "to develop products for agriculture"; in December 1982, invested $25 million in 3 biotech firms: Molecular Genetics, Inc. ($11.9 million–21% share); Chiron Corp. (10–20% share); & Native Plants, Inc. (10–20% share)	with equity investments, involved in both crop & livestock genetics; in December 1982 stated that it plans "to assume a leading role in cooperative programs aimed at penetrating major agricultural, health, and industrial markets"	is a major corporation engaged in aerospace, electronic data processing systems, aluminum, & other areas

Corporation with Fortune 500 Rank (1984)	Seed/Livestock, Food/Agriculture Subsidiaries, &/or Foreign Interests	Biotechnology Ventures; Investments; Facilities	Agricultural Genetics Research & Related Activities	General Information; Agricultural/Food Products
McCormick & Company, Inc. Hunt Valley, MD Fortune Rank: 353		in January 1985, made a $2.5 million joint venture with Native Plants, Inc., to use biotechnology to research "crops of mutual interest," including a wide range of seeds & plants used by McCormick in making food seasonings	McCormick has a long history of in-house research & plant breeding on crops such as garlic & onions & others that produce pepper, cinnamon, & vanilla	largest producer of seasonings in the United States
Merck & Company Rahway, NJ Fortune Rank: 110	company's MSD AGVET division produces livestock drugs & pesticides	has in-house biotechnology research program aimed at human & animal health applications	company's Hubbard Farms division involved in chicken & turkey genetics; $20 million annually on agricultural research, with some work in plant genetics; has financially supported university research on its ivermectin parasite drug for cattle, Ivomec, & disease control studies in rice & soybeans	4th largest U.S. producer of animal health products

Corporation with Fortune 500 Rank (1984)	Seed/Livestock, Food/Agriculture Subsidiaries, &/or Foreign Interests	Biotechnology Ventures; Investments; Facilities	Agricultural Genetics Research & Related Activities	General Information; Agricultural/Food Products
Mitsubishi Japan		in August 1984 acquired small interest (less than 5%) in Sungene Technologies (CA); Mitsubishi may distribute Sungene's products in Asia & become a collaborator in research		one of Japan's largest corporations
Monsanto St. Louis, MO Fortune Rank: 51	in 1969 acquired Farmers Hybrid Co., Inc., a cattle & hog breeding firm; in 1982, acquired DeKalb's hybrid wheat program & the Jacob Hartz soybean seed company of Stuttgart, AR	in 1972 invested in Genentech; also holds equity in Biogen, Collagen, & Genex & is involved with other companies & universities (e.g., Harvard, Washington, & Rockefeller Universities) in research contracts & joint ventures; in 1984 made a 3-year agreement with Biotechnica International to use genetic engineering to manufacture an undisclosed "pesticide candidate for use on field crops"	heavily involved in both crop & livestock genetics; $190 million R & D budget for ag biotechnology in 1984; nearing market with genetically engineered crop, livestock & microbial products; is perhaps U.S. leader in ag biotechnology research	4th largest U.S. chemical company; a leading producer & marketer of herbicides worldwide; also sells plant growth regulator for sugarcane, hybrid swine, & cattle breeding stock, feed supplements, & nitrogen-based fertilizer products

Corporation with Fortune 500 Rank (1984)	Seed/Livestock, Food/Agriculture Subsidiaries, &/or Foreign Interests	Biotechnology Ventures; Investments; Facilities	Agricultural Genetics Research & Related Activities	General Information; Agricultural/Food Products
Nabisco Brands, Inc. Parsippany, NJ Fortune Rank: 54	Nabisco merged with American Brands in 1981; owns production facilities in 35 countries & markets products in more than 100 countries	in February 1984, began a joint venture with the Cetus Corporation to apply enzymology & recombinant DNA technology to the processing of food & food ingredients; company maintains food research facilities in Fairlawn, NJ, & Wilton, CT, & will complete construction of a new Corporate Technology Center in East Hanover, NJ, in 1985		3rd largest U.S. food producer; largest U.S. baked foods producer; 5th largest U.S. breakfast cereal producer; is also major U.S. producer of cookies, crackers, nuts, margarines, pet foods, & consumer yeast
National Distillers & Chemical Corporation New York, NY Fortune Rank: 168	owns Almaden Vineyards (5,561 acres owned & leased in California & 2,800 acres in Brazil)	in 1978 invested $5 million in the Cetus Corp. (16% share); also includes research for improving microorganisms used in the fermentation of alcohol	biotech interests related to fermentation of alcohol	2nd largest U.S. distiller; 4th largest U.S. vintner; also produces industrial chemicals

Corporation with Fortune 500 Rank (1984)	Seed/Livestock, Food/Agriculture Subsidiaries, &/or Foreign Interests	Biotechnology Ventures; Investments; Facilities	Agricultural Genetics Research & Related Activities	General Information; Agricultural/Food Products
Nestlé, S.A. Basel, Switzerland	with the $3 billion acquisition of Carnation in 1984, acquired additional crop & livestock genetics capabilities	in April 1984 made joint research agreement with Calgene, Inc., to develop herbicide-tolerant varieties of soybeans	involved in both crop & livestock genetics; an R & D subsidiary, Nestec, has succeeded in regenerating whole soybean plants from cell cultures	world's largest food corporation
Occidental Petroleum Los Angeles, CA	in June 1981 paid $813 million to acquire nation's largest meat packer, Iowa Beef Processors; also owns the Hooker Chemical Co.	was involved in plant biotech research & U.S. seed business until 1983–84, when it sold its Zoecon & Ring Around Products subsidiaries to pay off debts for acquisition of Cities Services	involved in livestock genetics through the company's Shadow Isle, Inc., purebred cattle breeding subsidiary; has financially supported university research on the phosphorus requirements of broiler chicks, layers, & young turkeys	is the largest U.S. producer of phosphoric acid & 4th largest supplier of phosphate rock; 34% of corporate income from meat processing, feeds, & fertilizer
Olin Corporation Stamford, CT Fortune Rank: 178			has financially supported university research on fungicides, soil fertility, sulfur nutrition of legumes, & the use of a fungicide as a nitrification inhibitor	is a producer of ammonia & urea fertilizers; also working on a "nitrification inhibitor" called Dwell

Corporation with Fortune 500 Rank (1984)	Seed/Livestock, Food/Agriculture Subsidiaries, &/or Foreign Interests	Biotechnology Ventures; Investments; Facilities	Agricultural Genetics Research & Related Activities	General Information; Agricultural/Food Products
PepsiCo, Inc. Purchase, NY Fortune Rank: 40	owns Frito-Lay snack foods & Pizza Hut & Taco Bell fast food restaurants	launched a 5-year program in 1983 to use technological breakthroughs to offset cost increases due to inflation; its potato & corn research "has led to increased potato solids that yield more finished product per pound of raw material than ever before . . ." (1983 annual report)	Frito-Lay has 60 Ph.D.s engaged in raw vegetable crop research at R & D facility in Irving, TX, & experimental farms in Wisconsin; is using tissue culture & cloning to produce new varieties of potatoes	largest U.S. snack food producer; 2nd in soft drinks; 5th in fast foods
Pfizer New York, NY Fortune Rank: 101	acquired 3 U.S. seed companies between 1973–75; later consolidated these, then merged with DeKalb AgResearch Inc. in 1982; sold its H & N poultry genetics company to Tatum Farms in 1982	is a 30% owner of DeKalb-Pfizer Genetics, a joint venture in crop genetics; has in-house biotechnology program at Pfizer Central Research in Groton, CT	conducts animal health product research on cattle, swine, & poultry at its Terre Haute facility; & crop genetics research on alfalfa, soybeans, sorghum, sunflowers, & hybrid corn at DeKalb-Pfizer; investing about $25 million in ag research annually	largest U.S. producer of animal drugs; #2 in hybrid corn seed with DeKalb venture

Corporation with Fortune 500 Rank (1984)	Seed/Livestock, Food/Agriculture Subsidiaries, &/or Foreign Interests	Biotechnology Ventures; Investments; Facilities	Agricultural Genetics Research & Related Activities	General Information; Agricultural/Food Products
Phillips Petroleum Company Bartlesville, OK Fortune Rank: 17	owns American Fertilizer & Chemical Co., which operates farm stores	in 1981, invested $10 million in Salk Institute Biotechnology/Industrial Associates, Inc. (37% interest) for biotech research to improve oil & gas production & for agricultural applications	using biotechnology at its Provesta Corp. subsidiary to produce a single-cell protein feed product called Provesteen; M.I.T. researchers are testing this Phillips product for potential human consumption; in 1982, Phillips began research effort focused on the production of synthetic pheromones	8th largest U.S. oil company; produces & sells nitrogen fertilizers
Pillsbury Minneapolis, MN Fortune Rank: 94	acquired Green Giant in 1979 for $290 million; acquired agricultural division and bean seed subsidiaries of the Wickes Companies in 1982; acquired Haagen-Dazs in 1983 for $130 million			major U.S. food company; largest U.S. processor of dry edible beans, peas, & lentils (Wickes); 2nd largest U.S. flour miller; also involved in rice milling, sunflower seed crushing, & poultry-feed ingredients
Pioneer Hi-Bred, Intl. Des Moines, IA	owns 3 other U.S. seed companies & 1 microbial products company	opened new $2.7 million biotechnology research lab in 1982	using classical genetics & biotechnology in research on corn, sorghum, wheat, soy-	world's largest hybrid corn seed producer

Corporation with Fortune 500 Rank (1984)	Seed/Livestock, Food/Agriculture Subsidiaries, &/or Foreign Interests	Biotechnology Ventures; Investments; Facilities	Agricultural Genetics Research & Related Activities	General Information; Agricultural/Food Products
PPG Industries Pittsburgh, PA Fortune Rank: 89		in 1985, made a 15-year $120 million agreement with the Scripps Clinic & Research Foundation of La Jolla, CA, for basic biotechnology research on herbicides & other ag products	"The Scripps program will concentrate on plant science using knowledge of biology at the molecular level. PPG will translate this knowledge into safer, more effective agricultural chemicals to improve worldwide farm productivity." (PPG)	9th largest U.S. chemical company; sells about $100 million worth of ag chemicals annually; Genep corn herbicide, Bud Nip tobacco-shoot inhibitor; Cobra soybean herbicide, potash fertilizer, & others
Purex Lakewood, CA	sells some specialty food products such as Ellio's Pizzas, & distributes canned tomatoes, wine vinegar, & packs olive & vegetable oil through its Pope Products Division	holds a 16% interest in the Cetus Corp.		major producer of household laundry & cleaning products such as Purex & Brillo soap pads

Corporation with Fortune 500 Rank (1984)	Seed/Livestock, Food/Agriculture Subsidiaries, &/or Foreign Interests	Biotechnology Ventures; Investments; Facilities	Agricultural Genetics Research & Related Activities	General Information; Agricultural/Food Products
Ralston Purina Company St. Louis, MO Fortune Rank: 72	acquired ITT Continental Baking in 1984 for $475 million; owns Jack In The Box fast food restaurants & Chicken of the Sea tuna; owns soybean processing plants, feed mills, & grain elevators throughout U.S.		operates hog breeding facilities in Iowa & Indiana; poultry breeding & processing outside the U.S.; & owns a shrimp hatchery & research facility in Panama; also involved in the production of mushrooms & mushroom spawn, & the use of isolated soy protein in food products	world's largest producer of feeds for livestock; produces more than 350 basic feed formulations (with drug & antibiotic additives) for poultry, hogs, dairy & beef cattle, & pets; also produces & sells livestock insecticides; is also a major food company: 6th in breakfast cereals, 10th in fast food
Reynolds (R.J.) Industries Winston-Salem, NC Fortune Rank: 23	in 1984 acquired the Bear Creek Corp., a mail-order business that includes Jackson & Perkins, #1 in hybrid roses, & Harry & David, a mail-order fruit business; also acquired Heublein & Canada Dry in 1984, & owns the Del Monte Corp.		through its subsidiaries & research activities, company is engaged in the breeding of roses, grapes, & other fruits & vegetable crops; e.g., owns 10,500-acre banana plantation & 6,000-acre pineapple farm in Costa Rica	largest U.S. producer of cigarettes & little cigars; Del Monte is a leading producer & processor of pineapples, bananas, fruits, & vegetables; Heublein is 2nd largest U.S. vintner; 3rd in distilled liquor; 2nd in fast food

Corporation with Fortune 500 Rank (1984)	Seed/Livestock, Food/Agriculture Subsidiaries, &/or Foreign Interests	Biotechnology Ventures; Investments; Facilities	Agricultural Genetics Research & Related Activities	General Information; Agricultural/Food Products
Rhone-Poulenc Agrochemie Lyon, France	acquired Mobil's agrichemicals division in 1981, & a Diamond Shamrock pesticide facility in England in 1983	has June 1984 research contract with Calgene, Inc., to develop new varieties of sunflower tolerant to the herbicide Bromoxynil	has financially supported university research in the U.S. on livestock drugs, sorghum seed, soybeans, crops, herbicide use, & evaluation of tree fruits & berries	in the mid-1970s made an imitation, "spun fiber" meat product from soy, field beans, sunflower, & milk proteins; in 1984 expressed interest in building plants in the U.S. & France to produce lysine for use as a livestock feed additive
Rohm & Haas Philadelphia, PA Fortune Rank: 179	in 1983 formed Rohm & Haas Seeds, Inc.; in 1984 formed CR Seeds, a partnership with Coker Pedigreed Seed Co. to develop new varieties of wheat, soybeans, & oats; in 1985 acquired Ring Around Products soybean seed co. from Occidental Petroleum	in 1981 invested $12 million in Advanced Genetics Sciences (AGS); research with AGS includes hybrid seed activities; also financing research at Plant Genetic Systems in Belgium; R & H researchers in U.S. using tissue culture screening to find crop strains resistant to herbicide Blazer	working on chemical gametocides for hybridizing wheat called "HyBrex"; also soliciting arrangements with universities for use of wheat lines in conjunction with hybridizing program; other company-funded univ. research on herbicides, insecticides, & fungicides; one-third of company's R & D budget devoted to ag-related projects	major U.S. chemical company; produces & sells fungicides, insecticides, & herbicides worldwide

Corporation with Fortune 500 Rank (1984)	Seed/Livestock, Food/Agriculture Subsidiaries, &/or Foreign Interests	Biotechnology Ventures; Investments; Facilities	Agricultural Genetics Research & Related Activities	General Information; Agricultural/Food Products
Royal Dutch-Shell Netherlands/U.K. Fortune Rank: 13 (Shell Oil)	owns at least 8 U.S. seed companies & other major seed companies in U.K. & Western Europe; in joint venture for 100,000-hectare pine/eucalyptus forest project in Brazil; other forest interests in Chile & New Zealand	has plant genetics research contract with Celltech, an English biotech company, & also has equity in the Cetus Corp.; in October 1984 opened $9 million Plant Sciences & Microbiology Laboratory at Shell Development's (V.S.) Modesto, CA, research headquarters	company's Sittingbourne Research Centre is working with Nickerson Seed Co. subsidiary (U.K.) to develop plant growth regulator to facilitate hybrid cereal breeding; Shell Development Co. (U.S.) is studying the mechanics of herbicide resistance in plants; searching for a gene in corn that would resist new herbicide "Cinch" (Cinmethylin)	world's 2nd largest corporation; produces & sells pesticides & fertilizers worldwide

Corporation with Fortune 500 Rank (1984)	Seed/Livestock, Food/Agriculture Subsidiaries, &/or Foreign Interests	Biotechnology Ventures; Investments; Facilities	Agricultural Genetics Research & Related Activities	General Information; Agricultural/Food Products
Sandoz Basel, Switzerland	owns seed companies in the U.S. (6), Canada (2), U.K. (1), & the Netherlands (1), with subsidiaries in Australia, Germany, & Italy	in 1983 acquired the Zoecon Corp., a California company producing insect pheromones & conducting some plant biotechnology research	genetics research focused on soybeans & vegetable crops; has also financially supported university research in the U.S. on its herbicides & biological insecticides; in 1979, made a $100,000 research contract with Texas A & M scientists to study integrated pest management in cotton	world's 8th ranked drug company; also produces herbicides, fungicides, & biological insecticides

Corporation with Fortune 500 Rank (1984)	Seed/Livestock, Food/Agriculture Subsidiaries, &/or Foreign Interests	Biotechnology Ventures; Investments; Facilities	Agricultural Genetics Research & Related Activities	General Information; Agricultural/Food Products
Schering-Plough Kenilworth, NJ Fortune Rank: 193	in 1981 acquired Douglas Industries, Inc., a livestock research company	in 1979 invested $8 million in Biogen (30% share) with first rights to certain Biogen products in human & animal health markets; in 1981, built new lab in Elkhorn, NE, to provide for recombinant DNA vaccine research; in 1982, acquired DNAX, Inc., a biotech company working on immunology	biotech interests lie in livestock-related products such as antibacterial vaccines & new "drug delivery systems" (i.e., oral, infections, implants, etc.)	major U.S. drug company; produces livestock antibiotics, steroids, antifungals, analgesics, & vaccines
Seagram Company, Ltd. Canada	owns Paul Masson; acquired Taylor California Cellars & Sterling vineyards from Coca-Cola in 1984	in December 1984 became an 11.6% owner of Biotechnica International, Inc., & made a 5-yr. $10 million research contract with Biotechnica		world's largest producer & marketer of distilled spirits & wines

Corporation with Fortune 500 Rank (1984)	Seed/Livestock, Food/Agriculture Subsidiaries, &/or Foreign Interests	Biotechnology Ventures; Investments; Facilities	Agricultural Genetics Research & Related Activities	General Information; Agricultural/Food Products
Southwide Memphis, TN	acquired the Delta & Pine Land Co. (cotton seed) of Memphis, TN, in 1978 & the Greenfield Seed Co. of Harrisburg, AZ, in 1979		plant breeding research in cotton & soybeans	a leading holder of "cotton patents" under the Plant Variety Protection Act
SmithKline Beckman Corporation Philadelphia, PA Fortune Rank: 136	its Norden Laboratories subsidiary produces animal health products; in 1983, acquired VPI, Inc., of Omaha, NE, a manufacturer of vitamin & antibiotic feed additives & mixes	in 1981 initiated a $200 million in-house R & D program with emphasis on molecular biology; 1982–83 joint venture with Cetus through Norden Labs to produce diarrhea vaccine for pigs; completed construction of new molecular biology building in Upper Merion, PA, in 1983	livestock vaccines & other animal health products	5th largest U.S. producer of animal health products; largest producer of animal vaccines; also sells feed additives & growth promotants

Corporation with Fortune 500 Rank (1984)	Seed/Livestock, Food/Agriculture Subsidiaries, &/or Foreign Interests	Biotechnology Ventures; Investments; Facilities	Agricultural Genetics Research & Related Activities	General Information; Agricultural/Food Products
Staley (A.E.) Mfg. Company Decatur, IL Fortune Rank: 170	in 1983 formed a trading partnership with Continental Grain Co. to export corn gluten feed & soybean meal; company's Ging & Livergood subsidiaries sell seed & fertilizer to Illinois farmers	in 1984 agreed to acquire a one-third interest in Genencor, Inc., a biotech company that will research for Staley "new processes & new products in corn refining; also holds 40% share of BioTechnical Resources, Inc., of Manitowoc, WI, which is focusing on molecular biology & enzymology	is interested in biotechnology to produce products from corn, soybeans, & other renewable resources	8th largest U.S. producer of corn sweeteners; is also a major processor & supplier of vegetable oils & soy proteins

Corporation with Fortune 500 Rank (1984)	Seed/Livestock, Food/Agriculture Subsidiaries, &/or Foreign Interests	Biotechnology Ventures; Investments; Facilities	Agricultural Genetics Research & Related Activities	General Information; Agricultural/Food Products
Standard Oil of Ohio (SOHIO) Cleveland, OH Fortune Rank: 24	the company's Agricultural Products Division owns a chain of 130 mixing plants & outlets in the Corn Belt, which sell fertilizers, pesticides, & seed.	is building an in-house research group in plant biotechnology		produces ammonia in the U.S.
Stauffer Chemical Company Westport, CT Fortune Rank: 235	in 1978, acquired Blaney Farms, a corn seed company in Madison, WI, & Prairie Valley, a corn & sorghum seed company in Philips, NE; in 1980, acquired RBA, a Springfield, IL, seed co.	has active in-house biotechnology research program at its De Guigne Technical Center in CA	making a major research commitment in plant genetics research aimed at corn, sorghum, & sunflowers, including work to make herbicide-resistant crop strains	3rd largest U.S. producer of herbicides & insecticides; also sells nitrogen fertilizers

Corporation with Fortune 500 Rank (1984)	Seed/Livestock, Food/Agriculture Subsidiaries, &/or Foreign Interests	Biotechnology Ventures; Investments; Facilities	Agricultural Genetics Research & Related Activities	General Information; Agricultural/Food Products
Unilever London, England	acquired National Starch & Chem. Corp. in 1975 for $485 million; Lawry's Foods (Lawry's Seasoned Salt) for $66 million in 1979; a tapioca starch co. in Thailand in 1983; & the margarine division of Beatrice Foods in 1984; owns palm oil, rubber, copra, cocoa, & tea plantations in West & Central Africa, Colombia, the Solomon Islands, & Malaysia; owns "Nordsee," a European fishing co., & operates a salmon farming business in Scotland	has in-house research effort underway in plant biotechnology in England & Holland; has set up a company to sell cloned palm oil plant seedlings	using biotechnology to improve yields of oil & coconut palms	world's 4th largest corporation outside the U.S.; world's largest producer of margarine; sells Lipton Tea, Bird's-Eye frozen foods, dairy products, & others; produces & sells livestock feeds in Europe

Corporation with Fortune 500 Rank (1984)	Seed/Livestock, Food/Agriculture Subsidiaries, &/or Foreign Interests	Biotechnology Ventures; Investments; Facilities	Agricultural Genetics Research & Related Activities	General Information; Agricultural/Food Products
Union Carbide Danbury, CT Fortune Rank: 35	owns Soilserv, Inc., a subsidiary providing pesticide application services to growers in the Salinas Valley, CA	in 1981, built a new agricultural research facility at Research Triangle Park, NC	working on plant growth regulators; spends about $46 million annually on agricultural research; has also financially supported university research on its herbicides & insecticides	3rd largest U.S. chemical company; produces & sells insecticides, herbicides, & plant growth regulators
Uniroyal Middlebury, CT Fortune Rank: 164	heavily involved in fungicide seed-treatment business; in 1982 acquired: (1) the herbicide & fungicide business of Thompson Hayward Agricultural & Nutrition Co. (valued at $30 million), (2) the Gustafson Co., a producer of seed-treated chemicals & equipment, (3) Leffingwell, a company producing micronutrients; in 1983, acquired the pesticide business of the Olin Corporation	company has an in-house biotechnology capability & is making a major effort in producing new agricultural chemicals; in 1985, made a 4-year, $6 million research agreement with Biotechnica International "to apply genetic engineering & nitrogen fixation technology to increase yields of selected major food crops"	has financially supported university research in such areas as: herbicide use in cotton & peanuts, chemical control of soil-borne disease in peanuts, & the effect of Harvade, a plant growth regulator, on grain sorghum kernels	a leading U.S. producer of plant growth regulators; also produces & sells herbicides, miticides, & micronutrients

Corporation with Fortune 500 Rank (1984)	Seed/Livestock, Food/Agriculture Subsidiaries, &/or Foreign Interests	Biotechnology Ventures; Investments; Facilities	Agricultural Genetics Research & Related Activities	General Information; Agricultural/Food Products
United Brands, Inc. Boston, MA Fortune Rank: 117	sells Full O'Life seeds & horticultural products in U.S. & U.K.; owns the United Fruit Company (Chiquita, Chico, Fyffes, & Petite bananas); John Morrel & Co. (hogs, cattle, & lambs); the Numar Group (palm, cottonseed, & veg. oils); & Golden Sun Feeds (hog, cattle & pet feeds); owns 91,000 acres & leases 42,000 acres in Costa Rica, Honduras, & Panama for cultivation of bananas & oil palm		believed to hold one of the world's largest stocks of banana germplasm	one of world's leading producers & marketers of bananas; also involved in meat packing & processing; & gourmet foods

Corporation with Fortune 500 Rank (1984)	Seed/Livestock, Food/Agriculture Subsidiaries, &/or Foreign Interests	Biotechnology Ventures; Investments; Facilities	Agricultural Genetics Research & Related Activities	General Information; Agricultural/Food Products
Upjohn Kalamazoo, MI Fortune Rank: 166	owns the Asgrow Seed Co. of Kalamazoo, MI, Morrison Brothers Seed Co. of Spokane, WA, O's Gold Seed Co. of Parkersburg, IA, & 2 smaller seed companies; is also involved in poultry breeding	in 1983 doubled in-house biotechnology staff to 150 & initiated a joint venture with Amgen, Inc., to develop & commercialize a genetically engineered bovine growth hormone; in December 1984 signed a 10-year agreement with Molecular Genetics to distribute its animal health products in 57 countries, also made a 3-year research agreement with Biotechnica International in 1984 to make genetically engineered protein products in gram positive bacteria	interested in both crop & livestock applications of biotechnology; a leading holder of PVP patents & U.S. biotech patents; has also financially supported university research on poultry & agricultural crops	leading supplier of hybrid corn, soybean, & vegetable seed; 6th largest U.S. producer of animal health products; also produces some chemical & biological pesticides

Corporation with Fortune 500 Rank (1984)	Seed/Livestock, Food/Agriculture Subsidiaries, &/or Foreign Interests	Biotechnology Ventures; Investments; Facilities	Agricultural Genetics Research & Related Activities	General Information; Agricultural/Food Products
U.S. Steel Corp Pittsburgh, PA Fortune Rank: 15	USS Agri-Chemicals sells its own line of hybrid corn seed & other companies' oat, soybean, wheat, & grass seed; owns some 80 "farm service centers," which also sell pesticides, fertilizer, & animal feed supplements			nation's #1 steel producer & 19th largest corporation; acquired Marathon Oil Co. for $5.9 billion in 1982
Weyerhaeuser Co. Tacoma, WA Fortune Rank: 66	owns 3 U.S. nursery companies; also owns Oregon Aqua Foods salmon fish hatchery, capable of spawning & harvesting over 1.5 million fish annually	has in-house biotechnology program & a research contract with Cetus Corp. to convert forest-product residues into products such as fertilizers	genetic improvement of forest trees & genetic research on salmon	one of the world's largest lumber companies; owns more than 5 million acres in U.S.; largest lumber exporter

Sources: Data compiled by the Environmental Policy Institute, Washington, D.C., from 1979–1984 annual reports and 10-Ks of publicly held corporations; personal interviews with corporate officials; *Commercial Biotechnology: An International Analysis*, January 1984, U.S. Congress, Office of Technology Assessment; the 1978 *CDE Stock Ownership Directory of Agribusiness*, published by Corporate Data Exchange, Inc., NY; the *1976 Directory of Major U.S. Corporations Involved in Agribusiness*, by

A. V. Krebs, published by the Agribusiness Accountability Project/San Francisco Study Center, CA; *Directory of the 200 Largest U.S. Food and Tobacco Processing Firms, 1975*, by John M. Connor and Lays L. Mather, ESCS, U.S. Department of Agriculture and North Central Regional Project, NC-117, July 1978; and news reports in *The Wall Street Journal, The Washington Post, The New York Times, Business Week, Science, Forbes, Seed Trade News, Seedsmen's Digest, Seed World, Chemical Week, Feedstuffs,* and *Packer Weekly*.

Note: This compilation of corporate involvement in agricultural genetics research and biotechnology is not meant to be an exhaustive or complete listing. This table is only meant to provide a representative cross section of the kind of American and foreign companies now involved in agriculture and food production, as well as some measure of the breadth of their agricultural and food interests, both from a research and commercial standpoint. More detailed information on these and other companies is no doubt available.

INDEX